INTRACELLULAR PROTEIN CATABOLISM II

INTRACELLULAR PROTEIN CATABOLISM II

Edited by

VITO TURK
Department of Biochemistry
J. Stefan Institute
University of Ljubljana
Ljubljana, Yugoslavia

and

NEVILLE MARKS
New York State Research Institute
for Neurochemistry and Drug Addiction
Ward's Island, New York, USA

Associate Editors

Alan J. Barrett
Strangeways Research Laboratory, Cambridge
C BI 4RN, England

and

J. Frederick Woessner
Department of Biochemistry and Medicine
University of Miami School of Medicine
Miami, Florida, USA

PLENUM PRESS · NEW YORK and LONDON

Proceedings of the 2nd International Symposium on Intracellular Protein Catabolism held in Ljubljana, Yugoslavia, May 26-30, 1975

(Contains most of the papers presented at the symposium, in revised form)

Library of Congress Catalog Card Number 77-72034
ISBN 0-306-31037-6

Published in coedition by J. Stefan Institute, University of Ljubljana, Jamova 39, 61000 Ljubljana, Yugoslavia

and

Plenum Press, New York
A Division of Plenum Publishing Corporation
227 West 17th Street, New York, N. Y. 10011, USA

Printed by Študentski servis, Ljubljana, Yugoslavia

PREFACE

This book contains most of the papers presented at the Second International Symposium on Intracellular Protein Catabolism, which was held on May 26–30, 1975 in Ljubljana, Yugoslavia. The papers presented in this book cover a wide variety of areas related to protein turnover and responsible enzymes. The symposium attracted about one hundred participants, and the cordial atmosphere, the frank discussions and the exchange of ideas were a stimulant for all those attending.

During the last decade, investigations in the wide field of intracellular protein catabolism have been rapidly taken up in many laboratories, because the mechanisms and regulation of degradative processes are rather less studied compared to synthetic processes. In view of this increased interest, Prof. H. Hanson and his coworkers organized the First International Symposium and the papers presented there were published in the book Intracellular Protein Catabolism, edited by H. Hanson and P. Bohley. In the preface of this book they concluded: „We hope, however, the symposium as well as the book will be the beginning of a series of symposia and their publication concerning intracellular protein catabolism."

We would like to express our thanks for financial support provided by the Research Council of Slovenia. Special thanks are due to the organizing committee of the symposium, and to Dr. I. Kregar and particularly Dr. J. Babnik for their efforts in preparing manuscripts for the publishers.

Finally, it is a pleasure to acknowledge the cooperation of the Plenum Press and the J. Stefan Institute and to thank our contributors for their efforts and patience.

Editors

CONTENTS

AUTHOR INDEX

INTRACELLULAR PROTEIN CATABOLISM IN VITRO AND IN VIVO

H. Hanson, H. Kirschke, J. Langner, B. Wiederanders, S. Ansorge and P. Bohley

Physiologisch-chemisches Institut der Martin–Luther–Universität, 402 Halle (Saale), Hollystr. 1 (GDR)

It is a great honour and pleasure for the group at Halle and for myself to participate in this Second Symposium on Intracellular Catabolism as invited guests and speakers of the Yugoslav Committee. I wish to thank Dr. Turk and coworkers for their kind invitation to read the opening lecture of this Second Symposium on „Intracellular Protein Catabolism". I hope to meet you all again at Reinhardsbrunn Castle in May 1977. For I would like to invite you on behalf of the Biochemical Society of the GDR and the University of Halle to our III. Symposium of „Intracellular Protein Catabolism", which should be held on the occasion of the hundredth anniversary of Emil Abderhalden's birth.

Our First Symposium on Intracellular Protein Catabolism two years ago at Reinhardsbrunn Castle covered the fields of protein turnover **in vivo** as well as investigations of proteolytic enzymes in **vitro.**

I will embrace the opportunity to give a survey of the methods for comparison of **in-vitro**-experiments with intracellular protein catabolism **in vivo** and of some presuppositions for investigations in these fields. This synopsis will be restricted mainly to the intracellular protein catabolism of the rat liver and I will particularly discuss results from our Institute in Halle.

In the whole I would like to avoid superfluous overlappings with the review of Goldberg and Dice on intracellular protein degradation. (1)

In my opinion, there is still a gap between investigations of protein turnover in vivo and the descriptions of intracellular proteinases, especially for rat liver. Such a division, as we could feel two years ago on the occasion of our First Symposium, fortunately, no longer exists between the investigators of these two important fields, surely, as we will see in the next few days at this nice place. I think all the persons working on this subject agree in the following opinions: investigators of protein turnover in vivo get a lot of data concerning velocities of degradation in vivo of many isolated proteins, but don't know the molecular mechanisms and the intracellular proteolytic enzymes involved in the respective degradation process. On the other hand, investigators of protein degradation in vitro, mainly enzymologists, get a lot of data concerning isolated proteolytic enzymes, but don't know exactly the in-vivo-conditions for their action and even in many cases the substrate proteins degraded in vivo by their enzymes. The reason for this division is, I think, mainly a methodical one.

Therefore the aim of our work should be to connect the knowledge of protein turnover in vivo with investigations of intracellular proteolytic enzymes in vitro (and subsequently in vivo) in such a way as to arrive at some conclusions on the molecular mechanism of protein turnover in vivo. Clearly, all experiments in vitro are performed to get information concerning the in-vivo reactions.

Therefore, I think it is not sufficient, although necessary, to isolate a proteolytic enzyme and to characterize it in the usual manner. In the isolated enzyme it must be proved whether the activity in vitro is also the essential activity in vivo and, especially, whether this activity is important for the degradation of cell-derived substrates. For example, there are some intracellular peptidases, which degrade a row of model peptides, but hitherto intracellular proteins could not be found which were split by these enzymes (p.e. Cath. A, Cath. B2).

On the other hand, there are some intracellular proteinases which degrade intracellular proteins very easily but don't split any of the model peptides used, synthetic substrates and even extraneous proteins. The latest example is: the metal-dependent proteinase of the bovine lens, found and purified by Blow et al. (2) with a great specificity for splitting α_2 – crystallin, but without effect against hemoglobin, azocasein, γ– globulin and synthetic peptidase – substrates. In the opinion of these authors, this enzyme could account for the entire endopeptidase activity of the eye lens.

In the second place it should be proved whether the isolated proteinase is important for regulatory processes. Unfortunately, in mammalian cells there are numerous different proteolytic enzymes. For that reason it is necessary to find out which of such enzymes are the most important ones for the intracellular protein catabolism. Firstly, this is a question depending on the protein substrate used. In our opinion, one of the most suitable substrate proteins for rat liver endopeptidases are the short–lived proteins from rat liver cytosol, because they show a high turnover rate in vivo, but a low degree of autoproteolysis in vitro.

Depending on the comparatively small amount of endopeptidases in rat liver cytosol, the autoproteolysis of isolated cytosol fractions in vitro does not exceed 20 percent of the protein degradation measured in vivo. Therefore, we postulated an intracellular cooperation in vivo, in which cytosol proteins are substrates for particulate (especially lysosomal) protein degrading systems.

From the release of TCA-soluble nitrogen or radioactivity (after in vivo-labeling of proteins) we can (in contrast to other methods) exactly calculate the percentage of protein degraded in vitro. This permits the required comparisons with the known turnover rates in vivo. If we assume a weight ratio between cytosol proteins and lysosomal proteins of thirty to one for rat liver, then the activities detected in lysosomes in the presence of 5 mM reduced glutathione at pH 6.1 were found to be sufficient for degradation of more than 8 % of cytosol proteins per hour, and this value exceeds the normal rate of degradation in vivo also for short-lived proteins.

Which enzymes are particularly involved in this process? If we fractionate the lysosomal proteinases and determine the proteolytic activities against cytosol proteins at pH 6.1, we find about eighty percent of the total activity in the enzyme fractions with molecular weight form twenty – to thirty thousand (L 20), and only ten percent in the fractions with molecular weights form forty – to eighty thousand (L 60) containing the well-known cathepsin D. There are also other reasons to call in question an important role for cathepsin D in intracellular breakdown in vivo in rat liver, and I will discuss them later.

After subfractionation of the fraction L 20, Dr. Kirschke of our group obtained mainly four different endopeptidases, all of which are SH–proteinases. (3). The comparison between their in-vitro-activities and the in-vivo turnover rates of the substrate proteins from rat liver cytosol shows the most important role of cathepsin L (capable of degrading more than fifty percent of cytosol proteins in the same time as in vivo), as follows from comparisons with the turnover rates determined in vivo and of the cathepsins **B1** and **B3** (capable of degrading more than twenty percent of the cytosol proteins in the same time as in vivo). Therefore, I propose to investigate in the future especially the proteolytic enzymes cathepsin L, B1 and B3. The carboxypeptidases cathepsin A and B2 and the prolylcarboxypeptidase and also the lot of lysosomal aminopeptidases play a minor role in the degradation of **protein** substrates, but are necessary for the degradation of peptides deriving from the action of the endopeptidases in vivo, as well as in vitro, yielding amino acids.

The use of **short–** and **long-lived substrate** proteins in vitro allows one to prove whether the conditions chosen in vitro are a sufficient simulation of conditions in living cells, or not. Using such double labeled substrate proteins, Dr. Bohley from our Institute could already in 1971 show the selecitive degradation of short-lived proteins by **lysosomal** proteinases (4). The same holds true for other proteinases, as shown in 1973 by Dice et al (5). Our results (6,7) concerning the preferential degradation of in vivo short-lived proteins by lysosomal endopeptidases in vitro published in 1971 and 1972 were also fully confirmed in 1974 by Huisman (8) and Segal and coworkers (9). Moreover, in our studies in 1972 a preferential degradation of in vivo short-lived cytosol proteins was clearly demonstrated for the endopeptidases in L 20 in comparison to the endopeptidases in L 60 (containing cathepsin D).

On the other hand, the protein degradation in vitro in the microsomal fraction must be an entirely insufficient simulation of the in vivo process, because a selective autolytic degradation of the in vivo long-lived membrane proteins could constantly be observed, as shown in 1973 at the same time in our Institute (10) as well as by Betz and coworkers (11). One of the reasons for this behaviour in vitro may be an accelerated dissociation of apolar proteins from the endoplasmic membranes during the incubation, whereas in vivo the integrity of the endoplasmic membranes and therefore the protection of their apolar proteins should be maintained. The preferential degradation of apolar proteins in the cytosol will be discussed later. The manifold possibilities of **interactions** among all organelles important in vivo has often been overlooked in vitro. The special case of interaction between two organelles only was investigated last year by Ekkehard Schön from our Institute (12, 13). In these experiments, especially the in-vitro degradation of in vivo short-lived proteins (labeled with ^{14}C – or ^{3}H – leucine) was investigated. Labeled subcellular fractions from rat liver were incubated with each other at pH 6.9 in 5 mM reduced glutathione. The highest specific activity for the degradation of short-lived proteins of all fractions was found in the soluble lysosomal fraction. Short-lived cytosol proteins are particularly suitable substrates for all fractions, whereas lysosomal proteins are always degraded slowly. In comparison with the activities in lysosomes, the ability of the other fractions (especially of the cytosol and the microsomal fraction) to degrade the proteins from nuclei, mitochondria, lysosomes and themselves has always been low. These results are in agreement with the scheme for autophagy elaborated mainly from morphological observations: in secondary lysosomes parts of mitochondria, membranes, ribosomes and so on had been found. Nevertheless, these experiments give only hints but not evidence for the mechanisms of the in-vivo-process of protein catabolism. Mitochondria as well as endoplasmic membranes are not degraded as a whole in the lysosomes, and the reasons for the short half-lives of the outer mitochondrial membranes and of some proteins in the endoplasmic membranes are still unknown. Furthermore, it is a rather strong simplification to assume only interactions between two organelles. In fact, there are manifold possibilities for interactions among three or more organelles in intracellular protein catabolism in vivo. For example, the formation of primary lysosomes in the Golgi apparatus (the synthesis de novo or only the rearrangement) can be an important factor also for the energy requirement of intracellular proteolysis. Therefore, from the results obtained by E. Schön it follows that the lysosomes certainly play the most important role in the overall process of proteolysis, and the interactions among the other organelles are comparatively low. There are, however, possibilities of in-vivo interactions between proteinases in the cytosol in substrate proteins from mitochondria and also nuclei. It will be important to investigate subfractions from mitochondria and isolated membranes, Golgi apparatus and ribosomes as substrates and also to prove their proteolytic activities. Last year, Dr. Langner from our Institute found proteolytic activity in purified ribosomes (14). Ribosomal proteins are very easily degraded by this proteolytic activity and also by lysosomal and other endopeptidases. The mechanisms of ribosome catabolism in **vivo,** however, are still not known.

Now I will give some examples of hints concerning the regulation of intracellular protein catabolism in vivo obtained in different physiological conditions:

Over ten years ago, Schimke (15) published his classical experiments on synthesis rate and degradation rate of rat liver arginase under starvation, and following the change from 70 % to 8 % casein in the diet.

These experiments, I think, are an important milestone on our way, and therefore I would like to recall them to our minds. Schimke could show an inhibition of arginase catabolism in starvation and an extremely high activation of this process following a change from 70 % to 8 % casein in the diet. Concomitant changes in the synthesis rates for this enzyme resulted in variations of the amount of arginase in the rat liver. These results clearly demonstrated the importance of degradation rates for the regulation of intracellular enzyme concentrations.

Some further examples of changes in protein degradation in different physiological stages:

(Such changes are described for a number of single enzymes, but in some cases also for the total protein content of rat liver). For instance, the average half-lives of rat liver proteins are prolonged after hepatectomy, after treatment with adrenocorticotrophin, or as a result of aminoazodye feeding or also in aged rats (dr. B. Wiederanders will report on new results in this field tomorrow). It is very difficult to define at present the molecular mechanisms for such in-vivo-alterations as well as the decreased protein catabolism during stimulated growth. An important possibility for getting new results on protein catabolism in vivo, which allows conclusions on molecular mechanisms, in the use of specific inhibitors in vivo. Using leupeptin alone and leupeptin in combination with pepstatin, M. Miyamoto et cow. (16) found inhibitions of RNA-synthesis, DNA-synthesis and mitosis in regenerating rat livers, supporting their hypothesis that lysosomal proteinases may be involved in the initiation of cell proliferation. In our opinion, the enzymes inhibited in these in-vivo experiments by leupeptin must be particularly cathepsin L and also cathepsin B1 (compare the report of Dr. H. Kirschke concerning the action of leupeptin on the different lysosomal hydrolases in vitro). Leupeptin also inhibits proteolytic activities involved in early events in lymphocyte transformation (17). I think in the future we will succeed in bringing out many new results by use of leupeptin and other N-protected peptidealdehydes in vivo and in vitro.

Another important possibility for connecting in vivo and in vitro processes is the perfusion of rat liver. The stimulating role of glucagon (18) and the importance of lysosomes (19) have been confirmed last year using this method.

A further important connection between in-vivo and in-vitro conditions is the injection of labeled proteins in vivo and the investigation of their uptake and degradation in vivo and in vitro thereafter.

The pioneering work of Mego (20) yielded highly important results, and also the studies of Katayama and Fujita (21) on elastase, as well as of Davidson (22) on ribonuclease, describe in detail this process of heterophagy and subsequent degradation of extraneous proteins in secondary lysosomes. Such experiments, however, cannot give sufficient information about the molecular mechanisms of the catabolism of cell-derived substrate proteins, because the uptake into lysosomes and the preceding reactions of substrate proteins in the cytosol are very important steps for catabolism. In our opinion, the exploration of the primary reactions of intracellular catabolism in vivo is a very important question and will in future give new insights also into the regulation of this process. Since the specificity of proteolytic enzymes alone is generally insufficient to explain the selectivity of intracellular protein catabolism, there is a need to find other molecular mechanisms responsible for this selectivity. Therefore, in this connection let me please sum up the results in this field obtained by Dr. Bohley from our Institute during the last few years.

In the opinion of Schimke and coworkers, the most important property for the selective catabolism of short-lived proteins is their high molecular weight, implying more susceptible bonds, more different conformational stages, less stable native conformations, more errors in amino acid sequence. Confirming this result in 1971 at Varna, we found, however, also short—lived proteins with lower molecular weights (30—50000 daltons) in rat liver cytosol. Therefore, the size of substrate protein molecules is only one factor influencing catabolic rates. In 1968 Bohley (23) proposed a hypothetical mechanism for the selectivity of intracellular protein catabolism: different degrees of exposure of superficial hydrophobic areas of substrate proteins could cause different affinities for lysosomal membranes and for proteolytic enzymes.

Using double-labeled proteins from rat liver cytosol (long-lived proteins ^{14}C-labeled, short-lived proteins ^{3}H-labeled) Bohley found recently in phase-partition-experiments (water—toluene, water—benzene, water-butanol, water-isoamylalcohol and chloroform-water) a selective enrichment of short-lived proteins at the apolar phase, whereas long-lived proteins have preferentially been found in the aqueous phase in each partition experiment.

Moreover, short-lived proteins are always enriched in the flotating, fat-rich layer of the cytosol after high-speed centrifugations at 10^6 g min. In conclusion, not only a high molecular weight (24) but also a considerable proportion of apolar area at the surface is an important factor for the selective catabolism of short-lived proteins. This proportion of apolar area may also be the reason for the selective selfaggregation and sedimentation of these short-lived proteins, which occur under widely different conditions, as Dr. Bohley reported two years ago at our First Symposium, and last year at the FEBS-meeting in Budapest, and may, finally, enforce their uptake and degradation in lysosomes. The preferential degradation of these sedimented proteins by solubilized lysosomal endopeptidases was also shown last year.

From these results it seems necessary to find methods of proving **in vivo** possibilities for such preferential aggregation of the short-lived proteins and the importance of this process for rapid protein degradation in normal living cells.

Hitherto, dense aggregates of denatured proteins bound together by hydrophobic linkages have only been found in E. coli that are forced to produce large amounts of abnormal proteins, but similar inclusions have been reported in human diseases, where abnormal proteins accumulate intracellularly (e.g. Heinz – bodies in red cells), and possibly these proteins are also rapidly turning over (1). Recently important publications in a similar field have also appeared by Ballard and coworkers.

The degradation of phosphoenolpyruvate carboxykinase (GTP) in vivo and in vitro was investigated and yielded results leading to the following succession of reactions:

1. loss of catalytic activity
2. disappearance of immunological activity
3. finally a loss of solubility before
4. any evidence of proteolytic cleavage.

The authors propose that denaturation of the enzyme is the rate-limiting step in degradation in vivo. Moreover, Hadvary and Kadenbach (25) found in rat liver mitochondria and other fractions apolar proteins possessing particularly short half-lives in vivo. All these results are, I think, in agreement with the hypothesis made in 1968 by Dr. Bohley (23). Nevertheless, there are also other mechanisms for some individual proteins such as ornithinaminotransferase (26).

Generally, we propose to distinguish **more** and **less** stable conformations with respect to protein breakdown in vivo and in vitro:

As a rule in newly synthesized proteins the bulk of hydrophobic amino acids are buried (exceptions are membrane proteins and very short-lived proteins), and after

dissociation into monomers
dissociation from organelles
dissociation of substrates
dissociation of cosubstrates
splitting of disulfide bonds
splitting of **single** peptide bonds

and other changes, the proportion of superficial hydrophobic amino acids rises. Therefore, the protein molecule has more susceptible peptide bonds and also more possibilities for random aggregations and precipitation. These aggregates might be autophagized preferentially by lysosomes and proteolysis could follow. The process of **autophagy** itself however, is unfortunately not very clear. In **vivo** undoubtedly there occurs autophagy, as we know from morphological experiments. But all efforts to simulate autophagy in **vitro** have not been successful as yet. We hope the partial energy requirement of intracellular protein catabolism might also allow some first insights into the molecular mechanisms of autophagy. Interactions should be proved among organelles in autophagy (especially the role of resynthesis or rearrangement in the Golgi apparatus), but there should also be proofs for further causes of this energy requirement, for example:

1. resynthesis of short-lived proteolytic enzymes or activators such as glutathione
2. lysosomal ATPases (27, 28)
3. maintainance of intralysosomal effector concentrations in general
4. inactivation of inhibitors (29); certainly, Dr. Mego will bring more than theories to this problem.

Now I will bring to an end my synopsis. For this I will give a more hypothetical view of the succession of steps in intracellular protein catabolism in rat liver **in vivo**.

Please compare my statements with our present knowledge in this field and you will see that a lot of work has to be done in future.

The **primary** reaction of intracellular protein catabolism for many proteins may be the change from a more stable conformation (more buried apolar residues) to a **less** stable conformation (more superficial hydrophobic areas), followed by random aggregations and uptake in the lysosomes (autophagy). Certainly, some proteins can be taken up directly by lysosomes as well as debris from organelles. The change from more to less stable conformations may be regulated by substrates, cosubstrates and many effectors.

In the **lysosomes** (or other organelles), primarily **endopeptidases** are acting with different substrate and effector specificities. The actual activity of these endopeptidases depends on their synthesis and degradation, effector and inhibitor concentrations and may also be regulated. Thereafter **exopeptidases** act in lysosomes and also (to a great extent) in the cytosol to degrade peptides to amino acids. Regulations at the level of exopeptidases are also possible, but less probable, because the amount of exopeptidases exceeds by about ten fold the amount required for the in-vivo-turnover.

In all these processes, we have got clues from in-vitro-experiments in the past. Our problem in the future is to confirm or refute their validity in vivo, in the living normal cells. In my synopsis I hope you have seen a series of possibilities for confirming our present conception on intracellular protein catabolism in the living cells.

REFERENCES

1. Goldberg, A.L., Dice, J.F., Ann. Rev. Biochem., 43, 835 (1974)
2. Blow, A.M.J., van Heiningen, R., Barrett, A.J. Biochem. J. 145, 591 (1975)
3. Kirschke, H., Langner, J., Wiederanders, B., Ansorge, S., Bohley, P., Acta biol. med. germ. 28, (1972)
4. Bohley P., Langner, J., Ansorge S., Miehe, M., Miehe, C., Hanson, H., FEBS Meeting Varna, Abstr. 360 (1971)
5. Dice, J.F., Dehlinger, P.J., Schimke, R.T., J. Biol. Chem 248, 4220 (1973)
6. Bohley, P., Kirschke, H., Langner, J., Ansorge, S., Wiederanders, B., Hanson, H., in: Tissue Proteinases (Barret A.J., Dingle J.T., eds) North—Holland, Amsterdam, p. 187 (1971)
7. Bohley, P., Miehe, M., Miehe, C., Kirschke, H., Langner, J., Wiederanders, B., Ansorge, S., Acta Biol. Med. Germ. 28, 223 (1972).
8. Huisman, W., Dissertation, Groningen.
9. Segal, H.I., Winkler, J.R., Miyagi, M.P., J.Biol. Chem. 249, 6364 (1974).

10. Bohley, P., Kirschke, H., Langner, J., Wiederanders, B., Hanson, H., FEBS Meeting, Budapest, Abstr. 5256 (1974).
11. Betz, H., Gratzl, M., Remmer, H., Hoppe Seyler's Z. Physiol. Chem. 354, 567 (1973).
12. Schön, E., Diplomarbeit, Halle (1974).
13. Schön, E., IX. Jahrestagung der Biochemischen Gesellschaft der DDR, p. 90 (1974).
14. Langner, J., Bohley, P., Ribosomensymposium, Reinhardsbrunn (1974).
15. Schimke, R.T., J. Biol. Chem. 239, 3808 (1964).
16. Myamoto, M., Terayama, H., Ohnishi, H., Biochem. Biophys. Res. Comm. 55, 84 (1973).
17. Saito, M., Yoshizawa, T., Aoyagi, T., Nagai, Y., Biochem. Biophys. Res. Comm. 55,569 (1973).
18. Woodside, K.H., Ward, W.F., Mortimore, G.E., J. Biol. Chem. 249, 5458 (1974).
19. Neely, A.N., Mortimore, G.E., Biochem. Biophys. Res. Comm. 59,680 (1974).
20. Mego, J.L., in: Intracellular Protein Catabolism, (Hanson, H., Bohley, P. eds.) p. 30, Verlag, J. Ambrosius Barth, Leipzig, (1974/6)
21. Katayama, K., Fujita, T., Biochim. Biophys. Acta 336, 165 (1974).
22. Davidson, S.J., Song, S.W., Biochim. Biophys. Acta375, 274; 385, 163 (1975).
23. Bohley, P., Naturwiss. 55, 211 (1968).
24. Shimke, R.T., Munro, H.N., Mamm. Prot. Metabol. 4, 172 (1970).
25. Hadvary, P., Kadenbach, B. Europ. J. Biochem. 39, 11 (1973).
26. Katunuma, N., Kominami, E., Kobayashi, K., Banno, Y., Suzuku, K., Chichibu, K., Hamaguchi, Y., Katsunuma, T., Europ. J. Biochem. 52, 37 (1975).
27. Iritani, N., Wells, W.W., Arch. Biochem. Biophys. 164, 357 (1974).
28. Schneider, D.L., Biochem. Biophys. Res. Comm., 61, 882 (1974).
29. Saheki, T., Holzer, H., Biochim. Biophys. Acta 384, 203 (1975).

AN ENERGY REQUIREMENT FOR INTRALYSOSOMAL PROTEOLYSIS IN PHAGOLYSOSOMES AND TISSUE SLICES

John L. Mego and Roderick M. Farb

Department of Biology, University of Alabama University, Alabama 35486, U.S.A.

SUMMARY

A partial energy requirement has been demonstrated for intralysosomal proteolysis in a cell free system. This requirement is manifested only at alkaline pH. Divalent cations enhance the stimulatory effects of ATP in this system; however, Mg alone stimulates intralysosomal proteolysis in the absence of buffers and it appears to enhance proteolysis at pH 8 by membrane stabilization. The effects of Mg are abolished by high concentrations of monovalent cations. Monovalent cations have no effects on stimulation of intralysosomal proteolysis by ATP–Mg at pH 8. Preincubation of mouse kidney phagolysosomes in a solution of Pronase abolishes the stimulatory effects of ATP-Mg, Mg and pH 5 buffer. These results are consistent with the presence of a proton transport facilitating protein in the phagolysosome membrane which can function to translocate protons by facilitated diffusion in the absence of buffer but which requires ATP to transport protons into the phagolysosome at pH 8. Evidence is presented that a partial energy requirement also exists for intralysosomal proteolysis of intravenously injected protein in mouse liver slices. Results obtained with inhibitors of ATP formation, fluoride and chloroquine are similar to those obtained by Wibo and Poole in studies on intracellular proteolysis of endogenous proteins in rat fibroblasts. The results are also, for the most part, consistent with those obtained in the cell-free system and they suggest that intralysosomal proteolysis in intact cells is an energy requiring process. This energy requirement may be to maintain intralysosomal acidity.

In 1953, Simpson discovered an apparent energy requirement for the catabolic aspect of protein turnover in rat liver slices (1). This discovery has since been confirmed by a number of workers (2-4). An energy requirement for an exergonic process seems redundant but the involvement of lysosomes in intracellular protein catabolism would provide an answer to this apparent enigma. Some evidence for an involvement of lysosomes in intracellular endogenous proteolysis has recently been obtained (5). Not only might energy be required to incorporate proteins into lysosomes but also perhaps for maintenance of intralysosomal acidity. Although several reports have appeared in the literature that certain protein specific proteases may exist in the cell outside of lysosomes (6-8), most of the known intracellular protease activity is in lysosomes.

The acid pH optima of lysosomal hydrolases suggests that a mechanism might exist in the lysosome membrane to maintain intralysosomal acidity. This might be especially true in kidney lysosomes since a crude mouse kidney lysosomal cathepsin preparation exhibits a sharp peak of activity at pH 4.6 and falls off rapidly to approximately 20 % of this activity at pH 6.2 with serum albumin as substrate. A similar preparation from mouse liver has a somewhat broader peak of activity around pH 5, and at pH 6 there is still substantial activity present, although there is negligable activity at pH 7 (9). A new proton is generated for each carboxyl group liberated during proteolysis but this proton is taken up by the newly exposed amino group. The pH generated by amino acid liberation from a protein such as serum albumin is about 6.3. This pH might be expected to inhibit proteolysis in kidney phagolysosomes as much as 85 % and in liver phagolysosomes about 65 %.

In previous work, we presented evidence that intralysosomal proteolysis in a cell-free system is partially inhibited at pH 8 and this inhibition is reversed by additions of ATP during incubation (10). The primary function of ATP in this system appeared to be to stimulate intralysosomal proteolysis rather than to stabilize phagolysosomes. These observations were interpreted in terms of an energy-dependent system which functions to maintain intralysosomal acidity perhaps in the form of a proton pump.

MATERIALS AND METHODS

The materials and methods used in this research have been, described in previous publications (9-13). Mice were injected intravenously with denatured (formaldehyde treated) ^{125}I-labelled bovine serum albumin. The animals were sacrificed 30 min later and the kidneys and/or livers were removed and homogenized in cold 0.25 M sucrose. The homogenates were centrifuged, to sediment a particulate fraction containing phagolysosomes filled with labelled protein. Proteolysis in these phagolysosomes was measured by precipitation of undegraded protein with trichloroacetic acid. Routinely, only 0.25 M sucrose and 50 mM mercaptoethanol were present in the reaction mixtures. Other additions are described in legends to figures and tables.

Phagolysosome breakage rates were measured by diluting samples of suspensions 1:10 with ice-cold 0.01 M pH 8 buffer and in 0.01 M buffer-0.25 M sucrose. These dilutions were centrifuged and the radioactivities in the supernatants from samples diluted in sucrose were subtracted from the corresponding samples diluted in buffer. These differences were calculated as per cent of the total in the suspensions.

RESULTS AND DISCUSSION

Effects of Divalent Cations

In previous studies (10) we found that Mg^{2+} enhanced the effect of ATP on intralysosomal proteolysis at pH 8. Table 1 shows that Ca^{2+} and Mn^{2+} also enhanced the effect of ATP. Furtheromore, these divalent cations alone, without ATP, enhanced proteolysis in mouse kidney phagolysosome suspensions but not in suspensions of liver particles. This enhancement occurred at pH 8 or in the absence of buffers. Although the effect of Mg^{2+} at pH 8 appeared to be primarily on membrane stabilization (Fig. 1) the cation stimulated rates of intralysosomal proteolysis in the absence of buffer (Fig. 2). At pH 8, there was no apparent enhancement of intralysosomal proteolysis by Mg^{2+} (Fig. 3).

TABLE 1

Effects of Divalent Cations and ATP on Proteolysis in Mouse Liver and Kidney Phagolysosome Suspensions Incubated in Various Buffers

		Trichloroacetic acid-soluble radioactivity in 40 min incubation (% of the total)							
							ATP	ATP	ATP
	Buffer	Cont.	Mg	Ca	Mn	ATP	Mg	Ca	Mn
Liv	NH_4^+ 8	9.1	–	9.7	–	12.1	15.9	15.3	17.7
Liv	NH_4^+ 8	7.8	7.5	–	–	10.3	11.8	13.5	13.2
Liv	Bor. 8	15.5	15.6	–	–	16.0	20.3	–	–
Kid	None	20.4	32.8	26.2	28.1	–	–	–	–
Kid	None	18.5	25.5	–	–	13.6	32.1	–	–
Kid	TA 5	41.9	42.6	–	–	–	–	–	–
Kid	TA 8	8.5	12.9	–	–	9.3	22.6	–	–
Kid	NH_4^+ 8	7.5	–	10.4	–	11.6	22.7	22.9	–
Kid	Bor. 8	15.6	23.6	19.9	19.9	–	–	–	–
Kid	Bor. 8	9.2	17.0	–	–	24.4	–	–	–

Buffers were NH_4^+8, pH 8 ammonium chloride; Bor. 8, pH 8 sodium borate, pH 8; TA 5 or TA 8, Tris-acetate pH 5 or 8. All buffer concentrations were 0.025 M. Tissues (liver or kidneys) were homogenized in 0.25 M sucrose with 1 mM EDTA neutralized to pH 7. The 500-30,000 g fractions were not washed. Incubations were for 60 minutes at 35° in 0.25 M sucrose — 0.05 M mercaptoethanol and the above additions. Divalent cations were chlorides at concentrations of 4 mM. ATP additions were as described in Fig. 1.

The stimulatory effects of Mg^{2+} on intralysosomal proteolysis suggests that the cation may enhance proton uptake by facilitating the establishment of a Donnan equilibrium. The presence of large quantities of acidic lipoproteins has been established in lysosomes (14). If proton transport is facilitated by some mechanism in the lysosome membrane, proton concentration and hence a high buffering capacity might thus become established in the phagolysosome. The presence of an alkaline buffer (0.025 M sodium borate or Tris-acetate, pH 8) might be sufficient to lower intralysosomal acidity by stripping protons from acidic lipoproteins and by partially neutralizing intralysosomal acidity. Membrane stabilization, perhaps by prevention of thermally-induced phase transitions known to occur in membranes (15), might account for the enhancement of proteolysis at pH 8 by Mg^{2+}. Although Mg^{2+} seems unable to cope with pH 8 buffer in the establishment of intralysosomal acidity, additions of ATP effectively reversed the inhibitory effects of pH 8 buffer. Furthermore, higher rates of intralysosomal proteolysis were noted at pH 8 with ATP-Mg^{2+} than in the absence of buffer with only Mg^{2+} (Fig. 4). These experiments suggest that although Mg^{2+} may facilitate proton translocation by a passive process, ATP may energize a proton pump under alkaline conditions. The energized system should be more efficient than a passive process such as the Donnan equilibrium.

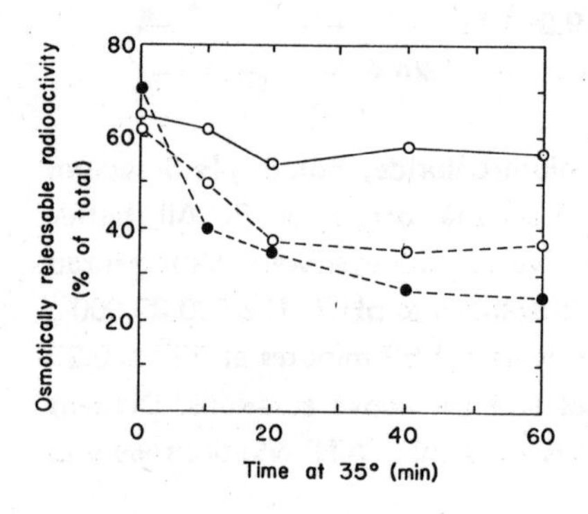

Figure 1. Effects of Mg and ATP on breakage of mouse kidney phagolysosomes at pH 8 and 35°. Incubations were carried out in 12 ml 0.25 M sucrose and ATP (●---●), 4 mM $MgCl_2$ (○——○) or no additions (○---○). Mercaptoethanol was omitted to inhibit proteolysis which was negligible in all the suspensions (about 5 % of the total radioactivity solubilized in trichloroacetic acid during 60 min incubation). ATP (disodium salt neutralized to pH 8) 0.12 M, was added in 0.1 ml aliquots at 0, 10, 20, 30, and 40 min. Breakage rates were determined as described in Materials and Methods.

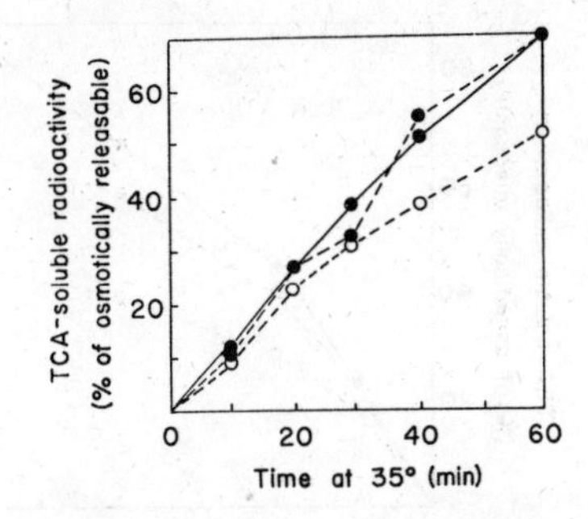

Figure 2. Effects of ATP-Mg (●——●), Mg (●- - -●) or no additions (○- - -○) on intralysosomal proteolysis in mouse kidney phagolysosome suspensions incubated in the absence of buffer. Proteolysis (TCA-soluble radioactivity) was calculated as a function of the radioactivity releasable by osmotic shock rather than as per cent of the total. Mg and ATP concentrations were the same as described in Fig. 1.

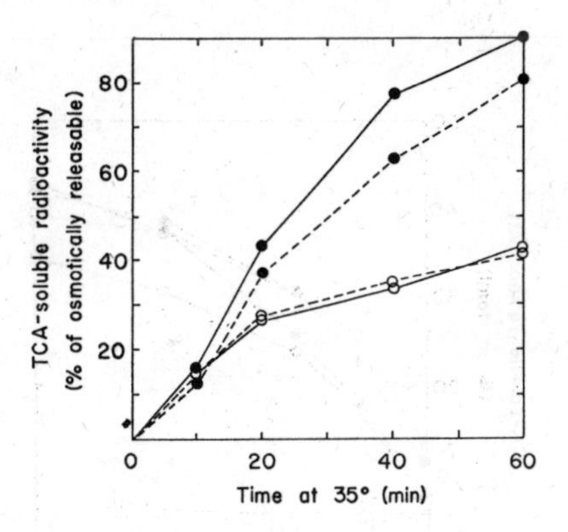

Figure 3. Effects of ATP-Mg (●——●), ATP (●- - -●), Mg (○——○) or no additions (○- - -○) on intralysosomal proteolysis in mouse kidney phagolysosomes incubated at pH 8. Conditions were the same as in the experiment illustrated in Fig. 2 except that 0.025 M sodium borate buffer, pH 8, was present in the incubation mixtures. Homogenization of the kidneys was performed without EDTA and the particles were unwashed; the suspensions therefore most likely contained endogenous Mg.

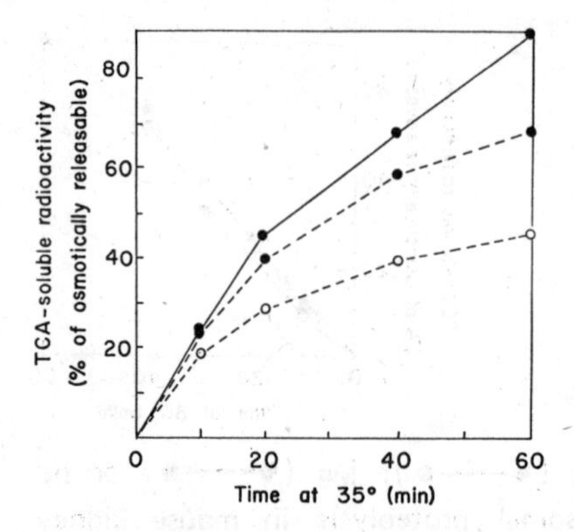

Figure 4. Intralysosomal proteolysis in mouse kidney phagolysosome supensions incubated at pH 8 with ATP-Mg (●——●), at pH 8 with no additions (o- - -o), and in unbuffered medium with Mg (●- - -●). Conditions were the same as described in the legends to Figs. 2 and 3.

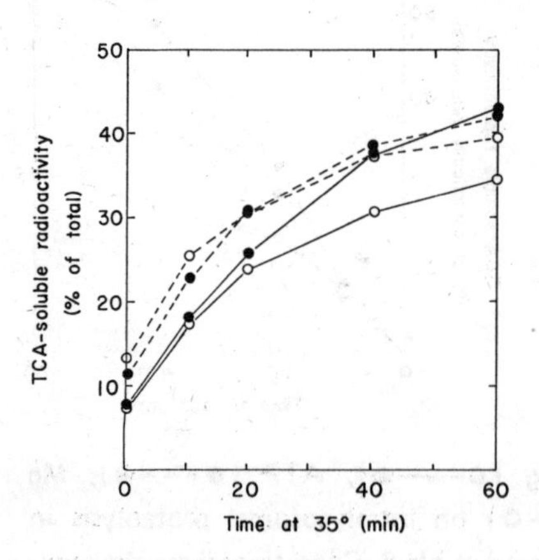

Figure 5. Proteolysis in mouse kidney and liver phagolysosome suspensions in the presence (solid lines) and absence (dashed lines) of 10 mM chloroquine diphosphate. Kidney suspensions: filled circles; liver suspensions: empty circles.

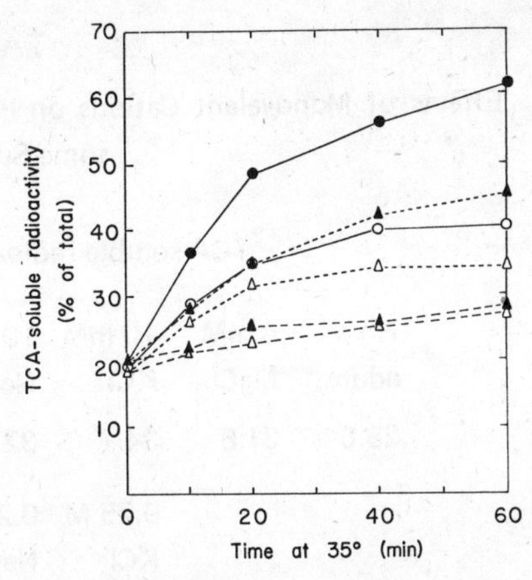

Figure 6. Inhibition of proteolysis in mouse kidney phagolysosome suspensions by neutralized chloroquine (pH 8) and reversal by ATP-Mg. $MgCl_2$ concentrations were 4 mM and ATP additions were as described in the legend to Fig. 1. Incubations were carried out at 35° in 0.25 M sucrose, 0.05 M mercaptoethanol and 0.01 M sodium borate, pH 8. Control (no chloroquine), solid lines; 0.1 mM chloroquine (–––), 0.01 mM chloroquine (........). Filled circles or triangles, ATP.

Effects of Monovalent Cations

If Mg^{2+} does facilitate the establishment of intralysosomal acidity by promoting a Donnan equilibrium, it should be possible to suppress this effect with monovalent cations such as K^+, Na^+, or Li^+. However, the phagolysosome is relatively impermeable to salts such as NaCl or KCl since these salts protect phagolysosomes from osmotic disruption (12). Furthermore, we found that low concentrations of monovalent cations (ca. 0.01 M) stimulated proteolysis in phagolysosome suspensions. This stimulatory effect was noted only in the absence of pH 8 buffer and it was additive to the stimulatory effects of Mg^{2+} (Table 2). High concentrations of monovalent cations (0.20 – 0.25 M), either in the presence of 0.25 M sucrose or alone abolished the stimulatory effect of Mg^{2+} both at pH 8 or in the absence of buffer. Furthermore, high salt concentrations reduced the stimulatory effect

TABLE 2

Effects of Monovalent Cations on Proteolysis in Mouse Kidney Phagolysosome Suspensions

TCA-soluble radioactivity, Δ/40 min

No addns.	4 mM $MgCl_2$	10 mM KCl	10 mM NaCl	10 mM LiCl	KCl Mg	NaCl Mg	LiCl Mg
25.6	31.6	34.7	32.7	28.4	39.0	38.0	32.5
		0.25 M KCl	0.25 M NaCl	0.25 M LiCl	KCl Mg	NaCl Mg	LiCl Mg
27.2	41.4	32.4	33.8	33.5	36.0	34.1	32.2

The data in this table represents two experiments. In each experiment, 500–30,000 g mouse kidney subcellular particles were divided into 8 parts and incubated at 35° in 12 ml 0.25 M sucrose – 0.05 M mercaptoethanol with the above additions. The figures represent the percent increase in trichloroacetic acid-soluble radioactivity during 40 min incubation.

---

of ATP-Mg^{2+} in the absence of buffer but had no effect at pH 8 (Table 3). Thus, monovalent cations suppressed the stimulatory effect of Mg^{2+} but had no effect on the enhancing effect of ATP on intralysosomal proteolysis. These findings are consistent with the hypothesis that Mg^{2+} promotes establishment of a Donnan equilibrium perhaps by facilitating proton diffusion into the phagolysosome and ATP-Mg^{2+} converts the system to an active proton pump.

The Effects of Chloroquine

There are a number of weak bases which accumulate in lysosomes (16). These substances have pKs around 8, and their accumulation appears to be a function of intralysosomal acidity. In the uncharged (non- protonated) form, they are able to diffuse into lysosomes where they become

TABLE 3

Effect of High Salt Concentrations on the Stimulation of proteoloyis in Mouse Kidney Phagolysosomes by ATP

TCA-soluble radioactivity, Δ/40 min

No buffer	$Mg2^+$	$ATP\text{-}Mg2^+$	% stimulation
0.25 M sucrose	28.5	39.7	39.3
0.25 M sucrose 0.20 M KCl	29.8	33.3	11.7
0.25 M sucrose 0.20 M LiCl	28.5	32.6	14.4
0.25 M KCl	31.7	35.7	12.6
pH 8 Borate			
0.25 M sucrose	16.0	31.7	98.1
0.25 M sucrose 0.20 M LiCl	10.3	20.9	102.9
0.25 M LiCl	10.0	21.8	118.0

Mouse kidneys were homogenized in 0.25 M sucrose in all cases. Incubations were carried out in 12 ml volumes at 35°. All samples contained 4 mM $MgCl_2$ and 0.05 M mercaptoethanol in addition to sucrose or salts. ATP additions were as described in the legend to Fig. 1.

---

protonated. Once inside in the protonated form, these substances cannot pass back through the lysosomal membrane. However, protonation inside the lysosome raises intralysosomal pH and further accumulation should cease unless more protons are pumped in. Wibo and Poole (17) have observed concentration ratios of 200 in rat fibroblasts exposed to 0.1 mM chloroquine. De Duve has postulated that a proton pump must be present in order to permit such extensive accumulations of chloroquine in lysosomes (16).

Incubation of mouse kidney phagolysosomes with concentrations of chloroquine diphosphate as high as 10 mM had no effect on intralysosomal proteolysis (Fig. 5). However, in the presence of 0.01 M. borate buffer, pH 8, neutralized chloroquine at concentrations of 0.1 mM completely inhibited intralysosomal proteolysis and 0.01 mM inhibited about 40 %. Additions of ATP reversed this inhibition by 0.01 mM chloroquine (Fig. 6). Furthermore, in the presence of ATP-Mg^{2+} five times higher concentrations of chloroquine (0.05 mM) were required to inhibit intralysosomal proteolysis to the same extent as 0.01 mM in the absence of ATP. Chloroquine has been shown by Wibo and Poole (17) to inhibit cathepsin BI which has been postulated to be the major protease in lysosomes involved in the degradation of albumin (18). We have also found that chloroquine inhibits hydrolysis of formaldehyde treated serum albumin by a crude rat kidney lysosomal enzyme preparation at concentrations of 0.01 mM. However, this inhibition was pH dependent. For example, 0.01 mM chloroquine had no effect at pH 5 but inhibited about 30 % at pH 6. The reversal of chloroquine inhibition of intralysosomal proteolysis by ATP therefore may have been to lower intralysosomal acidity to around pH 5. Chloroquine may have two effects in the isolated phagolysosome: to raise intralysosomal pH to inhibitory values and to directly inhibit cathepsin function at this elevated pH.

Preincubation With Pronase

Some attempts were made to determine the nature of the component or components in the phagolysosome membrane affected by Mg^{2+} or ATP. In these experiments, mouse kidney phagolysosomes were preincubated at 35° with 0.2 mg per ml „Pronase" (**Streptomyces griseus** protease). The suspensions were then washed free of the protease and incubated with ATP-Mg^{2+}, Mg^{2+}, monovalent cations, and in pH 5 buffer. Table 4 shows some of the results obtained in these experiments which indicate that Pronase abolished the stimulatory effects of ATP, Mg^{2+} and pH 5 buffer but not those of monovalent cations. This suggests that all the effects except those of monovalent cations might be mediated by proteins in the phagolysosome membrane. If these effects are mediated by the same protein, this would be consistent with the hypothesis that the protein mediates proton transport either by a facilitated diffusion or by an ATP-energized proton pump.

TABLE 4

Effects of Preincubation in Pronase on Proteolysis in Mouse Kidney Phagolysosomes

TCA-soluble radioactivity, Δ/40 min

Expmt. 1	No additions	5 mM $MgCl_2$	pH 8 borate	pH 8 borate 5 mM $MgCl_2$
Control	27.6	40.2	13.2	20.1
Pronase	24.6	23.8	10.5	11.3

Expmt. 2	5 mM $MgCl_2$	5 mM $MgCl_2$ pH 8 borate	ATP-Mg pH 8 borate
Control	26.6	15.1	32.2
Pronase	19.3	15.1	13.7

Expmt. 3	No additions	10 mM KCl	0.025 M pH 5 buffer
Control	30.6	36.4	36.2
Pronase	20.3	28.4	19.4

In each experiment, the 500-30,000 g mouse kidney subcellular fraction was divided into two parts. One part was incubated in 0.25 M sucrose – 10 mM phosphate buffer, pH 7, and the other part was incubated similarly in sucrose-buffer containing 0.2 mg per ml Pronase for 10 min at 35°. The suspensions were then centrifuged, divided into the number of samples to be incubated, centrifuged again and then suspended in 0.25 M sucrose-0.05 M mercaptoethanol with the above additions. Incubations were for 60 min at 35°. Samples were withdrawn at 0, 10, 20, 40 and 60 min. ATP additions were made as in Fig. 1.

Intralysosomal Proteolysis in Liver Slices

In the experiments described thus far, we have demonstrated a partial energy requirement for intralysosomal proteolysis in alkaline media in a cell-free system. Since the experimental system is one in which we are certain that proteolysis is occurring in lysosomes, and since we can measure this proteolysis in intact cells as readily as in the cell-free system, we considered it of interest to determine if intracellular proteolysis of formaldehyde-treated ^{125}I-labelled albumin also would exhibit an energy requirement. Livers from mice injected with ^{125}I-labelled albumin were sliced with a Stadie-Riggs tissue slicer and incubated in a modified (calcium-free) Krebs-Ringer solution. Samples were removed at intervals and homogenized in 5 % trichloroacetic acid. These were counted to determine the total and trichloroacetic acid-soluble radioactivity. Table 5 shows that additions of 2,4-dinitrophenol, sodium azide, sodium cyanide, fluoride and chloroquine to the incubation mixture produced an inhibition of this intracellular proteolysis. Partial inhibition by inhibitors of ATP formation is consistent with effects on intralysosomal acidity since pH 7-8 buffers never completely inhibited intralysosomal proteolysis in the cell-free system. The relatively larger inhibition by chloroquine is also consistent with results obtained in the cell-free system. The inhibitory effects of fluoride however

TABLE 5

Effects of Some Inhibitors on Intralysosomal Proteolysis in Mouse Liver Slices

Inhibitor	Concentration (mM)	TCA-Soluble Radioactivity Δ/60 min	Per cent Inhibition
2,4-dinitrophenol	5.4	24.7	38.4
Sodium Fluoride	10.0	16.6	58.6
Sodium Azide	100.0	22.1	44.9
Sodium Arsenate	1.2	39.6	1.2
Sodium Cyanide	1.0	28.3	29.4
Chloroquine	0.1	16.8	58.1
Ouabain	0.1	37.2	7.1
Valinomycin	0.005	37.7	6.0
Control	–	40.1	–

TCA-soluble radioactivity represents the per cent of the total radioactivity solubilized in trichloroacetic acid during 60 minutes incubation of the slices.

are difficult to interpret since this anion has no effect on intralysosomal proteolysis in suspensions of liver or kidney phagolysosomes. It is of interest to note, however, that Wibo and Poole also found fluoride and chloroquine to be effective inhibitors of intracellular degradation of endogenous proteins in fibroblasts (4,17). The results obtained in the studies summarized in this presentation therefore suggest that intracellular protein turnover may take place in lysosomes and the energy requirement for this process may be to maintain intralysosomal acidity.

ACKNOWLEDGEMENTS

Mrs. Jessica Hsia performed many of the experiments described in this communication. Some of the work was supported by Grant ES 00591 from the National Institute of Environmental Health Sciences, NIH. I am grateful to the University of Alabama for a travel grant to attend this Symposium.

REFERENCES

1. Simpson, M.V. J. Biol. Chem. 202, 143 (1953).
2. Brostrom, C.O. and Jeffay, H., J. Biol. Chem. 245, 4001 (1970).
3. Hershko, A. and Tomkins, G.M., J. Biol. Chem. 246, 710 (1971).
4. Poole, B. and Wibo, M., J. Biol. Chem. 248, 6221 (1973).
5. Neely, A.N. and Mortimore, G.E., Biochem. Biophys. Res. Commun. 59, 680 (1974).
6. Katunuma, N., Kominami, E. and Kominami, S., Biochem. Biophys. Res. Commun. 45, 70 (1971).
7. Duckworth, W.C., Heinemann, M.A. and Kitabchi, A.E., Proc. Nat. Acad. Sci. USA 69, 3698 (1972)
8. Mego, J.L. in: Intracellular Protein Catabolism, (Hanson, H., Bohley, P., eds.) Verlag J. Ambrosius Barth, Leipzig, p. 30 (1974)
9. Mego, J.L., Biochem. J. 122, 445 (1971).
10. Mego, J.L. Farb, R.M. and Barnes, Judith., Biochem. J. 128, 763 (1972).
11. Mego, J.L., Bertini, F. and McQueen, J.D., J. Cell Biol. 32, 699 (1967).

12. Mego, J.L. in: Lysosomes in Biology and Pathology (J. T. Dingle, ed.), Vol. 3, pp. 138-168. North Holland, Amsterdam-London (1973).
13. Mego, J.L. in Lysosomes in Biology and Pathology (J. T. Dingle, ed.), Vol. 3, pp. 528-537. North Holland, Amsterdam-London (1973).
14. Goldstone, A., Szabo, E. and Koenig, H., Life Sciences. 9, 607 (1970).
15. Trauble, H. and Eibl, H., Proc. Nat. Acad. Sci. USA 71, 214 (1974).
16. De Duve, C., de Barsy, Th., Poole, B., Trouet, A., Tulkens, P. and Van Hoof, F., Biochem. Pharmacol. 23, 2495 (1974).
17. Wibo, M. and Poole, B., J. Cell Biol. 63, 430 (1974).
18. Huisman, W., Lanting, L., Doddema, H.J., Bouma, J.M.W. and Gruber, M., Biochim. Biophys. Acta 370, 297 (1974).

INHIBITION OF CELL PROTEIN DEGRADATION BY MICROTUBULAR INHIBITORS

J.S. Amenta, F. M. Baccino*, M. J. Sargus

Department of Pathology, University of Pittsburgh, School of Medicine, Pittsburgh, Pennsylvania 15261 U.S.A.; *Instituto di Patologia Generale, Corso Raffaello, 30, I-10125 Torino, Italia

That the degradation of cell protein in eukaryotic cells involves more than simple hydrolysis of peptide bonds is now well-established. Simpson(22), using rat liver slices, first demonstrated an energy requirement for the degradation of cell protein, while later studies with this same system by Steinberg and Vaughan(23) showed that protein synthesis also appeared to be a necessary condition for the degradation of cell proteins. Although most workers(2),(9),(11),(27), using other experimental systems, have corroborated these requirements for energy and protein synthesis in degradation of cell proteins, other studies(1),(8),(20),(26) have suggested that the requirement for protein synthesis appears to be less than absolute.

A possible reason for these discrepant results is suggested by experiments which indicate that more than one mechanism may exist for cell protein degradation and that different mechanisms very likely have different sensitivities to various metabolic inhibitors. These workers, using both liver perfusion(15) and cell culture techniques(1),(6),(7),(20) have demonstrated that serum-deficient (SD) media markedly augment the degradation of cell proteins and that inhibitors of protein synthesis affect only this augmented proteolysis and are without effect upon the basal turnover rate of cell proteins(6),(7),(20),(26). In addition, there is strong evidence to support the hypothesis that this augmentation of cell protein degradation, induced by deficient media, involves the vacuolar system (17), though the involvement of lysosomes in the basal turnover of cell proteins still remains debatable (20).

Observations that microtubules are involved in the functioning of the vacuolar system (24), particularly in processes involving degradation of exogenous proteins,(4),(13),(18) suggested to us that microtubular function may also be involved in the degradation of cell proteins. Our hypothesis was that the SD-induced mechanism of cell protein degradation involved the functioning of the lysosomal system, including microtubules, and, as such, SD proteolysis should be inhibited by inhibitors of microtubules.

Our results indicate that microtubular inhibitors do inhibit cell proteolysis induced by serum deficiency, but have no effect on the basal turnover rate of cell proteins.

METHODS

Stock fibroblast cultures were derived from 18 day embryos, Fisher strain F-344 (Hilltop Laboratories, Scottsdale, Pennsylvania, U.S.A.). Fibroblasts subcultured from frozen Stock cultures were grown in 75 cm^2 Falcon flasks in Eagle's minimum essential medium (MEM) supplemented with 10 % fetal calf serum. ^{14}C-L-leucine (0.8 uCi/flask) and ^{3}H-thymidine (3.0 uCi/flask) were added four days prior to each experiment at the time of the subculture. Cultures were then placed in fresh cold chase medium for 24 hours prior to experiment. On the day of the experiment, cultures generally were in a sub-confluent stage, with about 500 μg of protein per flask.

For each experiment cells were washed twice with phosphate-buffered saline and placed in 10 ml of fresh unlabeled medium, containing either 10 % serum (basal) or 0 % serum (SD). In later experiments media with 1 % serum was used in place of the media with 0 % serum. Various metabolic inhibitors, as indicated in each experiment, were added at the same time. At intervals of 0,1,2,4 and 24 hours, 0.5 ml aliquots of media were removed for analysis. Approximately 10 mg of bovine albumin were added to the aliquots taken from the SD media before the addition of TCA to a final concentration of 8 %. Acid-solube ^{14}C and ^{3}H were measured in the clear supernatant, by liquid scintillation counting. At the end of each experiment the cell layers were washed twice with phosphate-buffered saline and solubilized in 4 ml of 0.1 N NaOH-0,4 % sodium deoxycholate (20). Total cell ^{14}C and ^{3}H were determined from an 0.4 ml aliquot; acid-soluble ^{14}C and ^{3}H were determined from an aliquot precipitated with TCA, as described for the SD media.

Total cell protein was determined by the method of Lowry (12). Values for total cell protein and total cell radioactivity were corrected for cell losses, based upon recovery of ^{3}H-thymidine in the cell DNA. All results are reported as % of total radioactivity.

To measure the rate of leucine incorporation into cell proteins, fibroblasts were grown in Eagle's MEM with 10 % calf serum for 5 days. On the day prior to the experiment, fresh medium was placed on the cell cultures. On the day of the experiment cultures were placed in 5 ml of fresh MEM, containing either 10 % or 1 % serum (SD) and 0,2 μCi/ml of ^{3}H-L-leucine. Cycloheximide or vinblastine were added at the same time, as described for each experiment. At 0.5, 1,2, and 4 hours flasks were processed by washing the cells twice with cold phosphate-buffered saline and the cell proteins precipitated by adding 10 ml of cold 5 % TCA. Cells were removed by scraping, the cell protein was centrifuged, and an aliquot of clear supernatant taken for measurement of the cell acid-soluble radioactivity. The cell protein was washed once with cold 5 % TCA, solubilized with 1 ml of 0.1 N NaOH; aliquots were taken for protein analysis and acid-insoluble radioactivity.

RESULTS

Effects of vinblastine on proteolysis

In fibroblast cultures the degradation rate of cell proteins was a function of the serum concentration in the incubation medium (Figure 1). In cells maintained in MEM with 10 % serum, the rate of cell protein degradation averaged approximately 1 % per hour during a four hour incubation period. In cells reincubated with serum-deficient medium, the average rate of protein degradation varied between 3 and 4 % per hour during the four hour experimental period. Vinblastine (10^{-5} M) had no detectable effect upon the basal rate of protein degradation but decreased

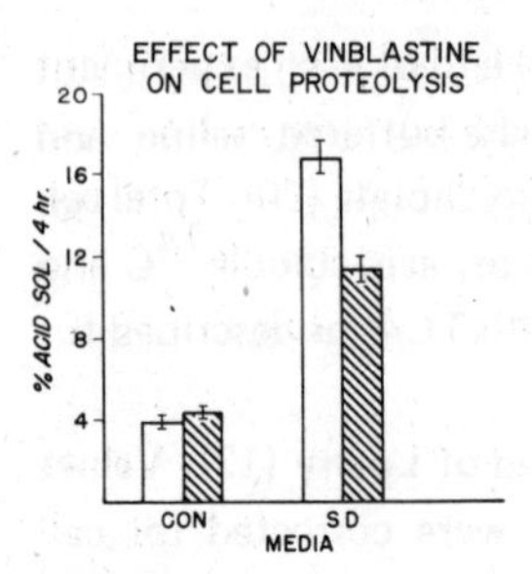

Figure 1 — TCA-soluble ^{14}C in media after 4 hour incubation. Control medium contains 10 % calf serum; SD medium contains no calf serum. Hatched bars, media with $10^{-5}M$ vinblastine. Each bar represents mean of 8 values; standard errors indicated in brackets.

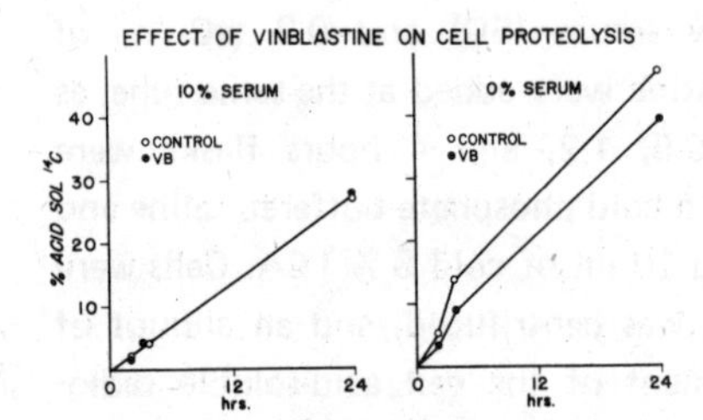

Figure 2 — TCA-soluble ^{14}C in media plotted against time of incubation in media containing serum as indicated. Vinblastine added at a concentration of $10^{-5}M$. Each point represents the average of at least four experiments.

---

the induced rate of degradation by approximately 50 %. Observed with phase contrast microscopy some of the fibroblasts in SD medium retracted into stellate and then rounded forms. Although this morphologic phenomenon was partially inhibited by vinblastine, we could detect no distinguishing qualitative effects of vinblastine upon these cells in SD medium.

The effects of vinblastine upon cell protein degradation were studied for a 24 hour period (Figure 2). Consistent with the results of the initial four hour study, we found no difference in the degradation rate of cell proteins when vinblastine was added to medium containing 10 % serum. Cells placed in SD medium showed an initial stimulation of cell protein degradation during the first four hour period as previously noted, but thereafter resumed a more normal rate of cell protein degradation of approximately 1.5 % per hour. The addition of vinblastine to SD medium inhibited this initial burst of cell proteolysis, but had very little effect upon the rate of proteolysis beyond the initial four hour period. Thus, placing cells in a serum-deficient medium apparently induces an acute proteolytic response (a wave of proteolysis) associated with a retraction and rounding of the fibroblasts. Vinblastine partially inhibits both the morphologic alterations and the acute proteolytic response.

The question arose: is this wave of proteolysis comparable to the induced proteolysis observed by other workers using rat hepatoma cells (8), fibroblasts (20) and liver (16) exposed to serum-deficient media. In a second series of experiments (Table I), flouride at 10^{-2} M was found to be a general inhibitor of cell protein degradation, reducing the basal degradation rate by 56 % and the SD-induced rate by 69 %. Chloroquine also inhibited both basal degradation, 22 %, and SD-induced degradation, 43 %, data consistent with observations of Poole and Wibo (20).

Cycloheximide, in contrast, inhibited only the induced proteolysis by 43 % and had no significant effect upon basal proteolysis, again in accord with other reported studies (7), (26). Insulin decreased the SD-induced

TABLE I

Effects of Metabolic Inhibitors Upon Degradation Rate of Cell Proteins in Fibroblasts

Inhibitor	% Inhibition** in Basal Medium	SD Medium
Na F (10^{-2}M)	56	69
Chloroquine (10^{-5} M)	22	43
Cycloheximide (4×10^{-4}M)	10*	43
Insulin (100 mU/ml)	5*	36
Vinblastine (10^{-5} M)	2*	38

* Differences not significant from control.

** Expressed as % inhibition of appropriate control without added inhibitor in basal or SD media. Concentration given in parentheses.

proteolysis by 36 %, but was without effect on the basal proteolysis, again in accord with reported literature (16). From these results, it appeared that the SD-induced proteolysis in these fibroblasts was subject to the same metabolic inhibitors noted by other workers in other cell systems and appeared to be the same phenomenon.

Effects of other microtubular inhibitors

Other inhibitors of microtubules had the same effect as vinblastine upon the degradation of cell proteins (Figure 3). Like vinblastine, vincristine and colchicine had no effect upon basal cell protein degradation but inhibited SD-induced proteolysis. Cytochalasin B, which affects microfibrils, showed a markedly different response, stimulating basal proteolysis slightly and inhibiting the induced proteolysis slightly, although in neither case were these effects found to be significant. Thus, microtubular inhibitors as a group inhibited only the SD-induced proteolysis, suggesting that the induced proteolysis requires the functional integrity of the microtubular system. The different response to cytochalasin B clearly distinguishes between the role of microtubules and microfibrils in this SD-induced proteolysis.

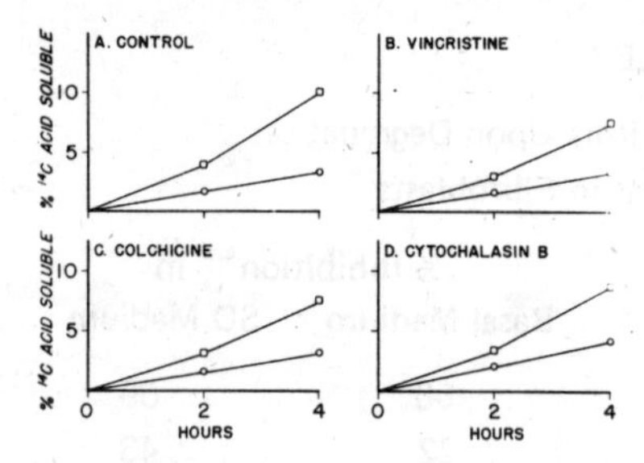

Figure 3 — Effect of vincristine, colchicine, and cytochalasin B on degradation of cell protein. Cells incubated in media containing (A) no inhibitor, (B) vincristine 10^{-5}M, (C) colchicine 2.5×10^{-5}M, and (D) cytochalasin B 1.2µg/ml. Acid-soluble ^{14}C measured in media at indicated times. Each point represents the average of four experiments. Circles are data from cells incubated in media containing 10 % serum; squares, stepdown media containing no serum.

Effects of combined inhibitors

A further experiment was performed to test the hypothesis that the inhibitory effects of cycloheximide and vinblastine were upon separate SD-induced proteolytic mechanisms. Under such circumstances, the combined effects of the inhibitors should be additive, reducing the SD-induced proteolysis rate almost to the level observed with basal medium. As shown in Figure 4, combining cycloheximide and vinblastine achieved an inhibition of the induced proteolysis no greater than that achieved with cycloheximide alone. As previously noted for cycloheximide and vinblastine individually, the combined agents had no effect upon the basal rate of proteolysis during this four hour experimental period. The results obtained with these inhibitors favor the hypothesis that the induced proteolytic mechanism is a single biochemical pathway, at least in part distinct from the basal mechanism of cell protein turnover, and that these two inhibitors both affect this induced pathway.

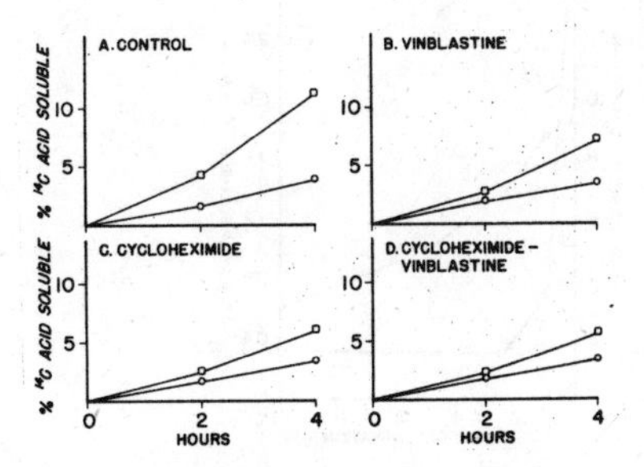

Figure 4 — Effect of vinblastine-cycloheximide on degradation of cell protein. Cells incubated in media containing (A) no inhibitor, (B) vinblastine 10^{-5}M (C) cycloheximide 4×10^{-4}M, and (D) both cycloheximide 4×10^{-4}M and vinblastine 10^{-5}M. Acid-soluble ^{14}C in media measured at indicated times. Each point represents the average of four experiments. Circles and squares as in Figure 3.

Effect of varying serum concentration on proteolysis

Morphologic observations and incomplete recoveries of ^{3}H-DNA in the cell layer indicated that some cells, when placed in medium containing no serum, tended to float free in the medium during the four hour experimental period. In addition, up to 3 % of the ^{14}C could be recovered in the acid-insoluble fraction of the media in some experiments, suggesting the possibility that some of the SD-induced proteolysis was in fact due to cell losses into the media, with subsequent destruction of the cell structure and hydrolysis of the proteins in the media. To test this possibility, we measured both TCA-soluble and TCA-insoluble ^{14}C in the media while progressively reducing the concentration of serum in the media (Figure 5). The SD-induced hydrolysis of cell proteins progressively increased as the serum concentration was reduced, while the acid-insoluble ^{14}C remained at negligible levels until a serum concentration of less than 1 % was reached. It is apparent from these data that the loss of cell protein into the medium, effected by lack of serum in the medium, is a different phenomenon from the increased degradation of cell protein induced by progressively lower serum concentrations. To check the possibility that cell proteins, released

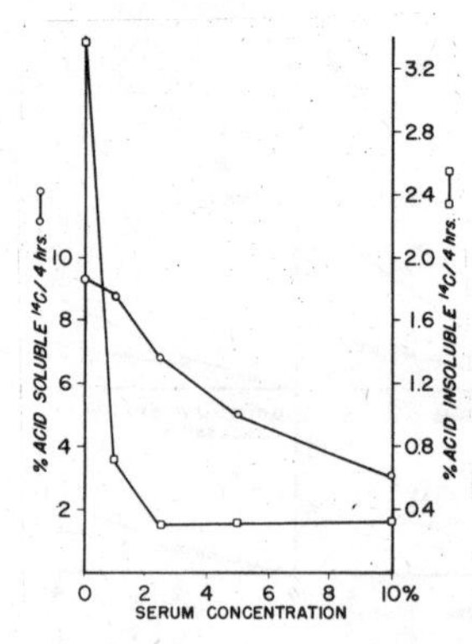

Figure 5 — Effect of varying serum concentration on hydrolysis of cell proteins and loss of intact cell proteins in media. Fibroblasts incubated for 4 hours in MEM containing 0-10 % serum concentrations as indicated. TCA-soluble ^{14}C (circles) and TCA-insoluble ^{14}C (squares) measured after 4 hours. Each point represents an average of three values.

from injured cells into the SD medium, were being degraded by hydrolases also released into this medium, we added ^{125}I-albumin to cultures incubated with both basal MEM and SD media. Release of acid-soluble ^{125}I in both groups averaged less than 0.1 %/ four hours, suggesting that the hydrolysis of proteins in the media was not a significant factor in the observed SD proteolysis.

Protein synthesis in SD media

Our experiments demonstrating an inhibition of the SD-induced proteolysis by cycloheximide suggested that protein synthesis is a necessary condition for this induced proteolytic mechanism. We considered the possibility that serum-deprived fibroblasts may not be synthesizing proteins (1) and that the cycloheximide effect was other than upon protein synthesis. Experiments were performed to measure the ability of these cells to incorporate ^{3}H-L-leucine into cell protein when incubated with SD media containing various inhibitors (Figure 6). We could detect only slight

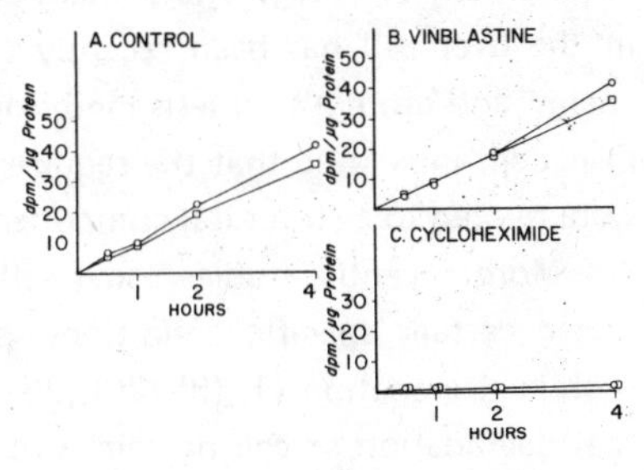

Figure 6 — Incorporation of ^{3}H-L-leucine into fibroblasts. Fibroblast cultures placed in either fresh basal media containing 10 % serum (circles) or transferred to SD media containing 1 % serum (squares). At t_o, ^{3}H-L-leucine 0.1 μC/ml was added to each flask along with (A) no inhibitor, (B) vinblastine 10^{-5}M, or (C) cycloheximide, 4×10^{-4}M. At indicated times cells in flasks were processed for total protein and total TCA-insoluble ^{14}C. Each point represents the average of two experiments.

differences (less than 10 %) in ^{3}H-L-leucine incorporation into cell protein in cells incubated in basal and SD media (Figure 6A). Vinblastine had no effect upon protein synthesis under either basal or SD conditions (Figure 6B), whereas cycloheximide, as expected, almost totally inhibited protein synthesis (Figure 6C). This data was consistent with the hypothesis that protein synthesis is necessary for SD-induced proteolysis and the effect of vinblastine is upon a separate step in this same biochemical pathway.

DISCUSSION

The mechanism by which cell proteins are degraded, although still unresolved, is clearly more complex than a simple hydrolysis of the peptide bonds. Simpson(22) first demonstrated that without a source of energy, the liver cell ceased hydrolysis of prelabeled cell proteins into amino acids, suggesting that protein turnover in the liver cell was an organized and finely regulated process. Consistent with this concept was the demonstration by Steinberg and Vaughan (23) that protein synthesis and protein degradation were somehow coupled together in the liver cell: inhibition of protein synthesis with amino acid analogues inhibited protein degradation. This requirement of energy and protein synthesis for general protein degradation in the liver cell has been recently extended by Brostrom and Jeffay (2): energy and protein synthesis inhibitors were most effective against the intact liver cell, indicating that the requirements for energy and protein synthesis were related to a structural component of the degradative process.

More recently studies using cell cultures have indicated that, at least under certain specific conditions, protein synthesis is not necessary for protein degradation (1),(8),(20),(26). And obvious alternative hypothesis is that degradation of cell protein involves more than one mechanism, at least one of which requires protein synthesis and at least one which does not. Evidence consistent with this dual mechanism hypothesis has been obtained by Tomkins and co-workers:(1) hepatoma cells show a marked increase in cell proteolysis when transferred from a medium containing 10 % serum to a medium containing 0 % serum. Significantly, cycloheximide and puromycin inhibited only the stepdown-induced proteolysis, while protein turnover in cells in complete medium containing 10 % serum and amino acids was not affected by these inhibitors. Additional data consistent with this hypothesis has also been obtained by Mortimer and co-workers, using a liver perfusion system (15),(16),(26). A significant increase in the rate of cell proteolysis occurs within 30 minutes after initiating perfusion, a phenomenon which could be partially prevented by adding either amino acids or insulin to the medium, suggesting that the original perfusate acted as a stepdown medium for the liver cells. Also, in accord with Tompkins'data on the hepatoma cell, a partial inhibition of this induced proteolysis in the perfused liver was effected when cycloheximide was added to the perfusate (26). These workers have also shown that the SD-induced proteolysis is associated with alterations in the lysosomes of the liver cell and suggest that the induced proteolysis involves an activation of the lysosomal system (17). It is highly probable that this augmented proteolysis in the liver perfusion system, as well as the stepdown-induced proteolysis observed in the hepatoma cell, is the same effect we have observed in our fibroblasts transferred to media containing 0 % or 1 % serum. Our data, showing that cycloheximide and insulin inhibit only the SD-induced proteolysis and are virtually without effect on the basal turnover rate of cell protein, are consonant with these data obtained with hepatoma cells and liver perfusion.

Poole and Wibo (20), using rat embryo fibroblasts in cell cultures, have demonstrated an induced proteolysis in cells placed in „conditioned medium", i.e. medium previously exhausted of certain specific factors by preincubation with another cell culture. Although the medium in this instance cannot be as well defined as in the instances cited above, it is apparent that placing the fibroblasts in this medium does induce a stepdown proteolytic mechanism, which very probably is the same mechanism we and others have observed in serum or insulin-deficient media. Using NaF, Poole and Wibo(20) have demonstated an energy requirement for all protein degradation and suggest a direct action of fluoride on proteolytic enzymes. We have been able to confirm this inhibitory effect of fluoride on all cellular protein degradation, both basal and induced, but have no further data to indicate whether this is secondary to a decrease in cellular ATP or due to a direct inhibition of lysosomal enzymes. Also consistent with the previously cited data with protein synthesis inhibitors is their observation that puromycin does not inhibit basal protein turnover in cultured fibroblasts. It would be of obvious interest to determine if protein synthesis inhibitors are effective in inhibiting proteolysis induced in cells exposed to conditioned media. Our observations that cloroquine inhibits both basal and SD-induced proteolysis corroborates the original observations of Poole and Wibo with this agent. Adding another dimension to this problem are their studies on proteins with fast and slow turnover rates: only the slow-turnover proteins are sensitive to an SD-induced augmentation of proteolysis, suggesting that these proteins may have steps in their normal degradative mechanism distinct from the pathway for degradation of fast-turnover proteins. In may be that the SD-induced mechanism is an acceleration of those degradative steps which are distinctive for slow-turnover proteins. We also have not been able to establish any SD-augmentation of proteolysis when measuring degradation of fast-turnover proteins (unpublished data). It would appear from all these studies that cells transferred to deficient media are able to activate a proteolytic mechanism which normally may be functioning at a relatively slow rate; this activated mechanism then becomes superimposed upon basal degradation of cell protein. Our findings that inhibitors of microtubular function inhibt only SD-induced proteolysis support the dual mechanism hypothesis. The failure of cycloheximide and vinblastine to produce any additive effect argues for a single rather than multiple SD-induced mechanisms.

Our data (Figure 2) demonstrates that this induced proteolysis is not a long term augmentation of the proteolytic rate, but rather a wave of proteolysis, probably involving a sudden burst of lysosomal activity, as suggested by Mortimer and co-workers (26). Since the roles of lysosomes and microtubules in the degradation of exogenous protein are well documented (4),(13),(18), and in view of the suggested role of lysosomes in SD-induced degradation of cellular proteins,(26) the inhibitory effects of microtubular inhibitors in SD proteolysis should not be surprising. We interpret these findings with microtubular inhibition as additional evidence in favor of an induced lysosomal mechanism for the SD-induced proteolysis. A recent study (19) by Pfeifer and Scheller, using quantitative morphology on rat kidney tubules supports the concept that autophagy is a low-level, but significant, factor in the degradation of cell proteins. An obvious possibility is that SD-induced proteolysis accelerates this mechanism in the cell, although we have as yet no direct evidence for this. Another possible mechanism, which would involve increased activity of the lysosomal system, has been suggested by the studies of Buckley (3): in damaged cells in culture lysosomes increase in size and show highly active processes extending and retracting into the cytoplasm. This phenomenon, though different from our usual concepts involving fusion of primary lysosomes with autophagic granules, could very well require microtubules for the increased lysosomal activity. This is not to suggest we exclude the possibility that lysosomes are involved in basal protein degradation in the cell, but rather that such a mechanism, if it exists, is at least in part different from the SD-induced mechanism. The distinctiveness of the vacuolar mechanism of the SD-induced proteolysis is strongly suggested by the specific inhibitory effects of microtubular inhibitors and inhibitors of protein synthesis.

We have considered the possibility that protein synthesis and microtubular function may be required for general degradation in cell protein, both basal and induced, and that it is only when general cell protein degradation is accelerated that these requirements become manifest. In this model a single pathway of protein degradation contains a preformed pool of protein necessary for general protein degradation and only after this pool is exhausted does the requirement for protein synthesis become evident. However, preincubation of our cell system with cycloheximide or vinblastine does not increase the effectiveness of the inhibition for either agent, as would be expected were a single exhaustible protein pool involved in single common degradative pathway.

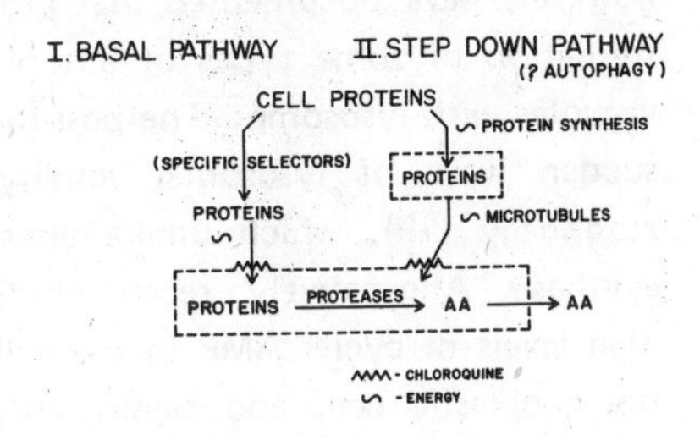

Figure 7 — Dual mechanism model for degradation of cell proteins. Involvement of vacuolar system indicated by dashed rectangles. Basal pathway (I) for protein turnover indicated on left; SD-induced mechanism on right (II). This particular model arbitrarily shows the final degradative step for both pathways to involve the lysosomes (20) and the inhibitory effect of chloroquine at the level of the lysosomal membrane.

At the present time, we favor the hypothesis that cell proteins are degraded by at least two degradative mechanisms in the cell (Figure 7). A basal mechanism, which obviously involves specific selective factors within the cytoplasm to account for the great diversity in turnover rates of proteins within the cells, appears to require energy and can also be inhibited by chloroquine. The final degradation of these proteins may occur in lysosomes as suggested by Poole and Wibo (20), but the evidence for this final step remains somewhat tentative at this time. More definitive is the evidence for a second pathway which can be induced by placing the cells in contact with media deficient in certain substances. For this second pathway, apparently protein synthesis and microtubular integrity are necessary for optimum functioning, with the data at the present time strongly suggesting involvement of the vacuolar system. Energy is also required for the functioning of this induced pathway, and like the basal pathway, chloroquinine is also inhibitory. The mechanism of chloroquine inhibition may be via inhibition of proteolytic enzymes(25) or possibly through some general alteration of cell or lysosomal membranes involved in the degradative processes (5). Whether this SD-induced pathway involves activation of autophagy, with fusion of autophagic vacuolar and primary lysosmes, or involves merely a burst of „amoeboid" activity of enlarging lysosomes cannot yet be ascertained.

Trump and co-workers (21), and Mortimore and co-workers (26) however, have documented that protein synthesis is not required for the formation of some types of autophagic vacuoles or the fusion of these vacuoles with lysosomes. The possibility remains that we are dealing with a sudden wave of lysosomal activity different from the usual form of autophagy (19), which unlike classical autophagy, may require protein synthesis. Alternatively, recent studies by Tomkins(10) have shown that high levels of cyclic AMP in the cell stimulate protein degradation in the non-neoplastic cell, and significantly, that cycloheximide reduces cyclic AMP levels in these cells along with protein degradation. Cyclic AMP has been shown to affect the functioning of the microtubule in the degradation of exogenous protein (4), suggesting that vinblastine directly and cyclohexi-mide indirectly may be inhibiting microtubular functioning. It is conceivable, therefore, that these inhibitors of protein synthesis are effective by altering the levels of intracellular cyclic AMP rather than by inhibiting a protein-requiring step in the degradative pathway. Of interest in this case would be to know what effect a sudden elevation in cyclic AMP has upon lysosomal and/or microtubular activity. Further studies are obviously necessary to determine the role of protein synthesis, if any, and of cyclic AMP in SD-induced proteolysis.

REFERENCES

1. Auricchio, F., Martin, D. Jr., Tomkins, G., Nat. 224,806(1969).
2. Brostrom, C. O. and Jeffay, H., J. Cell Biol. 245,4001 (1970).
3. Buckley, I., Lab. Invest. 29,411 (1973).
4. Ekholm, R., Eriscon, L.E., Josefsson, J. – O., and Melander, A., Endocrinology 94,641 (1974).
5. Federko, M.E., Hirsch, J.G., Cohn, Z.A., J. Cell Biol. 38,377 (1968); and J. Cell Biol. 38,392 (1968).
6. Gelehrter, Thomas D., and Emanuel, Janet R., Endocrinology 94,576 (1974).
7. Hershko, A., Tomkins, G.M., J. Biol. Chem. 246,710 (1971).
8. Jerrell, K.J. and Seglin, P.O., Biochim. Biophys. Acta 174,398 (1969).
9. Kenny, F.T., Sc. 156,525 (1967).
10. Kram, R., Mamont, P., and Tomkins, G.M. Proc. Nat. Acad. Sc. (USA) 70,1432 (1973).
11. Levitan, I.B. and Webb, J.E., J. Biol. Chem. 244,4684 (1969).
12. Lowry, D.H., Rosebrough, N.J., Farr, A.L., and Randall, R.J., J. Biol. Chem. 193,265 (1951).
13. Malawista, S.E., Blood 37,519 (1971).
14. McIlhinney, A. and Hogan, B.L.M., FEBS Letters 40,297 (1974).
15. Mortimore, G.E. and Mondin, C.E., J. Biol. Chem. 245,2375 (1970).
16. Mortimore, G.E., Neely, A.N., Cox, J.R., and Guinevan, R.A., Biochem. Biophys. Res. Com. 54,89 (1973).
17. Neely, A.N. and Mortimore, G.E., Fed. Proc. 32,299 (1973).
18. Neve, P., Ketelbant-Balasse, P., Willems, C., and Dumont, J.E., Exp. Cell Res. 74,227 (1972).
19. Pfeifer, U. and Scheller, H., J. Cell Biol. 64,608 (1975).
20. Poole, B. and Wibo, M., J. Biol. Chem. 248,6221 (1973).
21. Shelburne, J.D., Arstilla, A.U., and Trump, B.F., Am. J. Path. 73,641 (1973).
22. Simpson, M.V., J. Biol. Chem. 201,143 (1953).

23. Steinberg, D. and Vaughan, M., Arch. Biochem. Biophys. 65,93 (1956).
24. Stossel, T.P., New Eng. J. Med. 290,774 (1974).
25. Wibo, M., Poole, B., J. Cell Bioi. 63,430 (1974).
26. Woodside, K.H., Ward, W.F. and Mortimore, G.E., J. Biol. Chem. 249,5458 (1974).
27. Yagel, G. and Feldman, M., Exp. Cell Res. 54,29 (1969).

INCREASED DEGRADATION RATES OF CANAVANINE–CONTAINING PROTEINS IN HEPATOMA CELLS

F.J. Ballard and S.E. Knowles.

Division of Human Nutrition, Commonwealth Scientific and Industrial Research Organization, Kintore Avenue, Adelaide, South Australia 5000, Australia

SUMMARY

Reuber H35 hepatoma cells incorporate the arginine-analogue, canavanine into cell protein when arginine is omitted from the culture medium. Sequential labelling of arginine-containing proteins with ^{14}C-leucine and canavanine-containing proteins vith ^{3}H-leucine followed by a chase period has shown an accelerated degradation rate for the canavanine proteins. Further, the turnover rate of phosphoenolpyruvate carboxykinase (GTP) is also increased in cells incubated in the presence of canavanine. Although several proteolytic inhibitors and metabolic poisons inhibit the degradation of both normal and canavanine-containing proteins, a selective effect on the breakdown of normal proteins has been shown for insulin, puromycin, foetal calf serum and the trypsin inhibitor, tosyllysylchloromethoketone. We interpret these results as evidence for the existence of a separately regulated proteolytic system for the removal of aberrant proteins.

Abbrevitions used: PEPCK, phosphoenolpyruvate carboxykinase (GTP) (EC 4.1.1.32); PMSF, phenylmethansulphonyl fluoride; TAME, p-toluenesulphonyl-L-arginine methyl ester; TI, trypsin inhibitor; TLCK, N-tosyl-L-lysine chloromethyl ketone; TPCK, N-tosyl-L-phenylalanine chloromethyl ketone.

Alteration of the amino acid sequence of a protein can be produced by the replacement of certain amino acid analogues for the natural amino acid in a system where proteins are being syntesized. Proteins in which canavanine replaces arginine, or 6-fluorotryptophan replaces tryptophan have been produced in mammalian cells (1,2) while further substitutions have been introduced in various E. coli auxotrophs (3-5). Of particular interest is the finding that these altered or aberrant proteins are unstable to proteolytic attack **in vivo** or **in vitro** (1-6), since it implies that a few changes in the primary sequence of a protein result in alterations in overall structure that can be distinguished by a proteolytic system. In this report we describe experiments in which we have tested the action of several proteolytic inhibitors to determine whether the breakdown of normal and aberrant proteins are regulated in different ways.

RESULTS AND DISCUSSION

The degradation of normal, arginine-containing proteins that have been labelled with ^{14}C-leucine proceeds with a half life of approximately 15 h (Fig. 1a). However, canavanine-containing proteins in the same culture dishes are much less stable, so that 30 % of the ^{3}H-leucine is released to the medium in 2 h. These results and our previous experiments (1,2) show that the replacement of arginine by canavanine in proteins of Reuber H35 cells increase the degradation rate of the proteins by 3-5 fold. Although amino acid analysis of total proteins in cells incubated with canavanine shows only a 5 % replacement of arginine by canavanine (1), we consider that several arginine residues in each protein must have canavanine substituted in order for the protein to be degraded at an accelerated rate. This conclusion is based on experiments with ^{14}C-canavanine where the incorporation of label does not produce unstable proteins unless large amounts of unlabelled canavanine are included in the culture medium (F.J. Ballard & S. E. Knowles, unpublished experiments).

Since ^{14}C-canavanine is unsuitable for experiments on the regulation of aberrant protein degradation, we have used the methodology described in Fig. 1 and measured the effects of various agents on the breakdown of arginyl-protein and canavanyl-protein. The data have been expressed as the per cent inhibition of degradation over the first 2 h after removal of

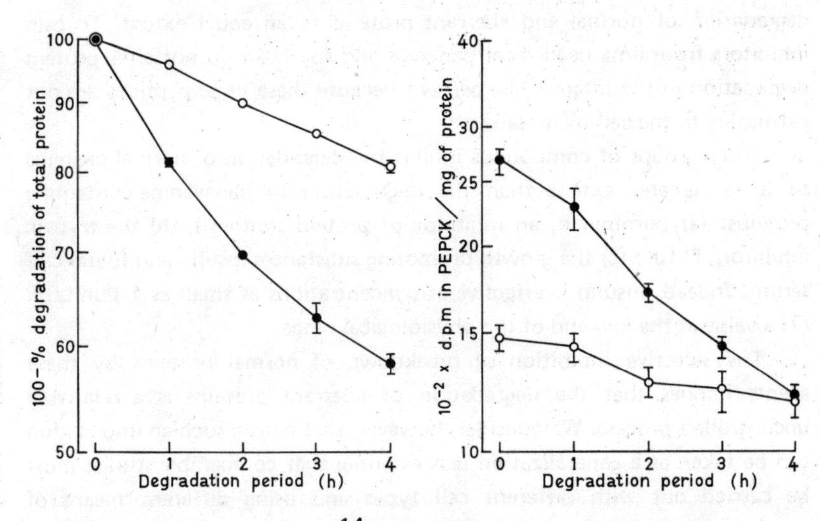

Fig. 1a Degradation of (^{14}C) leucine-labelled arginine- containing proteins (○) and (^{3}H) leucine-labelled canavanine-containing proteins (●) in Reuber H35 cells.

Fig. 1b Degradation of (^{14}C) leucine-labelled PEPCK (○) (^{3}H) leucine-labelled PEPCK (●) in the same cells. Details of the methods used for growing the cells and performing the experiments are given in previous publications (1,2). Briefly, dibutyrylcyclic AMP, theophylline and dexamethasone were added to confluent cultures in enriched Eagle's minimal essential medium together with (^{14}C) leucine, and the cells left for 15 h. This results in the induction of PEPCK and the labelling of normal proteins with (^{14}C) leucine. At the end of the incubation period the medium was replaced with a similar medium but with arginine replaced by canavanine and (^{14}C)-leucine replaced by (^{3}H)-leucine After incubation for 3 h to label canavanine-containing proteins, the medium was removed, the cells washed three times with Dulbecco salts solution and the degradation followed in Eagle's minimal essential medium with 2 mM leucine added as a chase. At the appropriate time cells were harvested, broken and the radioactivity in proteins, amino acids and PEPCK measured. Breakdown of total proteins is plotted as per cent radioactivity remaining in the proteins, while PEPCK radioactivity is expressed as total dpm per mg of protein. Values are means ± SEM three determinations at each time period.

^3H-leucine and canavanine from the culture medium (Table 1). It can be seen that metabolic poisons such as dinitrophenol, NaF or azide inhibit the degradation of normal and aberrant proteins to an equal extent. Trypsin inhibitors from lima bean, beef pancreas and soy bean do not alter protein degradation in the intact cells, perhaps because these large peptides are not permeable to the cell membrane.

Three groups of compounds inhibit the degradation of normal proteins to a far greater extent than the dègradation of canavanine-containing proteins: (a) puromycin, an inhibitor of protein synthesis; (b) the trypsin inhibitor, TLCK; (c) the growth promoting substances insulin and foetal calf serum. Indeed, insulin is effective at concentrations as small as 1 μunit/ml (7) a value at the low end of the physiological range.

The selective inhibition of breakdown of normal proteins by these agents implies that the degradation of aberrant proteins is a relatively uncontrolled process. We consider, however, that before such an implication can be taken as a generalization it is essential that comparable studies must be carried out with different cell types and using different means of producing error proteins. Although the present experiments are the first done with mammalian cells, research with E. coli has shown comparable differences in the degradation response of normal and error proteins towards both nutritional alterations (4) and proteolytic inhibitors (5). Moreover, a mutant E. coli strain has been isolated in which the rapid system for degrading aberrant proteins is absent (8). If the rapid, unregulated degradation of canavanine-containing proteins represents a special degradative process that occurs in E. coli and mammalian cells, then the retention of this process during evolution suggests that cells which have the ability to remove mistakes in protein synthesis and post-translationally modified proteins have a selective advantage in coping with unfavourable conditions.

In addition to measurements of general protein degradation we have also followed the turnover of an individual enzyme, PEPCK, in the Reuber cells. The degradation rate constant for canavanyl-PEPCK is 3 to 4 times as large as the rate constant for arginyl – PEPCK (Fig. 1b, Table 2). However, there is no consistent effect of insulin or foetal calf serum on the degradation of either arginyl – PEPCK or the canavanine-containing enzyme. Whether this result reflects a qualitative difference between regulatory effects on the degradation of different proteins must await further experiments.

TABLE 1

Inhibitor effects on the degradation of arginine-containing protein and canavanine-containing protein.

Agent tested	% inhibition of the degradation of arginyl-protein	canavanyl-protein
Amino acid mixture (6)	5 ± 4	1 ± 3
10 mM EDTA (4)	6 ± 4	3 ± 8
2 mM NaF (6)	19 ± 3	18 ± 1
0.1 mM dinitrophenol (4)	39 ± 7	33 ± 6
1 mM NaN_3 (4)	19 ± 3	14 ± 5
0.1 mM cycloheximide (5)	12 ± 11	7 ± 4
0.05 mM puromycin (5)	32 ± 5	8 ± 3*
1 mM PMSF (4)	−5 ± 11	−5 ± 4
1 mM TAME (4)	10 ± 3	9 ± 3
0.5 mg/ml lima bean TI (4)	−7 ± 5	−4 ± 8
0.5 mg/ml soy bean TI (4)	7 ± 2	5 ± 4
0.1 mg/ml beef pancreas TI (4)	−4 ± 3	−12 ± 4
0.1 mM TLCK (7)	30 ± 4	13 ± 2*
0.5 mM TLCK (4)	51 ± 9	21 ± 3*
0.05 mM TPCK (6)	17 ± 5	16 ± 3
0.1 mM TPCK (4)	55 ± 3	45 ± 8
10 % foetal calf serum (5)	30 ± 1	5 ± 3*
1 m unit/ml insulin (11)	26 ± 2	0 ± 3*

Cells were labelled as indicated in the legend to Fig. 1 and the various agents added for 2 h during the chase period. Inhibition of degradation is expressed as the mean percentage effect ± SEM over this 2 h period with the number of incubations noted in parentheses.

---

* Conditions where the inhibitory effect on degradation of canavanyl-protein is significantly less ($P > 0.01$) than for arginyl-protein.

TABLE 2

Degradation rate constants for PEPCK labelled in the presence of canavanine or arginine.

	Degradation rate constant for	
	arginyl-PEPCK	canavanyl-PEPCK
Control	0.055 + 0.011	0.206 + 0.012
Insulin, 1 m unit/ml	0.025 + 0.013	0.178 + 0.014
Control	0.076 + 0.024	0.260 + 0.017
Foetal calf serum, 10 %	0.113 + 0.020	0.266 + 0.15

Cells were labelled and incubated as described in the legend to Fig. 1. Insulin and foetal calf serum were added during the chase period which was for various times from 1 h to 6 h. The degradation rate constant is given as the regression co-efficient ± SE for the natural logs of the PEPCK specific radioactivities (dpm/mg protein).

REFERENCES

1. Knowles, S.E., Gunn, J.M., Hanson, R.W., and Ballard, F.J., Biochem. J. 146, 595-600 (1975).
2. Knowles, S.E., Gunn, J.M., Reshef, L., Hanson, R.W., and Ballard, F.J., Biochem. J. 146, 585-593 (1975).
3. Pine, M.J., J. Bacteriol. 93, 1527-1533 (1967).
4. Goldberg, A.L., Proc. Nat. Acad. Sci. USA 69, 422-426 (1972)
5. Prouty, W.F., and Goldberg, A.L., J. Biol. Chem. 247, 3341-3352 (1972).
6. Goldberg, A.L., Proc. Nat. Acad. Sci. USA 69, 2640-2644 (1972).
7. Ballard, F.J., In Report of the Ross Conference on the Consequences of Overnutrition, in press (1975).
8. Bukhari, A.I., and Zipser, D. Nature (London) New Biol. 243, 238-241 (1973).

STUDIES OF THE DEGRADATION OF PROTEINS IN ANIMAL AND BACTERIAL CELLS

Alfred L. Goldberg

Department of Physiology Harvard Medical School Boston, Massachusetts 02115

SUMMARY

Proteins within animal and bacterial cells vary widely in their rates of degradation, and these differences appear to reflect differences in protein conformations. For example, half-lives of intracellular proteins as well as serum appear to correlated with their inherent sensitivity to various proteolytic enzymes. We have also found that the half-lives of proteins are related to their net charge. In various tissues and in rat serum, proteins with acidic isoelectric points are degraded more rapidly that neutral or basic ones. Isoelectric point and subunit molecular weight are two independent parameters that influence degradative rates of proteins.

Protein breakdown in animal and bacterial cells also helps to protect the cell against accumulation of abnormal polypeptides. **E. coli** for example selectively degrade incomplete proteins that result from nonsense mutations or proteins containing amino acid analogs. Similarly mammalian reticulocytes rapidly degrade abnormal globins that contain valine or lysine analogs. This degradative process in mammalian tissues generally resembles that in **E. coli.** For example in **E. coli** and reticulocytes the degradative process requires metabolic energy.

In an attempt to define the mechanisms of protein degradation, we have investigated the rapid degradation of **X–90**, a fragment of β–galactosidase resulting from a nonsense mutation. A smaller polypeptide that contains the original amino terminal sequence has been found that is an intermediate in **X–90** degradation. When energy metabolism was inhibited, the loss of **X–90** and of the smaller fragments was also inhibited. Thus energy is required not only for the initial cleavage but also for subsequent proteolytic steps. The initial cleavage in the degradative process seems to be due to an endoproteolytic cleavage. **Deg T^{-}** mutants defected in degradation of abnormal proteins have a reduced capacity to carry out this reaction.

Proteins within animal and bacterial cells vary widely in their rates of degradation (1–3). These differences in protein half-lives must have important implications for the control of cell metabolism (2,3). For example, in rat liver those proteins with short half-lives tend to be rate-limiting enzymes in metabolic pathways (2). This rapid degradation insures that the intracellular levels of such enzymes can fluctuate rapidly in response to environmental changes (3). Thus protein half–lives appear to have evolved in part to regulate the flux of substrates through metabolic pathways (3). Such considerations also imply that the variations in the half-lives of different proteins are determined by differences in protein structure. In fact there is now appreciable experimental evidence (1,2) that rates of degradation must be encoded in the primary sequence of the polypeptide along with information for its catalytic and allosteric functions. However, the exact structural features of proteins that influence their degradative rates are still not clear.

Influence of protein conformation on degradative rate

Previous studies from this laboratory and others in both bacterial and mammalian cells have demonstrated a general correlation between the inherent sensitivity of proteins to proteolytic enzymes **in vitro** and their half-lives **in vivo** (1,2). For example, abnormal proteins which are degraded especially rapidly by animal and bacterial cells (see below) appear to be particularly sensitive to all proteolytic enzymes tested (4). Experiments by our group (4) in **E. coli** and by Dice and Schimke (5) in several mammalian tissues have also shown that normal proteins with short half-lives are on the average more rapidly digested by neutral endoproteases than more stable cell proteins.

Similar correlations between proteolytic sensitivity and half-life have also been obtained by several research groups using lysosomal enzymes (6,7). Since such correlations can be demonstrated with proteases of very different specificity (e.g. trypsin, subtilisin, chymotrypsin, or lysosomal enzymes), these findings do not provide specific insights into the selectivity of the responsible degradative enzymes. Instead these results suggest that degradative rates are determined by general conformational features (e.g. tightness of protein folding) that are recognizable by all classes of endoproteases (1,2).

Such model experiments were originally carried out with intracellular proteins. We have recently demonstrated a similar correlation between **in vivo** half-life and proteolytic susceptibility for serum proteins both in the rat and mouse (Fig 1) (8). These experiments utilized the double-isotope approach of Arias et al. (9) and gave analogous results with a wide variety of proteolytic enzymes, including both neutral and acidic proteases (8). Thus the rates of catabolism of serum proteins are also related to their native conformations which must determine their susceptibility to proteolytic attack. In accord with this conclusion, these correlations were abolished when the serum proteins were treated with the ionic detergent sodium dodecyl sulfate (Fig. 1), prior to addition of the protease. This treatment destroyed normal protein conformations and greatly increased proteolytic susceptibility of all fractions.

Two other features of protein structure have been found to correlate with rates of degradation. Work primarily by Schimke and colleagues has indicated that in mammalian tissues, polypeptides of large subunit size tend to be degraded more rapidly than proteins of lower molecular weight (1,2). Since the degradative rates of many specific proteins were reported not to obey this correlation, we recently carried out a statistical analysis of the published half-lives of rat liver proteins to test whether such a relationship might be demonstrable without the use of SDS-gel electrophoresis (10). This statistical survey demonstrated a highly significant correlation ($p<0.01$) between degradative rates and subunit molecular weight (10), although numerous exceptions to this correlation ($r = 0.66$) were found. Thus subunit size must be only one of several structural parameters that can influence intracellular half-lives (3).

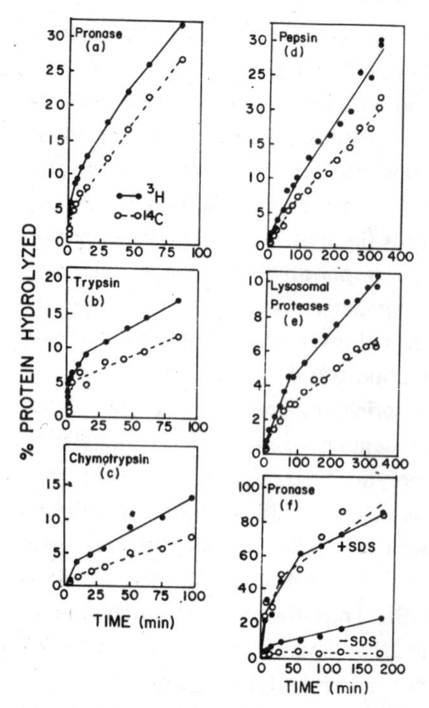

Fig. 1. Comparison of the susceptibility of stable and rapidly turning – over serum proteins to proteolytic enzymes. To label these fractions preferentially a male rat (125 g) was administered 100 μCi of 14C–leucine intraperitoneally. Four days later the same animal was injected with 500 μCi of 3H–leucine and killed 4 or 5 hours thereafter. Consequently, the 14C is found primarily in proteins with relatively long half lives, while 3H-proteins are enriched for more labile polypeptides. The serum was collected and was dialysed against phosphate-buffered saline (pH 7.2) (for experiments a-c and f) or 100 mM citrate buffer (pH 4). (for experiments d and e) to remove free amino acids. Each incubation mixture contained approximately 5 mg of serum protein and enzymes at the following concentrations: a, Pronase, 100 μg ml–l; b, trypsin, 100 μg ml–l; c, chymotrpysin, 1 mg ml–l; d, pepsin, 50 μg ml-l; e, crude lysosomal fraction 10 mg/ml; f Protase, 50 ug ml-l. Conformations of serum proteins were disrupted in (f) by addition of 1 % sodium dodecyl sulfate and 50 mM dithiothreitol and boiling for 5 min. (Pronase is at least partially active in SDS). Protein hydrolysis is expressed as radioactivity initially in protein that is converted to trichloroacetic acid-soluble form.

On account of these many exceptions to the size correlation, Dice and I investigated other structural features that might influence protein half-lives **in vivo** (11). We were able to demonstrate a marked correlation between the isoelectric point of proteins and their half-lives **in vivo** (Fig.2.) In these studies, rats or mice were injected initially with ^{14}C-leucine, and 4–10 days later with ^{3}H–leucine. After such a labelling procedure, a high $^{3}H/^{14}C$ ratio in a protein indicates a short half-life. The animals' tissues were homogenized and subjected to isoelectric focusing or ion exchange chromatography to separate proteins of different isoelectric points. These studies demonstrated that proteins with acidic isoelectric points tended to be degraded more rapidly than neutral or basic ones ($p<0.01$). Similar correlations were evident for soluble proteins in muscle, liver, heart, kidney, and brain of rats and mice. It is still unknown whether analogous relationships also hold for membrane or organelle-associated proteins in these tissues. No such influence of isoelectric point was demonstrated in control experiments, in which proteins received both isotopes at the same time (Fig. 2) In addition to these experimental studies, Dr Dice carried out a statistical survey of published literature on half-lives of rat liver proteins (11). This analysis for 23 proteins confirmed a strong correlation between isoelectric point and degradative rate ($\gamma = 0.824$, $p<0.01$).

Using analogous techniques, we have also demonstrated a very dramatic correlation between isoelectric point and degradative rates for serum proteins (Fig. 3). Traditionally it has been believed that extracellular proteins are degraded by very different mechanisms than intracellular enzymes and that the rate-limiting step in the degradation of extracellular proteins is their rate of pinocytosis. However, little precise information is available concerning the catabolism of such proteins. Our finding that the half-lives of serum proteins correlates with their isoelectric points and their inherent protease-sensitivities (8) raises the possibility that degradation of intracellular and extracellular proteins may involve common mechanisms.

Additional experiments have indicated that protein size and charge are independent factors influencing intracellular half-lives. This conclusion was based both on a statistical analysis of the published literature (i.e. by calculation of partial correlation coefficients) and experiments in which double-labeled liver proteins were subjected first to isoelectric focusing and then to SDS-gel electrophoresis. With both approaches, a correlation between protein size and rate of degradation was demonstrated for proteins of a given isoelectric point (i.e. size and charge must independently influence

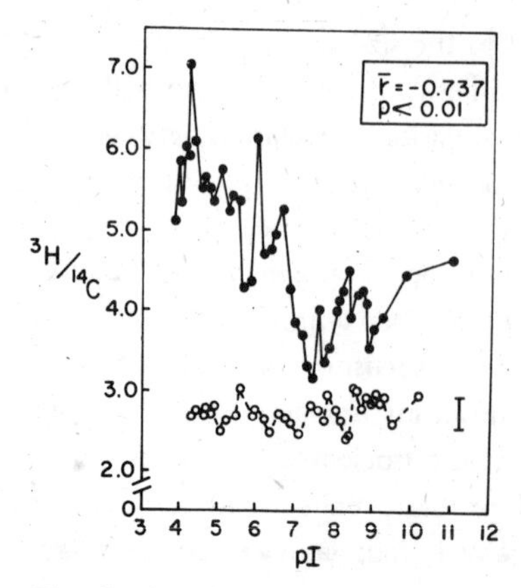

Fig. 2 Relative degradative rates of soluble proteins from rat liver with different isoelectric points separated by isoelectric focusing. The data represented by solid points are from an experiment in which a male rat received 100 μCi of 14 C-leucine 4 days prior to 500 μCi of 3H – leucine. Thus a high 3H/14C ratio indicates a rapid rate of degradation. The open points indicate the results of a control experiment in which the rat had been given 50 μCi of 14 C-leucine and 500 μCi of 3H-leucine 4 hr. before sacrifice.

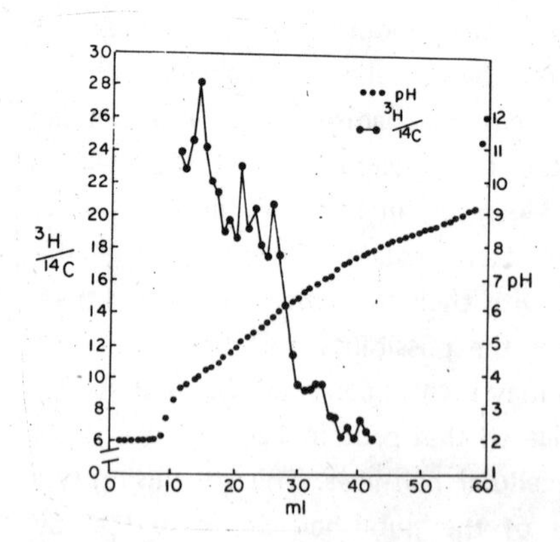

Fig. 3 Comparison of degradative rates of rat serum proteins with different isoelectric points. Double-labelled serum proteins were separated by isoelectric focusing as reported previously. A high 3H/14C ratio indicates more rapid turnover of the fraction. The correlation between 3H/14C and pH is highly significant (r = 0.957, P 0.02), pH•••; 3H/14C. •–•

degradative rates). In other words, if nature (or evolutionary pressures) would have favored a low degradative rate for a specific protein, it would have evolved to be both small and highly basic. It is of interest that the most stable polypeptides in mammalian tissues are the histones, which fit these criteria exactly.

The precise manner through which isoelectric point might influence protein half-lives is unknown. Possibly protein charge may influence degradative rates only indirectly. For example, acidic proteins may share certain structural properties (e.g. ease of denaturation) that lead to their rapid hydrolysis **in vivo** (see below). The proteins that are most rapidly degraded are the ones that are most negatively charged in environment of the lysosome. Exactly which environment is the site of the rate-limiting step in the degradative process remains to be established.

It is hoped that further understanding of the influence of size and charge on half-lives will provide valuable insights into the mechanisms and selectivity of the degradative process.

Degradation of abnormal proteins in E. coli and reticulcytes

The strongest evidence that protein conformations can influence intracellular half-lives comes from the study of proteins with abnormal tertiary structures. Both animal and bacterial cells have been found to degrade such polypeptides expecially rapidly (1,2, 12). Thus intracellular protein breakdown appears to serve as a sort of cellular sanitation system that helps prevent the accumulation of aberrant and potentially harmful polypeptides. The clearest evidence for this function of protein breakdown comes from studies in **E. coli.** A variety of evidence has demonstrated that such cells selectively hydrolyze incomplete proteins or may arise from mutations or defects in translation. For example, Goldschmidt (13) and Zabin and coworkers (14) showed that in **E. coli** nonsense fragments in β–galactosidase are degraded with half-lives as short as a few minutes, even though the wild-type enzyme appears completely stable (see below). Our own laboratory (15) and that of Pine (16) obtained similar conclusions by studying the fate of incomplete proteins induced with puromycin. This antibiotic is incorporated into growing polypeptides and leads to their premature release from the ribosome. Such puromycin-containing polypeptides are degraded far more rapidly than normal ones. Under the same conditions puromycin does not affect the catabolism of preexistent normal proteins. More recently a variety of mammalian cells, including reticulocytes, Hela cells, and rat hepatoma cell, have also been shown to hydrolyze selectively puromycin-containing polypeptides (3).

It is possible to induce bacteria or mammalian cells to make large amounts of complete polypeptides with abnormal conformations by exposing them to amino acid analogs (1,2). Many such compounds can be incorporated into proteins in place of the normal amino acids, although the resulting proteins tend to be rapidly degraded within cells (12, 14, 15). For example, bacteria auxotrophic for arginine can grow for several generations if provided with the analog, canavanine. However, canavanine-containing proteins are degraded up to 20 times more rapidly than those containing arginine (Fig. 4). The degradation of the aberrant proteins is a highly selective process, since normal arginine-containing polypeptides within the same cell are degraded at normal rates. Similar findings have been made with a large variety of amino acid analogs, although different analogs promoted the degradation of proteins to very different extents (Fig 4). Presumably these varying effects are related to the extent to which the presence of the analog in proteins prevents their assuming their normal conformation (1,2). In accord with this conslusion, we have shown that increasing the content of an analog within proteins increases rates of proteolysis.

The natural substrates of this system for degrading abnormal proteins probably include proteins that arise through spontaneous enzyme denaturation or that result from mutations and mistakes in gene transcription and translation. The frequency of synthetic errors in normal cells are unknown, and attempts to determine such rates may have been misleading because of the cells' ability to rapidly eliminate abnormal proteins. However, we (15) have studied average degradative rates in bacterial strains which produce frequent errors in protein synthesis as a consequence of a mutation in the ribosome (the „ribosomal ambiguity mutation" **ram1)** Such strains degraded proteins during growth 2–3 times more rapidly than normal controls (Fig. 5). Furthermore, revertant strains with more accurate protein synthetic machinery show normal rates of catabolism.

The chemical features that differentiate the abnormal polypeptides and lead to their rapid hydrolysis have not yet been defined. Probably the simplest model that can explain the selective degradation of aberrant proteins is that normal cell enzymes share conformational features that

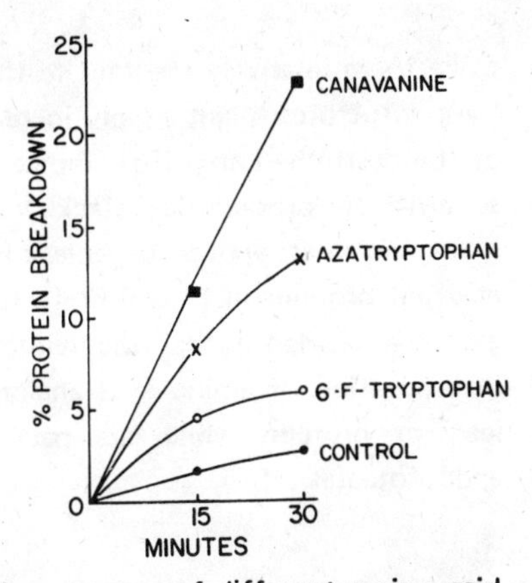

Fig. 4 Fate of proteins synthesized in the presence of different amino acid analogs. **E. coli A33,,** an auxotroph for both arginine and trytophan, was grown initially in medium containing these amino acids. The cells were then exposed for 5 min. to 3H-leucine in the presence of the required amino acids or one of their analogs. Canavanine is incorporated in protein in place of arginine, and azatryptophan and 5-fluorotryptophan in place of tryptophan. The labeled cells were then transferred to nonradioactive medium containing the required amino acids and the degradation of proteins made in the presence of the analogs or the natural amino acids'was measured in the normal fashion.

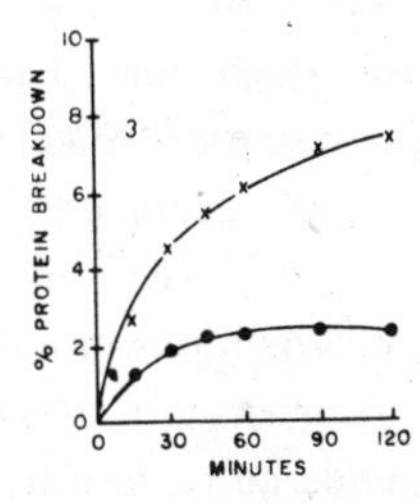

Fig. 5 Breakdown of proteins in **E. coli** carrying the **ram 1,**x — x,ribosomal ambiguity mutation) and wild type ●—● (ram +). Degradation of prelabeled proteins was measured during growth on glucose-minimal medium. This difference in degradative rates disappeared in revertants that lost the **ram** phenotype.

make them relatively resistant to these proteases (1,4), and deviations from these structures might simply increase degradative rates. The classic studies of Lindestrom–Lang first indicated that denatured proteins are more sensitive to proteolytic attack than native ones, and this simple generalization does appear capable of explaining the selective hydrolysis of aberrant proteins in **E. coli** (1,4). In accord with this idea, we have obtained extensive evidence that the failure to complete polypeptide chains, the incorporation of amino acid analogs, and mistakes in protein synthesis all lead to proteins which are particularly sensitive to attack by cellular endoproteases (4).

The degradation of abnormal proteins in reticulocytes

Mammalian cells also have this ability to hydrolyze rapidly abnormal proteins. For example, elegant studies by Cappechi et all. demonstrated the selective hydrolysis of mutant enzymes in mouse L cells (16), and several groups have reported rapid breakdown of analog-containing proteins in cultured cells (2). In order to learn more about this process in mammalian cells, we have investigated the fate of abnormal proteins in rabbit reticulocytes (17). Reticulocytes offer several unique advantages for the study of protein breakdown. 1) They synthesize one protein, hemoglobin, almost exclusively. 2) The structure of normal hemoglobin is very well characterized. 3) Pure preparations of reticulocytes can be prepared quite easily. 4) A large variety of human hemoglobin variants are known. Finally, although normal tetrameric hemoglobin undergoes little if any intracellular degradation, there is evidence that several human hemoglobin variants are rapidly degraded within reticulocytes (2).

We have investigated in depth the rapid breakdown of abnormal polypeptides in reticulocytes. Rabinowitz and Fisher (18) first showed that incorporation of valine analogs in reticulocytes leads to polypeptides, which undergo rapid intracellular degradation. Similar findings have been obtained by others for puromycil polypeptides in reticulocytes (3). Initially we surveyed the effects of several amino acid analogs in order to determine which ones promote protein degradation to the greatest extent (19). Incorporation of the lysine analog, β (aminoethyl) – cysteine, and the valine analog, t-aminochlorobutyric acid (ACB), were most effective in producing proteins that were rapidly degraded. For example, the ACB-containing protein was hydrolyzed to acid-soluble material with a half-live of 15–20 min (Fig. 6). The dramatic effects of these analogs presumably are a consequence of the large amounts of valine and lysine found in hemoglobin, or the particular importance that these residues play in its structure. Studies with SDS-gel electrophoresis indicated that the ACB-containing polypeptide are of similar size to normal globin, although it is not yet clear whether such analog—containing proteins from heme-containing tetramers like normal hemoglobin.

The degradation of such analog-containing proteins in reticulocytes shows a number of interesting features. Hydrolysis of these proteins is inhibited by the chloromethyl ketones (TPCK and TLCK), n—ethylmaleimide, and by iodoacetamide (17) These findings are consistent with the involvement of a sulfhydryl protease in the degradative process. Cycloheximide did not affect this process; thus, the degradation of abnormal proteins in reticulocytes, as in **E. Coli** is not dependent on concomitant protein synthesis (2). Finally experiments with metabolic inhibitors indicate that the degradation of these proteins appears to depend on high energy phosphates, in agreement with analogous results in other cells. (see Below).

Biochemical features of the degradative process

In **E. coli** , as in mammalian cells, inhibition of energy metabolism reversibly blocks protein degradation (2,12). For example, the catabolism of analog-containing proteins or nonsense fragments (7) can be inhibited almost completely if oxidative phosphorylation and glycolysis are simultaneously blocked. This inhibition of proteolysis can be completely reversed by removal of the inhibitors (Fig. 7), and thus these effects do not simply represent the irreversible killing of the cell. Such findings are particularly surprising on thermodynamic grounds, and no known proteolytic enzyme requires high energy cofactors. In fact a classical tenet of biochemistry is that biosynthetic processes require metabolic energy, while degradative processes produce high energy phosphates. Thus the biochemical basis for this energy requirement is an important unsolved problem.

Dr. Kenneth Olden in my laboratory has carried out a number of experiments to identify the biochemical processes that provide the energy for protein breakdown and to determine whether proteolysis requires high

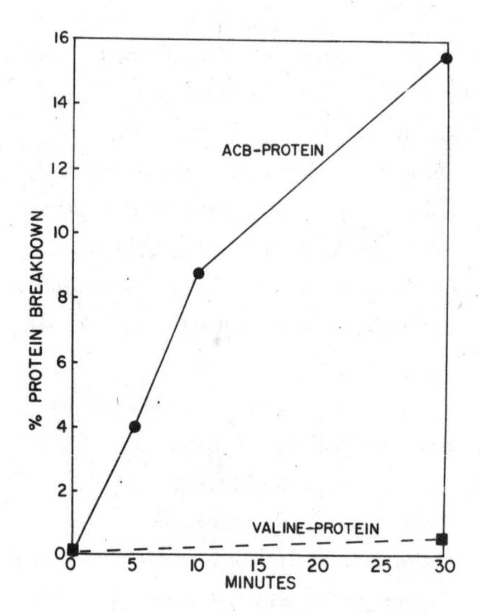

Fig. 6 Degradation of ACB—containing protein and valine-containing protein in rabbit reticulocytes. Breakdown of protein synthesized in the presence of the valine analog, (amino-chlorobutyric acid), or valine by reticulocytes isolated from phenylhydrazine-treated rabbits. To label proteins, the washed cells were suspended in Krebs—Ringer-phosphate buffer containing glucose, FeSO4 and plasma level amino acids minus valine. Either 1 mM valine or 1mM ACB was added to the media followed by 14C-leucine (0.1 uc/ml). After 15 min., the cells were washed in fresh medium lacking 14C-leucine but containing 10 mM nonradioactive leucine.

---

energy phosphates or the energy-rich state of the membrane. A number of processes in bacteria, such as uphill transport of amino acids, derive their energy directly from the high energy membrane state (20). Using various metabolic inhibitors and **E. coli** mutants deficient in the membrane Ca-Mg ATPase, it was possible to demonstrate that protein breakdown requires ATP or some related energy-rich compound rather than the energized membrane state. In typical experiments (12,21), ATP stores were completely depleted in the cell, and protein breakdown was blocked completely. This process was restored when such cells were allowed to generate ATP either by glycolysis or by oxidative phosphorylation.

However, when phosphorylation was prevented, but aerobic respiration permitted to occur (for example, by supply of arsenate), then protein breakdown did not occur. These studies and related ones thus indicated a strict correlation between the regeneration of ATP and the recovery of protein breakdown (12,21).

It is also noteworthy that protein breakdown requires a relatively low amount of ATP. To inhibit proteolysis, the ATP level must be reduced to below 10 % of the level found in growing cells. Similar findings have been made for the breakdown of normal proteins in starving bacteria (unpublished data), and proteins in mammalian tissues (2,22). Since the degradation of abnormal proteins is not affected by supplying chloramphenicol, high energy phosphates are not required for protein synthesis. The precise nature of this energy requirement remains a challenging unsolved problem, whose solution should provide important clues to the biochemical mechanisms of protein breakdown.

To learn more about the pathway for degradation of abnormal proteins and to define where in this process metabolic energy is required, Dr. Joel Kowit in my laboratory has investigated the fate of a specific abnormal polypeptide in greater depth (23). We chose to study the rapid degradation of an incomplete fragment of β–galactosidase that results from a nonsense mutation (X–90) close to its carboxyl terminus. The molecular weight of this abnormal polypeptide is very similar to that of the normal monomer of β–galactosidase (approximately 135,000MW). The study of such polypeptides offers important advantages for investigating the degradative process: 1) polypeptides lacking the carboxyl terminus are degraded with half-lives as short as several minutes; 2) they are inducible proteins; 3) it is possible to assay such incomplete polypeptides even though they lack enzymatic activity. Zipser and colleagues (24) have shown that these polypeptides when autoclaved (or treated with cyanogen bromide (14) release a polypeptide containing the amino terminus. This polypeptide (MW about 8,000), known as auto – α, can form active enzyme when mixed **in vitro** with an inactive fragment of β-galactosidase that lacks the amino terminus but contains a normal carboxyl end. This **in vitro** complementation assay for the amino terminal region has been used to monitor the intracellular fate of incomplete polypeptides (13,14,23–25). After removal of the inducer, the polypeptides containing the intact amino terminal region (Fig. 7) disappeared from the cell with a half-life of 17 minutes. This process, like the degradation of analog-containing proteins, can be inhibited by blockers of energy metabolism such as azide or cyanide, and this inhibition of degradation is fully reversible on removal of the inhibitor.

In an attempt to find intermediates in this degradative process, Dr. Kowit has combined the complementation assay with SDS -gel electrophoresis (25). In this procedure, aliquots of the induced culture were removed and the proteins separated on SDS-gels. The gels were then sliced, subjected to autoclaving to release auto – β peptides, incubated with the complementing polypeptide and assayed for auto– α activity. A typical experiment using this approach with mutant X–90 is shown in Fig. 8. In the induced cell, two different sized fragments of β –galactosidase containing auto– α activity were found. The larger polypeptide (A) is indistinguishable in size from normal β–galactosidase and presumably therefore represents the product of translation of the**X–90** gene. These findings are in accord with earlier observations of Goldschmidt (26), who suggested that the smaller polypeptide fragment (MW 98,000) might be an intermediate

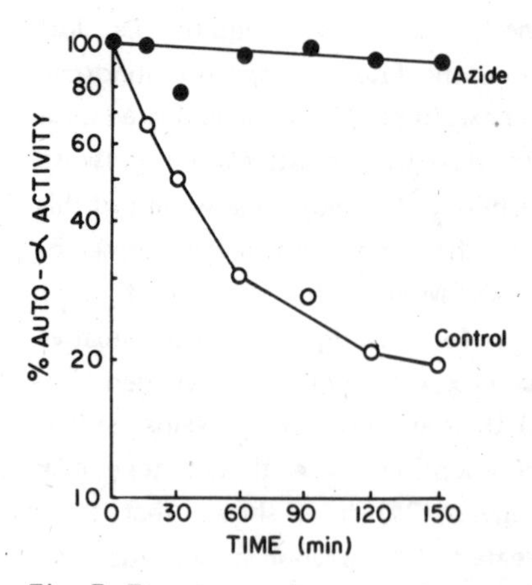

Fig. 7 Energy requirements for the degradation of nonsense fragments of β–galactosidase (X–90). Strain **X–90** was grown at 37°C on LB broth (20) in the presence of 5×10^{-4} M IPTG, harvested and washed by centrifugation in the cold. They were then resuspended (t=0) in fresh medium at 37°C in medium lacking the inducer but containing 100 ug/ml chloramphenicol. One half of the cells were treated with 25 mM sodium azide. At various times, aliquots were taken, chilled, centrifuged, and autoclaved. The auto α –polypeptides were taken measured by the complementation assay (24), and auto –α activity at the outset of the experiment is taken as 100 %.

produced in the degradation of the larger polypeptide. This hypothesis was tested by following the fate of the two polypeptides after removal of inducer. Careful kinetic analysis provided clear evidence that fragment A is indeed a precursor of fragment B. The level of fragment A decreased under these conditions with a half-life of only 6 minutes. The smaller fragment (B) initially increased almost 2-fold in the absence of further protein synthesis and subsequently disappeared with a half-life of 20 to 30 minutes. The disappearance of B thus correlated with the loss of auto – α activity from the intact cells (Fig. 7) (14).

These findings established a precursor product relationship between these two polypeptides. Only fragments A and B have been detected thus far by such techniques. Since no other polypeptides containing the α terminal region were detected on such gels, subsequent cleavage of fragment B must either be especially rapid or must occur within the alpha region. It is noteworthy that no other auto–α–containing polypeptides of size intermediate between A and B were detected. This finding suggests that the initial step in degradation of **X–90** involves an endoproteolytic cleavage

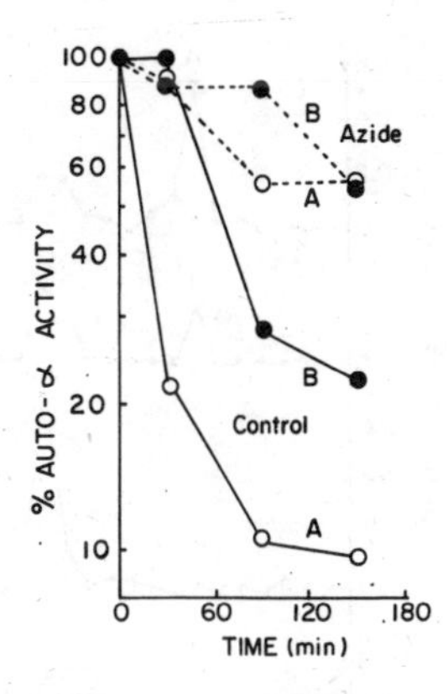

Fig. 8 Kinetic analysis of the degradation of the **X–90** polypeptide with acrylamide gels electrophoresis in the presence of sodium dodecyl sulfate (SDS). Cells were grown in the presence of IPTG as described above. At various times, aliquots were chilled and centrifuged. The cell samples were dissolved in 1 % SDS and electrophoresed to separate polypeptides (26). The gels were washed in trichloroacetic acid, neutralized, and sliced. After autoclaving to release the auto –α fragment, the slices were assayed by complementation (24).

rather than an exoproteolytic attack. If an exoprotease were responsible for the conversion of A to B, then the production of polypeptides of intermediate size would be expected. Proof that an endoproteolytic cleavage occurs would require the demonstration of the missing carboxyl-terminal polypeptide whose anticipated MW would be approximately 30,000. Because this polypeptide cannot be detected by the complementation assay, we have been attempting to identify and isolate it with anti $-\beta-$ galactosidase antibodies, although thus far without success.

Bukhari and Zipser (25) have isolated mutants designated **deg–T$^-$** that have a reduced ability to degrade incomplete fragments of β –galactosidase. The nature of this mutant and the biochemical basis of the defect are still unknown. Our studies of **deg–T$^-$** cells carrying the **X–90** mutation showed increased levels of peak A, and very little peak B. Thus in the mutant, the initial cleavage of A to B is reduced for reasons that remain to be established.

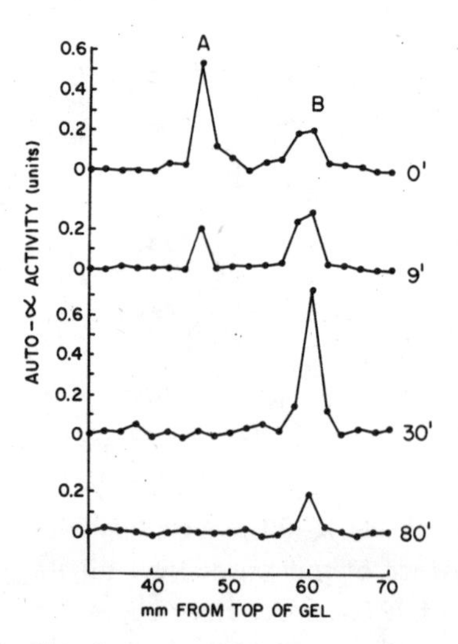

Fig. 9 Effect of azide on the degradation of polypeptides in strain **X-90** Aliquots from the experiment in Fig 7 were analyzed on SDS -gels to assay the auto $-\alpha$ activity in each peak at 0–time and 90 min. The levels of each fragment at time o were defined as 100 % and were calculated from the area under the curve determined as in Fig. 8.

The demonstration of an intermediate in protein breakdown (fragment B) enabled us to test whether metabolic energy is required either for the initial steps in the breakdown of this protein (i.e. the conversion of peak A to B), for the subsequent disappearance of fragment B, or in later steps in the conversion of polypeptides to free amino acids. When induced cells of strain **X–90**, growing on succinate, were washed free of inducer and treated with azide, the disappearance of auto $-\alpha$ activity was blocked almost completely. Simultaneously the loss of both peaks A and B were dramatically reduced (Fig. 9.). (No additional polypeptides appeared to accumulate under these conditions). These results clearly indicate that energy is required for multiple steps in the degradation of a single protein, both for the initial cleavage of peak A and for the subsequent disappearance of the intermediate.

We believe these studies to be highly promising in that they represent the first identification of intermediates in the degradative process. It is hoped that further investigations of this sort will define other intermediates and other properties of the degradative system. It is hoped that this approach will lay the basis for subsequent cell-free studies, which will be necessary to actually elucidate the responsible biochemical mechanisms. In any case, the information obtained on intact cells should be very helpful in helping us identify the responsible degradative enzymes and reconstructing this process in cell-free preparations.

ACKNOWLEDGMENTS

This work has been made possible by research grants from the National Institute of Neurological Disease and Stroke and the Muscular Dystrophy Association of America. Alfred L. Goldberg holds a Research career Development Award from the National Institute of Neurological Disease and Stroke.

I am grateful to Mrs. Elsa Fox and Miss Claire Kollin for their assistance in the preparation of this manuscript.

REFERENCES

1. Goldberg, A.L., and Dice, J.F., Jr.,Ann. Rev. Biochem. 43, 835–869 (1974)
2. Goldberg, A.L., and St. John, A.C.,Ann. Rev. Bicohem. 45, 747–803 (1976)
3. Schimke, R.T. in Mammalian Protein Metabolism, Munro, H. N., ed., Vol. IV, p. 177, Academic Press, New York (1970)
4. Goldberg, A. L.,Proc. Nat. Acad. Sci. USA 69, 2640-2644 (1972)
5. Dice, J.F., Jr., and Schimke, R.T.,J. Biol. Chem. 247, 98 (1972)
6. P. Bohley et al.,Acta Biol. Med. Ger. 28, 321 (1972)
7. Segal, H.L., Winkler, J.R., and Miyagi, M.P.,J. Biol. Chem. 249, 6364 (1974)
8. Dice, J.F., Jr., and Goldberg, A.L.,Nature 262, 514–516 (1974)
9. Arias, I.M., Doyle, D., and Schimke, R.T.,J. Biol. Chem. 244, 3303 (1969)
10. Dice, J.F., and Goldberg, A.L.,Arch. Biochem. Biophys. 170, 213–219 (1975)
11. Dice, J.F., and Goldberg, A.L.,Proc. Nat. Acad. Sci. 72, 3893–3897 (1975)
12. Goldberg, A.L., Olden, K., and Prouty, W.F. in Intracellular Protein Turnover, Schimke, R.T., and Katenura, A., eds., pp. 17–55, Academic Press, New York (1975)
13. Goldschmidt, R.,Nature 228, 1151 (1970)
14. Lin, S., and Zabin, I.,J. Biol. Chem. 247, 2205 (1972)
15. Pine, M.J.,J. Bacteriol. 93, 1527 (1967)
16. Capecchi, M.R., Capecchi, N.E., Hughes, S.H., and Wahl, G.M.,Proc. Nat. Acad. Sci. 71, 4732–4736 (1974)
17. Etlinger, J.D., and Goldberg, A.L.,Proc. Nat. Acad. Sci. (in press) (1976)
18. Rabinowitz, M. and Fisher, J.M.,Biochim. Biophys. Acta 91,213 (1964)
19. Olden, K. Etlinger, J.D., and Goldberg, A.L. In preparation.
20. Berger, E.A.,Proc. Nat. Acad. Sci. 70, 1514–1518 (1973)
21. Olden, K., and Goldberg, A.L. Submitted for publication.
22. Hershko, A., and Tompkins, G.M.,J. Biol. Chem. 246, 710–714 (1971)
23. Kowit, J.D., and Goldberg, A.L. Submitted for publication.
24. Morrison, S., and Zipser, D.,J. Mol. Biol. 50, 359 (1970)
25. Bukhari, A.I., and Zipser, D.,Nature New Biol. 243, 238–241 (1973)
26. Goldschmidt, R. Dissertation, Columbia University (1970)

PROTEIN DEGRADATION IN NEUROSPORA CRASSA

F.A.M. Alberghina and E. Martegani

Istituto di Scienze Botaniche dell'Universita – Centro di Studio del C.N.R. per la Biologia Cellulare e Moleculare delle Piante, 20133 Milano, Italy

SUMMARY

The metabolic stability of proteins has been determined in **Neurospora** mycelia in a variety of nutritional conditions. In exponentially growing mycelia the rate of protein degradation increases by decreasing the growth rate and it is pronounced only in ethanol growing cells.

During a shift–down transition of growth protein degradation is enhanced. 2–deoxyglucose prevents protein degradation in glucose starved cells.

In **Neurospora crassa** different rates of exponential growth are obtained at 30°C by changing the nutrients available to the culture (1). Characteristic rates of synthesis of rRNA, tRNA and proteins are measured in each of the different conditions of exponential growth (2).

The initial aim of this paper was to complete the analysis of the biochemistry of cellular growth in**Neuspora** by evaluating the rate of protein degradation in different conditions of exponential growth and to establish its relevance for the dynamics of cellular growth (3). Experiments were also performed on the regulation of protein degradation in **Neurospora** and the results are given in the present report.

MATERIALS AND METHODS

Conidia of **Neurospora crassa** (wild type strain 74 A St. Lawrence) were used to inoculate as described before (1) liquid Vogel's mineral medium to which, unless otherwise stated, one of following carbon sources was added: glucose (2 % w/v), glycerol (2 % w/v), ethanol (2 % w/v). Incubation was at $30^{o}C$ in a shaken water bath and growth was monitored as absorbance at 450 nm (A 450 nm) as previously indicated (1).

Proteins were labelled with l–^{14},COOH–leucine during exponential growth. To a 200 ml culture (A 450 nm: 0.100) radioactive leucine was added to final concentration of 5.10^{-7}M (specific activity 10 Ci/mole) for cultures in glucose or 6.10^{-8}M (specific activity 80 Ci/mole) for cultures in glycerol or ethanol. When all the radioactivity had been incorporated into hot TCA precipitable material (after 30 minutes in glucose or after 45 minutes in the other media) ^{12}C– leucine to a final concentration of 10^{-4} M was added to prevent reutilization of the radioactive leucine.

Protein degradation was assayed by measuring at various time intervals either the radioactivity retained in the proteins, as hot TCA precipitable material (5) or the radioactivity released in the TCA soluble fraction. The radioactivity released in the TCA soluble fraction was assayed as follows: 1 ml aliquots of the culture were added to 1 ml of cold 20 % trichloroacetic acid. After 30 minutes in the cold, the suspension was centrifuged at 3500 rpm for 15 minutes. An aliquot of the supernatant (0.5 ml) was counted in a Packard (mod. 3320) scintillation counter. The agreement between the two measurement was very satisfactory for the first 2–3 hours, then the radioactivity found in the TCA soluble fraction was less than that expected on the basis of its disappearance from the hot TCA precipitable material. Thus for long time kinetics the determination of the radioactivity in the TCA precipitate was preferred.

RESULTS AND DISCUSSION

Protein degradation during different exponential growth.

Fig. 1 shows the metabolic stability of **Neurospora** proteins in three different conditions of exponential growth. Protein degradation is almost undetectable in glucose ($\mu = 0.51$), while it is sizable both in glycerol (μ =

0.26) and in ethanol (μ = 0.13). The kinetics of protein degradation in glycerol indicated the presence of two components, a fast and a slow decaying one, while such biphasic kinetics is not detectable in ethanol or in glucose. The half–lives calculated from the slopes of the best fitting straight lines of the semilogarithmic graph of Fig. 1 are 57 hours for the cells in glucose and 24 hours for the cells in ethanol, the duplication times being respectively about 2 hours in glucose and about 8 hours in ethanol. So even assuming that all the cellular proteins are subjected to degradation, no more than 2 % of the total proteins are degraded during one duplication time in glucose and no more than 20 % in ethanol.

To estimate the half-life and the relative amount of the rapidly decaying component in glycerol, we made use of the calculations elaborated for a similar problem in **E. coli** (4). From the values of Fig. 1, we calculated the half-life of the fast decaying component as 45 minutes, of the slow decaying one as 60 hours, and the amount of the rapidly degrading proteins about 4 % of the total proteins. So even by assuming that the two slow and fast components account for all the proteins in cells growing in glycerol, no more than 8 % of the total proteins should be subjected to degradation in one duplication time (about 4 hours) in glycerol.

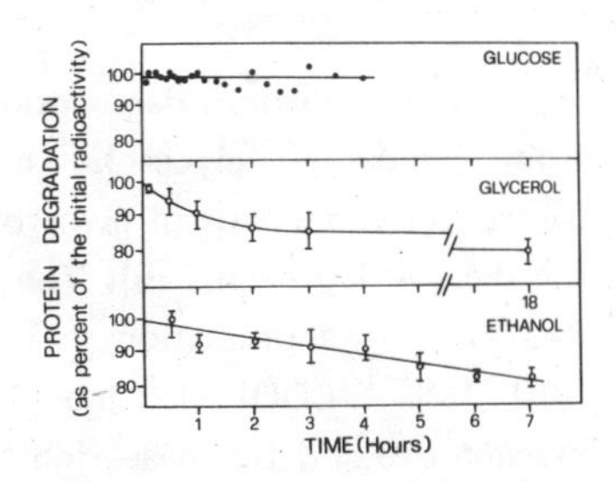

Fig. 1 — **Kinetics of protein degradation in cultures growing exponentially in glucose glycerol and ethanol.** Cultures growing exponentially in the different media were labelled with I–^{14} COOH–leucine as described under Methods. At the zero time 10^{-4}M^{12}C–I–leucine was added and thereafter the radioactivity incorporated into the hot TCA precipitable material was determined. The data are the average of four determinations and the standard deviations are given.

Regulation of protein degradation.

About 25 % of the radioactive proteins made during exponential growth in glucose are rapidly broken down during a previously described (5) shift–down transition from glucose to glycerol (Fig. 2). It appears as an adaptive response which may facilitate the rearrangement of cellular structures (for example mitochondria) which are very different in the two nutritional conditions (6).

The effect of glucose starvation on protein degradation in **Neurospora** was then assayed. Fig. 3 shows that a very marked protein degradation is observed in glucose starved **Neurospora** cells, as it has been in bacteria (7) (8). The addition of 3–O–methylglucose (3MG) to the glucose starved cells does not modify in a significant way the rate of protein degradation, while

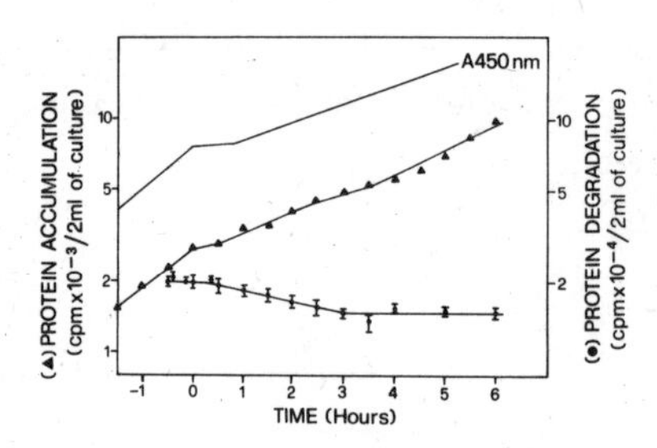

Fig. 2. – **Protein degradation during a shift down transition of growth from glucose to glycerol.** The culture is growing on limiting glucose (60μg/ml) and excess of glycerol: at the zero time glucose is exhausted and the diauxie lag begins (5). The growth is measured as A 450 nm (——). Protein accumulation (–▲–) is determined by adding 5.10^{-4}MI–^{14}COOH–leucine (0.05 Ci/mole) to the culture at the time of inoculation, and by measuring the radioactivity incorporated into the hot TCA precipitable material. To determine the stability of the proteins, during the diauxie, at time –60 min. the culture was labelled with 6.10^{-8}MI–^{14}COOH–leucine (spec. activity 80 Ci/mole). At – 30 min. 10^{-4}M^{12}C–l–leucine was added and at different times thereafter the radioactivity incorporated into hot TCA precipitable material was determined (–●–). The data are the average of six determinations and standard deviations are indicated.

the addition of 2-deoxyglucose (2DG) almost completely blocks it (Fig. 3). In **Neurospora** 3 MG is a substrate for the glucose transport system, but it cannot be phosphorylated and so is not metabolized (9). Instead 2 DG is partially metabolized, taken up and phosphorylated by hexokinase (10), and in yeast it is the origin of UDP and GDP derivates (11). In our experimental conditions, 2 DG (100 μg/ml) does not support growth and protein synthesis in glucose starved cells for at least 3 hours and it does not inhibit the growth rate of cells growing exponentially on glucose (200 μg/ml, initial concentration) (results not shown). These findings do not yet exclude the possibility that the block by 2DG on protein breakdown might be aspecific, for instance by interfering with the energy metabolism which migt be required for protein degradation (12) (13), but if proved to be specific, it may offer a lead to an understanding of some aspects of the regulation of protein degradation.

Finally, the addition of cyclic AMP (3.10^{-4}M) was found not to influence the rate of protein degradation in **Neurospora** cells growing exponentially on glucose (results not shown). No link between cyclic AMP and the rate of degradation has been yet found (8).

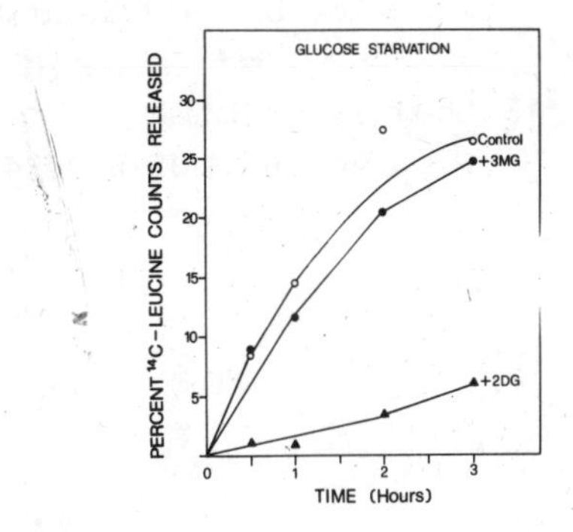

Fig. 3 – **Effect of 2–deoxyglucose and 3–0–methylglucose on protein degradation during glucose starvation.** The culture growing in limiting glucose (60μg/ml) was labelled as described under Methods with l–^{14}COOH–leucine. At the moment of glucose exhaustion (zero time) 10^{-4}M ^{12}C–l–leucine was added together with 3–O–methylglucose or 2-deoxyglucose 100 μg/ml. The TCA soluble radioactivity was determinated at different times, and to each value the zero time value was subtracted. The data are expressed as % of the radioactivity present in the TCA precipitable material at zero time.

REFERENCES

1. Alberghina, F.A.M.,Arch. Mikrobiol. 89, 83–94 (1973).
2. Alberghina, F.A.M., Sturani, E., Golke, J.R., J. Biol. Chem. (in press).
3. Alberghina, F.A.M. Bio System (in press).
4. Nath, K. and Koch, A.L., J. Biol. Chem. 245, 2889–2900 (1970).
5. Sturani, E., Magnani, F., Alberghina, F.A.M., Biochim. Biophys. Acta, 319, 153–164 (1973).
6. Alberghina, F.A.M., Trezzi, F., Chimenti Signorini, R.,Cell Different. 2, 307–317 (1974).
7. Willetts, N.S., Biochem. Biophys. Res. Commun. 20, 692–696 (1965).
8. Goldberg, A.L., Howell, E.M., Li, J.B., Martell, S.B. and Prouty, W.F. Federation Proceedings 33, 1112–1120 (1974).
9. Scarborough, G.A., J. Biol. Chem. 245, 1694–1698 (1970).
10. Sols, A., Heredia, C.F. and Ruiz–Amil, M., Biochem. Biophys. Res. Commun.,2, 126–129 (1960).
11. Biely, P. and Bauer, S., Biochim. Biophys. Acta 156, 432–434 (1968).
12. Holzer, H., Betz, H., Ebner, E.,Current Topics in Cellular Regulation (Horecker, B.L. and Stadtman, E.R., eds) Vol. 9, pp. 103 – 156, Academic Press New York (1975).
13. Shechter, Y., Rafaeli–Eshkol, D. and Hershko, A.,Biochem. Biophys. Res. Commun. 54, 1518–1524 (1973).

STUDIES ON THE MECHANISM OF ENDOCYTOSIS IN FIBROBLASTS WITH THE USE OF SPECIFIC ANTIBODIES

by P. Tulkens [(+)], Y.J. Schneider [(++)] and A. Trouet

International Institute of Cellular and Molecular Pathology and Universite Catholique de Louvain avenue Hippocrate 75 B – 1200 BRUSSELS (Belgium).

Endocytosis is one of the mechanisms trough which foreign material is taken up by living cells and transferred to their lysosomes (de Duve et al., 1974). The succession of events involves an invagination of the plasma membrane, the formation of a closed vesicle, called a phagosome, and the subsequent fusion of this vesicle with lysosomes, leading to the formation of a digestive vacuole (Jacques, 1969).

In this paper we describe experiments in which the functional relationships between plasma membrane and lysosomes has been analyzed on cultured fibroblasts by means of specific antibodies. Cells were incubated with labelled IgG directed against plasma membrane, lysosomal membrane and lysosomal soluble constituents. After various times, cells were fractionated and the subcellular distribution of the IgG were determined and compared with that of marker enzymes of plasma membrane, lysosomes and other subcellular components. Using the antibodies as a specific probe, it

---

(+)**Charge de Recherches** of the Belgian **Fonds National de la Recherche Scientifique.**

(++)**Boursier** of the Belgian **Institut pour l'encouragement de la Recherche Scientifique dans l'Industrie et l'Agriculture.**

becomes possible in this way to follow the fate of different antigens or structures related with plasma membrane and lysosomes during endocytosis.

Plasma membranes(MP) were isolated from rat liver by the method of Song **et al.** (1969) and re-purified after addition of digitonin (Remacle, J., unpublished results; de Duve, 1971). 5'Nucleotidase and alkaline phosphodiesterase I were purified 35 fold with a 15 % yield. Assay for contaminants revealed only trace amounts of glucose 6-phosphatase and undetectable levels of marker enzymes of other subcellular constituents.

Lysosomes were isolated from livers of rats injected with Triton WR–1339, according to the flotation method of Trouet (1974). Lysosome soluble (LS) and insoluble (LI) constituents were separated after osmotic disruption.

Antisera were obtained from rabbits. IgG were purified according to Sober **et al.** (1956) and labelled with $^3H/^{14}C$ acetic anhydride (Klinman and Karush, 1967) and/or fluorescein isothiocyanate (The and Feltkamp, 1969).

Fibroblasts, obtained from rat embryos, were incubated in the presence of antibodies for various times. They were thereafter collected and fractionated by isopycnic centrifugation according to Tulkens **et al.** (1974). Under those circumstances, plasma membrane (5'Nucleotidase) and lysosomes (cathepsin B1, cathepsin D, N–acetyl–β–glucosaminidase) display very distinct equilibration patterns.

1.PROPERTIES OF THE ANTISERA

IgG directed against rat liver plasma membranes (αMP) inhibit strongly both liver and fibroblast 5'Nucleotidase. Further, they are specific for rat liver plasma membrane and show little cross reaction with other liver subcellular components (Hauser et al 1973); they do not inhibit lysosomal enzymes.

IgG directed against lysosome soluble constituents of liver (αLS) inhibit fibroblast cathepsih D, DNase, N-acetyl–β–glucosaminidase and precipitate acid phosphatase (Tulkens **et al.** 1970; Trouet, 1969); they do not inhibit liver or fibroblast 5'Nucleotidase; they do not react with isolated liver plasma membranes (Hauser **et al.** 1973).

IgG directed against lysosome insoluble constituents of liver inhibit fibroblast 5'Nucleotidase, but to a much lower extent than α MP. They cross-react only slightly with liver plasma membrane and with lysosome soluble constituents; they do not inhibit lysosomal cathepsin D.

2. FATE OF CONTROL IgG: INFLUENCE OF TYPE OF LABEL

In the course of our study on the uptake and fate of specific IgG, we made the rather unsuspected observation that the type of label (fluorescein or $^{3}H/^{14}C$-acetyl, both binding to ϵ–amino group of the lysyl residues) influences impressively the level of accumulation of the IgG by fibroblasts, and their subcellular distribution.

During a 24 hours incubation, fibroblasts concentrate fluorescein – IgG about 20 to 40 fold from the culture medium. By comparison, acetyl – IgG are concentrated only 1 or 2 times. This higher accumulation is somewhat related to the degree of labelling, but is already apparent with IgG of low and uniform fluorescein/protein molar ratio. It is therefore not the result of a preferential uptake of highly labelled and partially denatured molecules (The and Feltkamp, 1969).

The type of label influences still more dramatically the intracellular fate of the IgG. When fibroblasts are incubated 30 hours with fluorescein – IgG and are thereafter fractionated, the distribution pattern of the fluorescence follows closely that of the lysosomal enzymes (Fig. 1) and differs clearly from those of 5'–Nucleotidase, inosine diphosphatase (plasma membrane) and the other subcellular constituents detected in these cells (Tulkens **et al.**, 1974). By immunodiffusion, fluorescent material was shown to react with anti – IgG goat antiserum. If cells are incubated 30 hours with ^{3}H–acetyl-labelled IgG, more than 65 % of the cell-bound radioactivity is found associated with 5'–Nucleotidase and inosine diphosphatase (Fig. 1). By gel electrophoresis and gel chromatography this radioactivity was conclusively shown to be linked to intact IgG.

These observations do not result from different susceptibilities of acetyl or fluorescein – IgG to lysosomal hydrolases. In our experimental conditions, no product of digestion of the IgG molecules could be observed in the culture medium; further, acetyl-labelled IgG were shown **in vitro** to be very resistant to proteolysis by purified lysosomes, and to the same extent as reported for unlabelled IgG (Fehr **et al.** 1970). Selective reincorporation of the label in lysosomes (fluorescein) or plasma membrane (radioactive acetate) could also be excluded.

If IgG are simultaneously labelled with fluorescein and acetic anhydride, they behave esentially like fluorescein IgG and the distribution patterns of both fluorescence and radioactivity coincide with those of lysosomal hydrolases (Fig. 1).

Before reaching lysosomes, fluorescein – IgG molecules bind to the plasma membrane and are only thereafter transferred to the vacuolar apparatus. As illustrated in Fig. 2, the distribution of double-labelled IgG

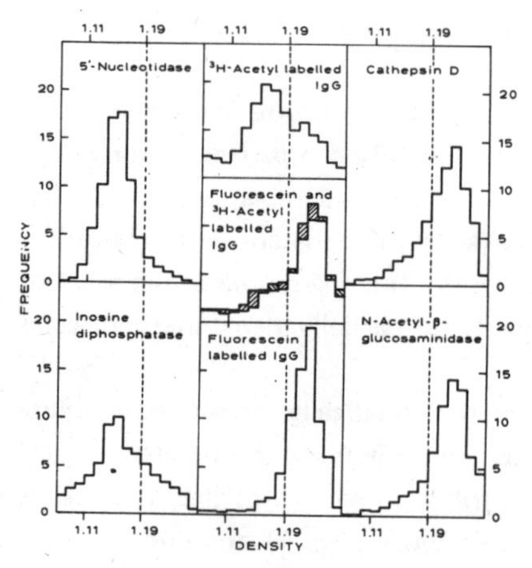

Fig. 1. Isopycnic centrifugation of post-nuclear supernates from homogenates of fibroblasts incubated 30 hrs with 100 ug/ml of ^{3}H–acetyl–labelled IgG, fluorescein-labelled IgG or ^{3}H–acetyl– and fluorescein-labelled IgG. The experimental density distribution pattern of each type of IgG is represented in the central column of the diagram. Double-labelled IgG were assayed both by fluorescence and radioactivity measurement; discrepancies between the two determinations are represented by the shaded part of the diagram; such discrepancies are not significant. The left and the right columns show the mean density distribution patterns of plasma membrane (left) and lysosome (right) marker enzymes. These distributions fall into the range of those observed for control rat fibroblasts by Tulkens **et al.** (1974)

after short exposure to cells is more related to 5'Nucleotidase than to N—acetyl—β—glucosaminidase. After 3 hours of incubation, the distribution pattern becomes clearly bimodal and after 34 hours of incubation, it follows closely that of lysosomal enzymes. However, it should be noticed that the amount of cell bound fluorescein — IgG increases tremendously between 1 hour and 34 hours of incubation. This increment concerns nearly exclusively the lysosomal part of the distribution pattern, so that, after prolonged incubation, the association of fluorescein — IgG to plasma membrane could no longer be satisfactorily demonstrated. We therefore conclude that both acetyl — and fluorescein — labelled IgG bind to the plasma membrane, but the latter ones are immediately transferred to lysosomes where they accumulate, whereas the former ones remain essentially attached to their membrane binding sites.

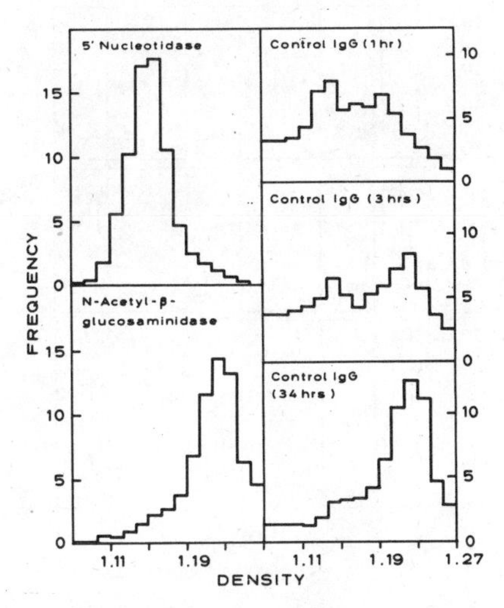

Fig. 2. Isopycnic centrifugation of post-nuclear supernates from homogenates of fibroblasts incubated with ^{3}H—acetyl— and fluorescein-labelled IgG for various times. Left and right diagram show the mean density distribution pattern of plasma membrane and lysosome marker enzymes. The central diagram shows the density distribution patterns of IgG after 1 hour, 3 hours and 36 hours of incubation with fibroblasts. Note that the absolute amount of cell bound to IgG was about 3 and 10 times higher after 3 hours and 36 hours than after 1 hour.

3. UPTAKE AND LOCALIZATION OF SPECIFIC IgG

Anti plasma membrane IgG

When cells are incubated with acetyl-labelled αMP IgG, they accumulate them at a level about 4 to 5 times higher than control acetyl– IgG. After 24 hours incubation, these αMP IgG are found, for the largest part, associated with the plasma membrane (Fig. 3). To evaluate the rate of interiorization of the surface antigens during the culture, cells were incubated 24 hours with αMP acetyl-IgG, washed and reincubated in fresh medium. At various intervals, they were collected and fractionated. After 6 days of incubation, 50 % of the radioactivity still remained linked to the plasma membrane. At the concentrations of αMP IgG used in our experiments, endocytosis was not

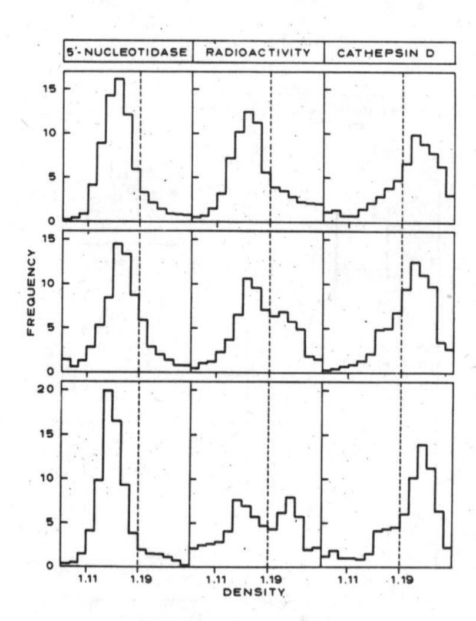

Fig.3. Isopycnic centrifugation of post-nuclear supernates from homogenates of fibroblasts incubated with antiplasma membrane ^{3}H–acetyl–IgG (50 ug/ml). **Upper diagram:** cells incubated 24 hours in presence of antibody; **middle diagram** cells incubated 24 hours in presence of antibody and 24 hours in new fresh medium, containing no antibody; **lower diagram:** cells incubated 24 hours in presence of antibody and 144 hours in new fresh medium. The total amount of radioactivity recovered in cells remained nearly constant throughout the experiment, including the wash-out period.

impaired; indeed a normal cell uptake and lysosomal accumulation of fluorescein-control IgG could be demonstrated during the incubation in the presence of αMP IgG and during the wash – out period. In view of the data of Fig. 2, the amount of αMP IgG still associated with the plasma membrane 6 days after its labelling seems therefore higher than expected, should endocytosis involve the plasma membrane in a random fashion.

This is still more evident if the fate of fluorescein-labelled αMP–IgG is studied. After 36 hours of incubation, they could, both biochemically and morphologically, be clearly localized on the plasma membrane, where they displayed some degree of patching. Under the same circumstances, control fluorescein – IgG are only observed as discrete intracellular granules which, by cell fractionation, were shown to be lysosomes (Fig. 1).

Anti lysosome IgG

Fibroblasts incubated with acetyl–IgG directed against soluble lysosome constituents (αLS) accumulate these immunoglobulins 3 to 4 times more than control acetyl – IgG. After 6 hours incubation the αLS IgG display a typical plasma membrane like distribution (Fig.4). However, if the incubation is prolonged up to 36 hours, the distribution pattern of the antibodies shifts to a bimodal one, a part of which (35 %) still coincides with the distribution of plasma membrane, the second part (65 %) overlapping the pattern of the lysosomal enzymes. The fate of acetyl IgG directed against insoluble lysosome constituents is very similar to that of αLS IgG; after 36 hours incubation, they distribute themselves on plasma membrane (1/3) and in lysosomes (2/3).

Uptake and localization of mixtures of IgG

We observed definite differences in the level of accumulation and the fate of IgG according to their label and immunological properties. To establish whether this behaviour was specific for each type of IgG or resulted from a general cell reaction to a given type of antibody, we undertook a systematic study where IgG from different sera (control, αMP, αLS,αLI) and with different labels (^{3}H or ^{14}C acetyl, fluorescein) were mixed at concentrations varying from 10 to 100 mg/ml, in various combinations. In no instance was the fate of one type of IgG significantly influenced by the other types of IgG present in the culture medium. The fate of each IgG appears thus to be the result of an individual interaction of the fibroblasts with that IgG.

DISCUSSION

In the course of this study, we came to the observation that in cultured fibroblasts, both the type of label and the immunological properties influence markedly the fate of immunoglobulins in respect to endocytosis. Such a result is far from being surprising as endocytosis was already shown to be a mechanism endowed with great specificity according to the biochemical nature of the molecule (Cohn and Parks, 1967; Morell **et al.** 1968; Morell **et al.** 1971; Bach **et al.** 1972).

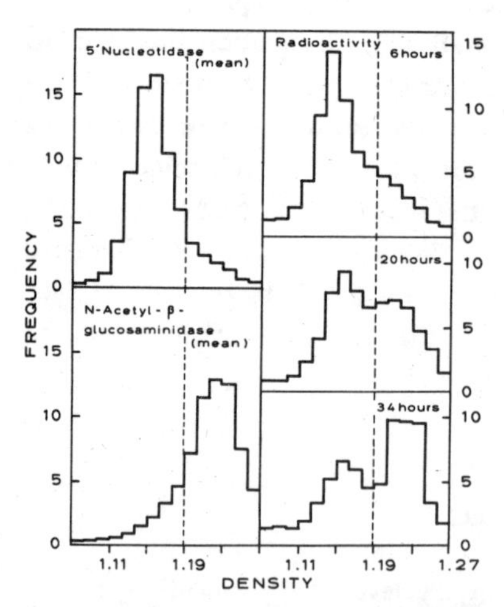

Fig. 4. Isopycnic centrifugation of post-nuclear supernates from homogenates of fibroblasts incubated in presence of ^{3}H–acetyl– LS IgG. Left part of the diagram shows the mean density distribution patterns of plasma membrane and lysosome marker enzymes. Right part of the diagram shows the density distribution patterns of the IgG after various times of contact with cells, respectively 6 hours (upper), 20 hours (middle) and 34 hours (lower).

It may be reasoned that each type of IgG reacts with particular sites of the cell surface and subsequently follows the fate of those sites during cell culture. We could indeed demonstrate that all IgG display some binding to the pericellular membrane during the first moment of contact.

Provided this binding is tight and little reexchange occurs, IgG could be considered as a probe reagent for the site they react with. It should, however, be stressed that the nature of these sites cannot be further specified, pending complementary biochemical and morphological observations.

Two models may explain our data. The first one would represent the cell surface as constituted by at least three types of independent sites. The **first type** would consist of the cell membrane antigens, recognized by the αMP IgG. These sites remain rather stable at the cell surface and are little involved in the phenomenon of endocytosis. Eventually, they may move on the cell surface giving rise to some **patching** as observed in the fluorescence microscope. The **second type** of sites would be able to bind fluorescein – IgG. Present in small amounts compared to the sites binding αMP IgG, they would be essentially involved in transporting their ligand towards the lysosomes. These sites may be present on cell types other than fibroblasts (Nairn **et al.,** 1958) and may also react with other fluorescein – labelled protein (Fiume **et al.,** 1971). The **third type** of sites would bind αLS and αLI IgG. They would behave like fluorescein – IgG sites, but less effectively. It is also possible that some of them remain on the cell surface like the αMP IgG sites. The initial binding of αLS or αLI to the cell surface may indicate that some lysosome antigens are present on the pericellular membrane; but it cannot be excluded, at present, that αLS and αLI IgG react with some secreted lysosomal enzymes which, in human fibroblasts, were shown to bind to a special membrane receptor and proceed thereafter to lysosomes (Bach **et al.,** 1972). Finally, control acetyl IgG would bind losely to sites little involved in endocytosis and whose nature cannot be further characterized. This model suggests that the pericellular membrane is a highly organized structure, only some portions of which are involved in endocytosis and subsequent fusion with lysosomes. The work of Tsan and Berlin (1971) on the fate of membrane transport systems in marcrophage during endocytosis points in the same direction, as well as our own work on endocytosis of Triton WR – 1339 in hepatoma tumour cells (Lopez–Saura **et al.,** 1972). It would also explain the rather different chemical and enzymological compositions of plasma and lysosomal membranes (Thines–Sempoux, 1973).

However, contradiction is offered by the work of Hubbard and Cohn (1975), who presented evidence that iodinated membrane proteins are found in phagolysosomes in an amount compatible with a random interiorization of plasma membrane during phagocytosis.

An alternative model may therefore be considered. The whole cell surface would be involved in endocytosis in an undifferenciated fashion, but a selection between the various membrane parts would take place at the level of the fusion between the phagosome and the lysosome. Only a part of the phagosomal membrane would finally becomes a constituent of the lysosomal membrane, the remaining part being, in some way, restituted to the plasma membrane.

The fate of each IgG will therefore depend on the fate of the antigen or site to which it is bound. Accordingly, the surface antigens to which αMP are linked would belong almost exclusively to fragments of the phagosome membrane restituted to the plasma membrane. Conversely, the antigens and sites binding αLS, αLI and fluorescein IgG would be directed to lysosomes either by membrane fusion or after hydrolytic detachment of these sites from the phagosomal membrane followed by their reassociation to lysosome constituents. The accumulation of αLS, αLI and fluorescein IgG in lysosomes could also be due to the detachment of these IgG from their primary binding sites and their subsequent association, during the fusion process, to higher affinity lysosomal antigens or binding sites. The same may be true for the iodinated surface proteins described by Hubbard and Cohn (1975).

Both models of endocytosis are still, for the present, largely hypothetical. But the experiments on which they rely stress the importance of the selectivity of this phenomenon. All together, they may shed new light on the precise mechanism of endocytosis. The uncovering of this aspect of cell physiology could also prove of particular importance in the rational design of a therapy based on the endocytotic properties of diseased cells (de Duve et al., 1974).

ACKNOWLEDGEMENTS

This work was supported by the Belgian **Fonds National de la Recherche Scientifique Medicale.**

REFERENCES

BACH, G., FRIEDMAN R., WEISSMANN B., and NEUFELD E.F., **Proc. Natl Acad. Sci. U.S.A.** 69: 2048–2051 (1972).

COHN, Z.A. and PARKS E., **J. Expl. Med. 125** : 213–232 (1967)

de DUVE, C. **J. Cell Biol. 50** : 20D–55D (1971)

de DUVE, C., de BARSY Th., POOLE B., TROUET A., TULKENS P., and VAN HOOF F., **Biochem. Pharmacol. 23**: 2495–2531 (1974).

FEHR K., LO SPALLUTO J. and ZIFF M., J. Immunol. 105 : 973–983 (1970)

FIUME, L., CAMPAOELLI–FIUME G. and WIELAND Th., **Nature New Biology. 230** : 219–220 (1971)

HAUSER P., REMACLE J., TROUET A., and BEAUFAY H., **Arch. Intern. Physiol. Biochim. 81** : 186–187 (1973).

HUBBARD, A.L. and COHN Z.A., **J. Cell Biol. 64** : 461–479 (1975)

JACQUES, P.**In** „Lysosomes in Biology and Pathology". (J. T. Dingle and H.B. Fell, editors), North–Holland Publ. Co., Amsterdam.**2** 395–420 (1969)

KLINMAN, N.R. and KARUSH F., **Immunochimie. 4** : 382–405 (1967)

LOPEZ–SAURA, P., TULKENS P., and TROUET A., **Arch. Intern. Physiol. Biochim. 80** : 977–978 (1972)

MORELL, A.G., IRVINE, R.A., STERNLIEB I., SCHEINBERG I.H., and ASHWELL G., **J. Biol. Chem. 243** : 155–159 (1968)

MORELL, A.G., GREGORIADIS, G., SCHEINBERG, I.H., HICKMAN, J., and ASHWELL' G.,**J. Biol. Chem. 246** : 1461–1467 (1971)

NAIRN, R.C., CHADWICK C.S. and Mc ENTEGART, M.C., **J. Pathol. Bacteriol. 76** : 143–149 (1958)

SOBER, H., GUTTER F.J., WYCKOFF M.M., and PETERSON E.A., **J. Am. Chem. Soc. 78** : 756 (1956)

SONG, C.S., RUBIN W., RIFKIND A.B. and KAPPAS K., **J. Cell Biol. 41** : 124–132 (1969)

THE, T.H. and FELTKAMP T.E.W., **Immunology. 18** : 875–881 (1969)

TROUET, A. 1969. Caracteristiques et proprietes antigeniques des lysosomes du foie. Thesis, Universite Catholique de Louvain.

TROUET, A. **In** „Methods in Enzymology" (S. Fleischer and L. Packer, editors) Academic Press, Inc., New York. **31** : 323–329 (1974).

TSAN, M.F. and BERLIN, R., **J. Expl Med. 134** : 1016–1035. (1971)

TULKENS, P., TROUET, A., and VAN HOOF, F., **Nature** (Lond.). **228** : 1282–1285 (1970)
TULKENS, P., BEAUFAY, H., and TROUET, A.,**J. Cell Biol. 63** : 383–401 (1974).

SPECIFICITY OF BREAKDOWN BASED ON THE INACTIVATION OF ACTIVE PROTEINS AND PEPTIDES BY BRAIN PROTEOLYTIC ENZYMES

Neville Marks

New York State Research Institute for Neurochemistry and Drug Addiction Ward's Island, New York 10035

The peptide nature of brain regulatory hormones poses a number of intriguing questions concerning the role of intracellular catabolism in endocrine control. Two distinct processes can be discerned: 1) formation of an active species from an inactive protein or polypeptide precursor, and 2) their subsequent inactivation by a separate set of hydrolases. The coupling of two such processes appears to form an integral part of the mechanisms regulating intracellular catabolism; thus a study of the specificity of the enzymes involved could help reveal some of the pathways involved. It is the purpose of this short account to focus attention on particular brain enzymes that are directly involved in this function with emphasis placed on contributions from the authors' laboratory.

Brain, like other tissues, is characterized by a spectrum of proteolytic hydrolases, only a few of which have been studied in detail and characterized (1-3). In several cases these brain enzymes show unexpected specificities as illustrated by the limited cleavage of biologically active materials (Tables 1 and 2). Of the active peptides, the recently discovered hypothalamic fractors are of particular interest, since several of them are characterized by the presence of N-(pyroglutamyl) or C-(glycinamide or other acylated amides) protected groupings, which could imply the existence in tissues of enzymes with novel specificities (Table 1). These groupings are present in a very large number of tissue components, but their role in breakdown is a relatively unexplored topic. Currently, there is considerable interest in hypothalamic peptides, since they have powerful pharmacological

TABLE 1

Cleavage of Active Peptides by Brain Proteinases

A) Neutral proteinase (cathepsin M)

i) LH-RF (luliberin)

pGlu-His-Trp-Ser-Tyr-Gly-Leu-Arg-Pro-Gly.NH_2 (4,8,9)

⇡ ↑ ⇡ ↑

ii) Kinin-9 (Bradykinin)

Arg-Pro-Pro-Gly-Phe-Ser-Pro-Phe-Arg (11,12)

↑

iii) Substance P

Arg-Pro-Lys-Pro-Gln-Gln-Phe-Phe-Gly-Gly-Leu-Met. NH_2 (13)

⇡ ⇡ ↑

iv) Somatostatin

Ala-Gly-Cys-Lys-Asn-Phe-Phe-Trp-Lys-Thr-Phe-Thr-Ser-Cys (14)

⇡ ↑ ⇡

B) Tryptic-like (pars intermedia)

ACTH .... α -MSH (1-13); ACTH-like (CLIP) (15)

C) Kallikrein

Bradykininogen .... kinins and protein (1)

D) Renin

Hypertensinogen .... proangiotensin + protein (1)

E) Cathepsin D (brain)

Myelin basic protein .... Phe-Phe linkages in positions 43-44 89-90 (bovine material) (16, 17) Fig. 2

Established points of cleavage (↑), postulated points of cleavage (⇡). The numbers in parentheses are references.

effects with possible clinical applications (4-6). Their use in man is limited, however, by their extremely short biological half-lives, and this has led to an extensive and costly search for longer acting derivatives (7,8). As will be discussed, a knowledge of biodegradation can provide a rational basis for the synthesis of longer acting derivatives with greater clinical potential. The

TABLE 2

Cleavage of Active Peptides by Exopeptidases

A) N-terminal (arylamidase)

i) Angiotensin-II

Asp-Arg-Val-Tyr-Ile-His-Pro-Phe (20)

↑ ↑ ↑ ↑ ↑

ii) Melanostatin (MIF)

Pro-Leu-Gly. NH_2 (21)

↑ ↑

B) C-terminal (lysosomal carboxypeptidase A)

i) Asp-Arg-Val-Tyr-Ile-His-Pro-Phe (18)

↑ ↑

ii) Encephalitogenic peptides (Phe release)

Phe^{44} ----------Val-His-Phe^{89}

↑

C) C-terminal (carboxamide peptidase)

i) Oxytocin

Cys-Tyr-Ile-Gly-Asn-Cys-Pro-Leu-Gly.NH_2 (19)

↑ ↑

ii) Arg-vassopressin

Cys-Tyr-Ile-Gln-Asn-Cys-Pro-Arg-Gly.NH_2 (19)

↑ ↑

See Table 1 for details. The data on encephalitogenic peptides represents previously unpublished findings.

survival of a peptide during transport to its target site is an important consideration and may be of equal importance to the traditional approach of preparing analogs based on structure-receptor relationships. The present description is limited largely to specificity studies rather than details of purification procedures and methods of assay, since these have been documented extensively elsewhere (1-3, 33, 34).

Hormonal Inactivating Enzymes Active at Physiological pH.

Cathepsin M (brain neutral proteinase): It was established over a decade ago that brain is characterized by a soluble endopeptidase differing markedly in its properties from those in other organs (1,25,26). This enzyme can be purified by simple procedures resulting in a 10-40 fold purification. Brain can be extracted with hypotonic buffers, or with 10 volumes of 0.32 M sucrose (in which it is more stable) followed by chromatography on DEAE-cellulose (1, 13, 14, and unpublished findings). This procedure has been repeated by other laboratories (1, 12, 27), and although differences remain concerning its properties, there is general agreement that a very active endopeptidase exists in brain. The term **'cathepsin M'** is proposed for this enzyme for descriptive purposes, and an improved purification procedure will be described shortly. The existence and specificity of brain cathepsin M can be demonstrated using a number of newer (biologically active) substrates containing blocked end-groups as described below.

LH-RF (Luliberin): This decapeptide is characterized by an N-pyroglutamyl and C-glycinamide group. It is present in the hypothalamus and can directly stimulate the anterior pituitary to release luteinizing hormone (LH) and follicle stimulating hormone (FSH) (5,6). Analogs have been prepared with the objective of inhibiting the release of LH and FSH for improved fertility or possibly for contraceptive purposes. In a typical experiment, we incubated 50 nmoles of this substrate with suitable aliquots of crude and partially purified cathepsin M, and we have shown that there is a time-dependent release of all amino acids (8). Results for a brain homogenate from rat are presented in Table 3: this experiment demonstrates the utility of this substrate for detecting endopeptidase activity even in the presence of other enzymes. It can be seen that there is a preferential release of internal amino acids Tyr, Gly, and Leu as compared to His (adjacent to pyroGlu), and of Gly.NH_2 on the C- terminal. Our conclusion that the preferential release of internal amino acids was evidence for a rate-limiting cleavage by cathepsin M, was confirmed using a partially purified preparation (28). Further evidence was provided by the use of analogs which prevented cleavage at the internal site thus blocking the secondary release of amino acids (8). Thus if Gly in position 6 was replaced by D-Ala there was no significant release of Tyr and Leu (Table 3); the release, however, of Gly. NH_2 implies the existence of a second inactivating enzyme specific for C-terminal amides (carboxamide peptidase, see below). This second enzyme can be blocked by substituting Gly. NH_2 with ethylamide ($C_2H_5.NH_2$) (Table 2). It is striking that analogs which resisted digestion to a greater degree: D-Ala^6-(LH-RF) was 6-fold, and D-Ala^6-(LH-RF)-ethylamide was 50-80 fold more potent than the native analog, in terms of LH and FSH release **in vivo** (Table 3). The inactivation of LH—RE by an endopeptidase was confirmed by the studies of Koch et al (9) using a crude soluble preparation obtained from rat hypothalamic tissue; they succeeded in isolating one of the intermediate peptidyl products pyroGly-His-Trp-Ser-Tyr-Gly from the digest mixture confirming the internal cleavage site (Table 1).

Conclusions: 1) brain contains an endopeptidase active at pH 7.6 that cleaves an internal bond of the decapeptide LH-RF (termed cathepsin M); 2) a new method of assay is described based on the preferential release of internal amino acids from an N- and C-blocked peptide. On the basis of studies with analogs the substrate LH-RF-ethylamide is ideally suited for assay of cathepsin M (see Table 3); 3) assays can be performed in the presence of exopeptidases (they are required for the secondary cleavage of the split fragments in order to detect free amino acids-this can represent a less arduous approach than isolation of intermediate peptidyl products); 4) analogs can reveal the presence of other (novel) inactivating enzymes (see carboxamide peptidase below); 5) studies on biodegradation in relation to **in vivo** activity provide a rational approach for the preparation of longer acting derivatives.

Kinin-9 (Bradykinin and analogs): There are three hypotensive peptides known collectively as kinins, bradykinin (kinin-9), kalliden (kinin-10), and Met-Lys-bradykinin (kinin-11) (see review 1). Kinins themselves are formed in plasma by the action of a proteolytic enzyme, kallikrein, on precursors known as kininogens (Table 1); kallikrein, or its suggested name kininogenin, (E.C.3.4.21.8) is a serine proteinase. The release of kallikrein from a proenzyme and of kinins from protein and polypeptide precursors further illustrates the fine control exercised by intracellular catabolic processes. In the case of kinin-9 itself, this polypeptide contains a N-and C-terminal Arg,

TABLE 3

Cleavage of LH-RH Analogs by Rat Brain Homogenate

Analog	nmoles per cent released								
	His^2	Ser^4	Tyr^5	Gly^6	Leu^7	Arg^8	Pro^9	$Gly.NH_2{}^{10}$	Activity
LH-RH	20	30	45	46	50	50	30	15^+	100
$ethylamide^{10}$	18	41	26	38	40	30	tr	$(0)^+$	
$D\text{-}Trp^3$	0	tr	5	38	43	38	23	29	
$3\text{-}Me\text{-}His^2$	0	0	22	18	28	17	15	18	
$D\text{-}Ala^6$	5	tr	tr	(0)	0	18	15	15	670
$D\text{-}Ala^6$, ethylamide	5	tr	tr	tr	0	tr	0	$(0)^+$	8,000

The incubation mixture contained 50 nmoles of each analog and 0.15 ml of a 1:10 rat brain homogenate in a volume of 0.5 ml of 10 mM Tris-HCl containing 0.5 mM Clelands'reagent and incubated for 3 h at 37°. Enzyme action was terminated by addition of 3 % sulfosalicylic acid. Amino acids and Gly. NH_2 was determined on an autoanalyzer. Results are the means of 3-5 experiments agreeing within 5 %.
The brackets represent substituted amino acids. Trp and pGlu were not determined.

tr = trace quantities.

+**In vivo** ovulatory activity compared to LH-RH as 100.

and it is not as suitable a substrate for the detection of endopeptidases because tissues are known to contain exopeptidases (in some cases referred to as bradykininases) that remove either the N-terminal or the C-terminal group. The discovery that kinin-9 was inactivated also by cathepsin M was accidental, and arose from the finding of a preferential release of selected internal amino acids in a series of time-related studies on breakdown using crude brain extracts (11). These results were verified using a partially purified enzyme preparation which pointed to the Phe^{5}-Ser^{6} bond as the site of cleavage (11) (see Table 1). Subsequently, Carmargo et al (12) also purified cathepsin M by a similar procedure and were able to separate the intermediate peptidyl products Arg-Pro-Pro-Gly-Phe and Ser-Pro-Phe-Arg. Kinin-9, despite the presence of a free N-terminal Arg, is not inactivated by a brain aminopeptidase (purified on the basis of Leu-Gly-Gly as the substrate) or an 'arylamidase' (Arg-β-naphthylamide). It is evident that in brain extracts there may be a specific exopeptidase inactivating kinin-9 (based on the observed release of Arg); but it is premature to label this as a bradykininase. There are many candidates for 'bradykininase' in the literature, all with different catalytic mechanisms, and this term would be of little value. For example, tissues are reported to contain a peptidyl dipeptidase (E.C. 3.4.15.1) removing a C-terminal peptide from kinin-9 and also from proangiotensin (22), but there are no studies as yet on its presence in brain.

Conclusions: 1) Kinin-9 and its analogs can be inactivated by a neutral proteinase (called tentatively cathepsin M since it has identical properties to the enzyme that inactivates LH-RF and hydrolyzes proteins). 2) Brain contains a second enzyme that removes N-terminal Arg, but is distinct in its properties from other described exopeptidases.

Substance P: Gut-contracting substances have been known for over 30 years, but their structures have been clarified only recently. Substance P is a unidecapeptide characterized by a C-terminal methionamide (29). The potent effects of this peptide combined with its differential distribution in the nervous system has led to suggestions that it may play a role in neurotransmission (10). A function of this nature would necessitate rapid inactivation at its site of action. Early studies indicated that brain tissue could inactivate substance P, but the purity of the substrate employed at that time did not permit characterization of the enzyme(s) involved (see review, 1). In recent studies, we showed that incubation of this unidecapeptide with crude or partially purified extracts of rat brain led to the preferential release of internal amino acids, implying the existence of a rate-limiting cleavage by cathepsin M (13) (Table 1). In addition, there was a slower release of N- and C-terminal groups, indicating the possible presence of other enzymes with the potential for inactivation. At long periods of incubation with crude extracts we also observed the release of free Met. Studies with model peptides revealed that deamidation occurred only after release of methionamide from peptide linkage (13). Thus, with the dipeptide Leu-Met. NH_2, the cleavage products at short incubation periods were Leu, Met. NH_2, and a trace of free Met. The formation of Met from methionamide was confirmed by incubation of this residue with brain extracts; this required extended incubation periods (13). Deamidation is a known property of some aminopeptidases, but there have been surprisingly few studies on their specificity using polypeptidyl amides (21, 32). This may be an important pathway for regulation of breakdown since there are many examples in nature of C-terminal amides; examples include gastrin, secretin, α-MSH, cerulein, calcitonin, LH-RF, TRH, and MIF, to mention just a few (see review, 1). Our experience with substance P, LH-RF, and MIF suggests that deamidation does not play any major role in the inactivation process for these hormones. This does not exclude, however, the possibility that deamidation could play a role in the inactivation of other hormones with different C-terminal amides.

The enzyme inactivating substance P by internal cleavage has been purified approximately 10-fold using the simple extraction and column chromatography procedures described above (13). The major peak of activity was coincident with enzyme degrading hemoglobin and histones, and exhibited the properties associated with cathepsin M. Further evidence that internal cleavage preceded release of free amino acids was provided by a bioassay procedure utilizing contraction of guinea-pig ileum.

Conclusions: 1) Substance P is a good substrate for cathepsin M, with cleavage at more than one internal site (as postulated in Table 1). 2) Cathepsin M preparations contain an aminopeptidase contaminant capable of releasing Arg at lower rates. 3) Preparations contain a C-terminal cleaving enzyme, resulting in the appearance of Met. NH_2 and traces of Met.

Somatostatin: This cyclic tetradecapeptide, which occurs in the hypothalamus, can inhibit the release of growth hormone from the pituitary, and also glucagon and insulin from the pancreas (6,35). Its use as a therapeutic agent, however, is severely limited by its short biological half-life (ca. 4 min) (14). Incubation with brain extracts or a partially purified cathepsin M preparation led to a preferential release of internal amino acids, suggesting that there was a rate-limiting internal cleavage similar in character to that of the three hormones described above (see also Table 1). A slower release of Ala implied that N-terminal inactivation was not a major route for inactivation. Conclusive evidence for the action of cathepsin M was supplied by the use of an analogs with a blocked N-terminal, des Ala-Gly-(Ac-Cys^3)-somatostatin. Incubation of this analog with enzyme led to a preferential release of internal amino acids (14). No differences were observed in the rates of breakdown of the cyclic versus linear forms of this hormone or its analogs. The enzyme inactivating somatostatin was purified 10-fold by the method described above for substance P and bradykinin (14).

Conclusions: 1) Somatostatin can be inactivated by a neutral endopeptidase (cathepsin M), 2) Cathepsin M preparations contain an aminopeptidase cleaving the Ala-Gly bond. 3) The presence or absence of a disulfide bridge does not affect rates of breakdown (in contrast with oxytocin, below).

Oxytocin and vassopressin: These hormones are characterized by a hexapeptidyl ring attached to a tripeptidyl alicyclic tail (Fig. 2). Both hormones are resistant to breakdown on prolonged exposure to crude brain extracts, and purified brain aminopeptidases (1,19). Incubation, is accompanied, however, by the release of C-terminal fragments notably Leu-Gly. NH_2 or Arg-Gly. NH_2 and Gly. NH_2. This mechanism of inactivation is in contrast to the action of 'oxytocinase' of sera that can remove N-terminal residues from oxytocin (see review, 1). Product-precursor studies showed that dipeptide release preceeds the formation of glycinamide implying that cleavage occurred at the -Cys^6-Pro^7-bonds (Table 2). This enzyme is novel in its specificity and acts on polypeptidyl amides and has been termed carboxamide peptidase (see review, 1). It can be differentiated on the basis of its specificity from chymotryptic-like enzymes and from brain cathepsin M (see above for LH-RF). It is, for example, incapable of cleaving analogs with a free C-terminal carboxyl group (arginine vasopressionoic acid). Studies with analogs show that the ring structure prevents the action of other enzymes. The removal of the ring to yield linear peptides leads to rapid degradation by a brain arylamidase (19) as illustrated by the following two materials:

S-bz-Cys-Tyr-Ile-Gln-Asn-S-bz-Cys-Pro-Leu-Gly.NH_2

Tyr-Ile-Gln-Asn-S-bz-Cys-Pro-Leu-Gly.NH_2

Replacement of the disulfide ring by other substituents yielding cyclic structures also prevented the action of aminopeptidases; (1,6)-aminosuberic acid-lysine vassopressin (replacement of the terminal amino group by hydrogen and the disulfide by an ethylene bridge).

The role of the disulfide bridge in breakdown is an important topic since it is a characteristic of a very large number of proteins and polypeptides. In the case of oxytocin there is evidence that the ring structure confers a stable conformation which is unfavorable to the action of aminopeptidases and proteinases (19). It is notable that in the case of somatostatin, which has a disulfide bridge spanning 12 residues, that no hindrance to cathepsin M was observed with purified preparations. In the case of insulin, tissues are known to contain transdehydrogenases capable of cleaving the ring structure to yield linear polypeptides; these in turn are better substrates for breakdown. Thus in the case of some hormones the mechanism of breakdown may be complex and require two distinct catalytic mechanisms: scission of the ring structure followed by the action of proteolytic enzymes.

The enzyme(s) releasing the C-terminal fragments of oxytocin and vassopressin have been purified from rat brain cytosol fractions (4). On elution from DEAE-cellulose columns two peaks of activity were obtained; one gave a higher release of the dipeptide as compared to glycinamide. Since the dipeptide is an excellent substrate for brain arylamidases, the release of glycinamide may be a consequence of contamination by this aminopeptidase (3).

Conclusions: 1) Cyclic nonapeptides oxytocin and vassopressin are stable in the presence of cathepsin M and brain aminopeptidases. 2) The major route for inactivation in brain is by the C-terminal release of a dipeptide. 3) The formation of glycinamide is a consequence of breakdown by a brain arylamidase.

2. Lysosomal Proteinases

Cathepsin D: It is well documented that lysosomal enzymes are greatly elevated in tissues subjected to degeneration, including diseases associated with demyelination, experimental allergic encephalomyelitis (EAE) and multiple sclerosis (see review 1). Destruction of myelin can be mediated only by degradative enzymes, but the source of these enzymes, the manner of their activation or release, have not been established. In an attempt to answer some of these questions, a model disease state, EAE, has been used for a number of years. It involves the induction of demyelination by injection of a purified myelin component (basic protein) into suitable donor animals under defined conditions (see review 1). Basic proteins, it might be noted, are just one of many different types of protein in myelin, as illustrated by separation on SDS-acrylamide gels (Fig. 1 a,b). In man and ox there is only one basic protein which can be extracted with acid solvents and purified; it is a protein with 170 residues and 18.000 daltons (16,17, Fig. 2) It is suitable as a substrate for proteinases for the following reasons: it is highly soluble at low pH; it is a linear polypeptide lacking cysteine; and it contains an N-protected (N-acetyl-alanine) grouping, and thus is not a substrate for aminopeptidases.

The mechanism by which a myelin component can induce demyelination is a subject of active research. It has been established that EAE is dependent on sensitization of thymus-derived lymphocytes (T-cells) by basic protein and this disorder falls within the class of a cell-mediated (hypersensitive), autoimmune disorders, with similarities to multiple sclerosis. It is not clear how a membrane-bound antigen can be released and interact at the relevant sites, although recent studies with cathepsin D provide suggestive evidence for a possible mechanism. Incubation of myelin membranes with a purified cathepsin D from brain does lead to large breakdown of the membrane-bound basic proteins, as shown by their change in density after separation on SDS-gels as compared to untreated controls (Fig. 1 a,b). In separate studies with purified bovine basic protein, we were able to demonstrate a limited cleavage with the production of only three polypeptidyl fragments (Fig. 3) (16,17). These were separated by preparative electrophoresis on acrylamide, and their structure and end-groups were determined, with results showing cleavage at only two sites, the Phe-Phe linkages at positions 43-44 and 89-90 (Fig. 2). The discovery of a limited cleavage is consistent with the finding of split fragments of basic protein on extraction of crude homogenates with dilute acids. For a period there was a considerable controversy concerning the structure of basic protein which can now be ascribed to proteolysis during extraction procedures. One of the fragments (Phe^{44}-Phe^{90}) produced by cathepsin D is identical to that isolated from bovine spinal cord and known to be highly encephalitogenic in rabbits (17). Another fragment contains the active encephalitogenic sequence (Phe-Ser-Trp-Gly-Ala-Gly-Gly-Gln-Lys), active in guinea pigs (Fig. 2). The results with catheptic fragments support the

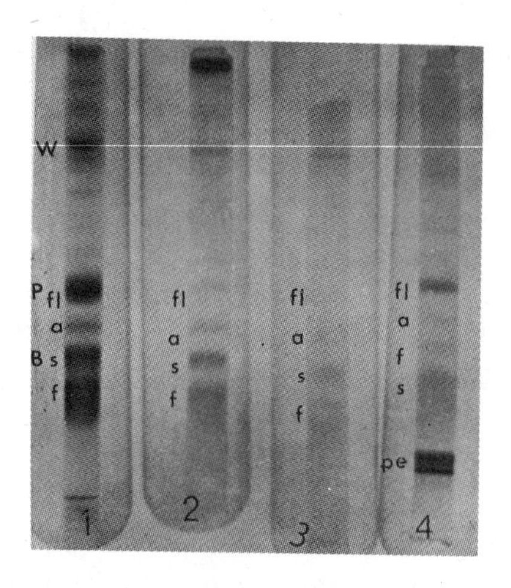

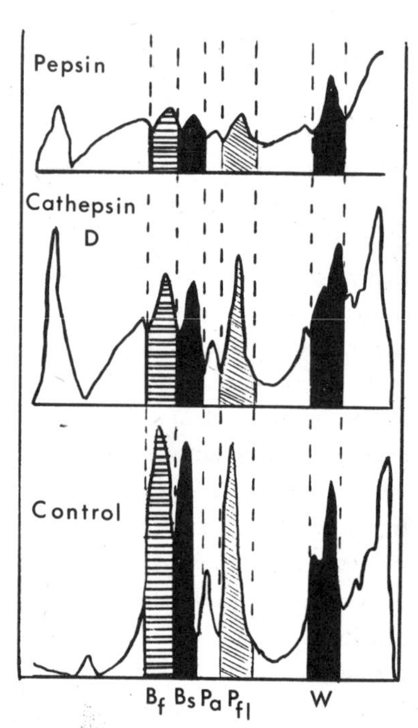

Fig. 1 a,b. Proteins present in purified rat myelin membranes after electrophoresis in 15 % bis-acrylamide gels containing 1 % w/v sodium dodecyl sulfate (SDS). Gels were 10 cm in legth and electrophoressed for 18 h at 45 v constant voltage and subsequent stained with acid Fast Green. The first gel represents myelin incubated in 50 mM sodium citrate buffer in the absence of added enzyme for 5 h at 37°; the entire incubation mixture was lyophilized prior to defatting with ether/ethanol and placing on the gel (100 ug protein). The second gel represent the change on incubation with calf brain cathepsin D for 5 h at 37° at the same pH (enzyme to substrate ratio 1:30), and the third gel represents the effect of pepsin (1:300), and the fourth gel the effect of trypsin (1:300). The identification of the bands is shown in Fig. 2 b which represents a densitometric scan prepared on a Gilford sphectrophotometer fitted with a linear attachment. (For full details see Marks, Grynbaum and Lajtha, Neurochemical Research,1, in press, 1975).

Abbreviations: B_f – fast basic protein (18,000 daltons)

B_s – slow basic protein (approx. 16.500 daltons)

Pa – agrawal proteolipid (also known as DM-20 25,000 daltons)

P_{fl} – Folch-Lees proteolipid (26,000 daltons)

W – Wolfgram component (High molecular weight 100,000 daltons);

Pe – split peptides of less than 10,000 daltons.

hypothesis that this type of cleavage is involved in the etiology of some demyelinating states. The release of smaller (toxic) peptides that are more diffusable from a membrane might account for interaction at the relevant immunological sites.

The highly selective cleavage of basic protein by cathepsin D as compared to the multiple products obtained with pepsin, is somewhat puzzling, since many more potential sites exist in this relatively large protein. Despite the absence of cysteine there may be peculiar structural features that prevent access to most of the available peptidyl bonds. Among the unusual features of basic protein are: N-protected group N-acetyl alanine, only one Trp residue, the presence of a tripoline bridge, and the presence of methylated Arg (Fig. 2). Currently we have prepared a series of

Fig. 2. Amino acid sequence of myelin basic protein from bovine spinal cord. T-tryptic peptides; P, peptic peptides and D (↓), points of cleavage by cathepsin D (Phe-Phe bonds at positions 43-44 and 89-90). This figure is based in part on that in **Multiple Sclerosis,** edited by E.J. Field, T.M. Bell, and P.R. Carnegie, North Holland, Elsevier, 1972.

hexapeptides based on the structure of the sensitive peptidyl bonds. Preliminary studies show that Gly-Arg-Phe-Phe-Gly-Gly is a substrate for cathepsin D and pepsin, yielding Phe-Gly-Gly as one of the split products. This approach may hold value in defining the catalytic sites of cathepsin D as compared to pepsin, since other hexapeptides were resistant to the lysosomal enzyme but degraded by pepsin. Studies show that degradation of basic protein by cathepsin D is inhibited by pepstatin, a pentapeptide (36); this or other synthetic peptides may offer promise in blocking tissue damage.

Cathepsin A: Along with cathepsin D, there is a large increase in lysosomal carboxypeptidase A during experimental demyelination, and in diseased tissue from patients with multiple sclerosis. We have now purified this enzyme from brain about 200-fold, using a modification of the methods of Weinstock and Iodice (30) and Kawamura et al (31). The enzyme can be extracted easily from frozen calf brain with 5 volumes of acetate buffer pH

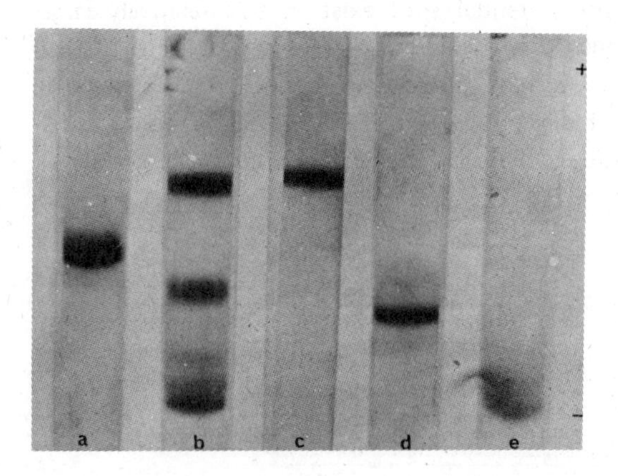

Fig. 3. Comparison of the cleavage products obtained from purified basic protein (a) after digestion with purified bovine acid proteinase of brain for 3 h (b) with the separated split products (c, sequence 44-89), (d, sequence 1-43), and (e, sequence 90-170). Electrophoresis was performed on 10 % acrylamide gels 0.6 x 6.5 cm at pH 8.7 (buffer 25 mM Tris-glycine) using a constant current of 1 mA per gel for the first 5 minutes followed by 2.5 mA per gel for 50 minutes. Gels were stained with 1 % amide black in 7.0 % acetic acid for 1 h and then destained by repeating washing in the same solvent. See Fig. 2 for structure of basic protein.

5.5 containing 0.5 % desoxycholate followed by heat treatment, precipitation with NH_4SO_4, and chromatography on DEAE-Sephadex in the presence of 100 mM sucrose and 20 mM KCl. Further purification can be achieved by concentration in an Amicon filter unit followed by gel filtration on Sephadex-G200. The enzyme is stable when frozen for several months in the presence of sucrose and salt. Two peaks of activity were eluted from DEAE-Sephadex both with very high M.Wt. (in excess of $10x10^3$) (Fig. 4). Enzyme from the first peak is devoid of aminopeptidase activity and preferentially degrades Z-Gly-Tyr and Z-Phe-Tyr, as compared to other N-protected dipeptides (18). Brain enzyme is similar in properties to kidney but different from liver in being inhibited completely by low concentrations of DFP and PCMB (31,37). This may indicate that there is more than one form of lysosomal carboxypeptidase A. Erdos and Yang (23, 24), observed the presence of a carboxypeptidase A-like enzyme in kidney cortex and leucocytes specifically degrading Z-Pro-Gly and now termed prolycarboxypeptidase).

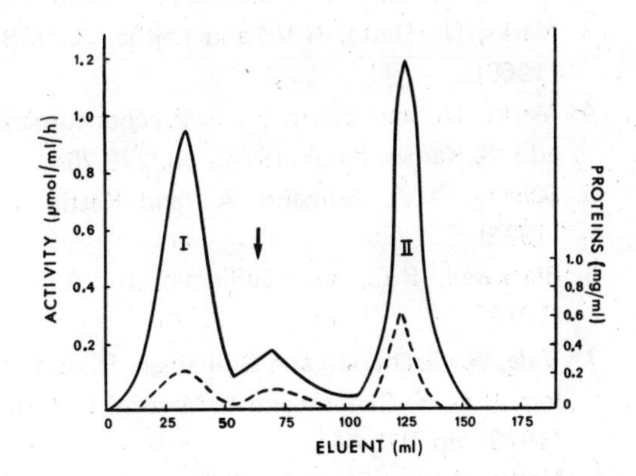

Fig. 4. Separation of cathepsin A activities on passage through a DEAE Sephadex A-50 column. The column was equilibrated with 10 mM acetate buffer pH 5.5 containing 100 mM sucrose and 20 mM KCl as stabilizing agent. Peak I activity appeared without application of a gradient, peak II appeared on addition of 0.42 M KCl as denoted by the arrow. Activity is expressed as u moles of tyrosine released from the Z-Gly-Tyr per ml per h when incubated under the standard conditions (18).

Our best evidence to date that the brain enzyme is a carboxypeptidase has come from studies on the inactivation of angiotensin-II. Incubation led to a release of Phe with time with only trace levels of the adjacent residues (18). On aging our preparations for several months, only the C-terminal Phe was detected. Purified brain carboxypeptidase A does not attack myelin basic protein (since the C-terminal groups are Arg-Arg, see Fig.2). It is therefore evident that cathepsin D action preceeds that of lysosomal carboxypeptidase A, as confirmed by studies **in vitro** (unublished findings). This provides yet another example of a sequential breakdown that may have biological significance in degenerative processes.

REFERENCES

1. Marks, N., and Lajtha, A., in Handbook of Neurochemistry, (A. Lajtha, ed.), Plenum Press, New York (1971), Vol. 5A, pp. 49-139.
2. Marks, N., Int. Rev. Neurobiol. 11, 57-90 (1968).
3. Marks, N., Datta, R.V., and Lajtha, A., J. Biol. Chem. 243,2882-2889 (1968).
4. Marks, N., and Stern, F., in Psychoneuroendocrinology, (N. Hatotani, ed.). S. Karger, Basel (1974), pp. 276-284.
5. Schally, A.V., Arimann, A., and Kastin, A.J., Science 179, 341-350 (1973).
6. Blackwell, R.E., and Guillemin, R., Ann. Rev. Physiol. 35, 357-390 (1973).
7. Vale, W., Grant, G., and Guillemen, R., in Frontiers of Neuroendocrinology (Eds. F. Ganony and L. Martins) Oxford Univ. Press, New York (1973), pp. 375-441.
8. Marks, N., and Stern, F., Biochem. Biophys. Res. Comm.61, 1458-1463 (1974).
9. Koch, Y., Baram, T., and Chobsieng, P., Biochem. Biophys. Res. Comm. 61, 95-103 (1975).
10. Takahashi, T., and Otsuka, M., Brain Research 87, 1-11 (1975).
11. Marks, N., and Pirotta, M., Brain Research 33, 565-567 (1971).
12. Camargo, A.C.M., Shapanka, R., and Greene, L.J., Biochem. 12, 1838-1844 (1973).

13. Benuck. M., and Marks, N.,Biochem. Biophys. Res. Comm. 65, 153-160 (1975).
14. Marks, N., and Stern, F., FEBS Letters, 55, 220-224 (1975).
15. Scott, A.P., Ratcliffe, J.G., Rees, L.H., Landon, J., Bennett, H.P.J., Lowry, P.J., and McMartin, C., Nature 244, 65-67 (1973).
16. Marks, N., Benuck, M., and Hashim, G., Biochem. Biophys. Res. Comm. 56, 68-74 (1974).
17. Benuck, M., Marks, N., and Hashim, G.A., Eur. J. Biochem. 52, 625-631 (1975).
18. Grynbaum, A., and Marks, N. (1975), J. Neurochem. in press.
19. Marks, N., Abrash, L., and Walter, R., Proc. Soc. Exp. Biol. Med. 142, 455-460 (1973).
20. Abrash, L., Walter, R., and Marks, N., Experienta, 27, 1352-1353 (1971).
21. Marks, N., and Walter, R., Proc. Soc. Exp. Biol. and Med. 140, 673-676 (1972).
22. Elisseeva, Y., and Orekhovich, V., Dookl. Akad. Nauk SSSR 153, 954 (1963).
23. Yang, H.Y.T. and Erdos, E.G., Excerpta Medica Inter. Cong. Series – Immunopathology of Inflammation. 229, 146-148 (1970).
24. Yang, H.Y.T., Erdos, E.G., and Levin, Y., Biochim. Biophys. Acta 214, 374-376 (1970).
25. Marks, N., and Lajtha, A., Biochem. J. 89, 438-447 (1963).
26. Marks, N., and Lajtha, A., Biochem. J. 97, 74-83 (1965).
27. Riekkinin, P.J., and Rinne, U.K., Brain Research 9, 126-135 (1968).
28. Benuck, M., and Marks, N., Brain Research, (1975) in press
29. Leeman, S.H., and Mroz, E.A., Life Sci. 15, 2033-2044 (1974).
30. Weinstock, I.M., and Iodice, A.A., in Lysosomes in Biology and Pathology (Eds. J.T. Dingle and H.B. Fell) North Holland Publ. Amsterdam, London (1969). Vol. 1, pp. 450-468.
31. Kawamura, Y., Matoba, T., Hata, T., and Doi, E., J. Biochem. 76, 915-924 (1974).
32. Smith, E.L., in Methods in Enzymology, Vol. 2, pp. 83-93 (1955).
33. Marks, N., Stern, F., and Lajtha, A., Brain Research 86, 307-322 (1975).
34. Serra, S., Grynbaum, A., Lajtha, A., and Marks, N., Brain Research 44, 529-592 (1972).

35. Koerker, D.J., Ruch, W., Chideckel, E., Palmer, J., Goodner, C.J., Ensinck, J., and Gale, C.C., Science 184, 482-483 (1974).
36. Marks, N., Grynbaum, A., and Lajtha, A., Science 181, 949-951 (1973).
37. Taylor, S.L. and Tappel, A. L., Biochim. Biophys. Acta 341, 112-119 (1974).

DETERMINATION OF THE AVERAGE DEGRADATION RATE OF MIXTURES OF PROTEIN

Peter J. Garlick*, and Robert W. Swick

Department of Nutritional Sciences, University of Wisconsin, Madison, Wisconsin 53706, USA

Rates of breakdown of proteins can be determined from the decay of label in protein after injection of a non-reutilisable precursor. With homogeneous proteins the decay is exponential which, plotted on semi-log graph paper, gives a straight line whose gradient is the fractional rate of breakdown. With mixtures of protein, such as occur in whole tissues, the semi-log plot of radioactivity against time is curved. The apparent rate of breakdown, therefore, varies with the duration of the experiment, becoming slower as the period of measurement becomes longer (1-4). We have attempted to obtain a unique value for the average rate of turnover of rat liver protein from the decay of label in protein after $^{14}CO_3^=$ injection.

Research supported jointly by NIH grant AM-14704, UW Graduate School and MRC of Great Britain

Present address: Department of Human Nutrition (Clinical Nutrition and Metabolism Unit) London School of Hygiene and Tropical Medicine, Keppel Streeet, London WC1E 7HT, England

METHODS

Female rats weighing about 240 g were injected with $^{14}CO_3^{=}$ and the liver protein isolated and counted as described previously (2). Animals were fed a diet containing 12 % casein and did not grow appreciably during the 41 days of the experiment. In a similar group of rats the rate of liver protein synthesis was determined from the incorporation of label into protein and from the specific activity of the tissue-free amino acid during continuous infusion of ^{14}C-tyrosine into a tail vein for periods of 6 or 10 hr as described previously (5-6).

RESULTS AND DISCUSSION

The decay of label in liver protein over a 41 day period is shown in Fig. 1. Peak labelling occurs at about 1 hr after injection. The apparent turnover rate calculated by assuming linear decay between 1 hr and 3 days is 53 %/day. By comparison, the rate calculated from the decay over 30 days is 17 %/day. It is therefore clear that it is not valid to calculate the turnover rate from the decay over an arbitrarily chosen time interval. We have estimated the average rate of turnover of liver protein from this curve by three methods: (a) by dividing the curve into 3 exponentials; (b) by stochastic analysis and (c) by an approximate method which involves measurement of the time at which 10 % of the initial radioactivity remains.

Firstly it was assumed that the proteins could be divided into three components with fast, medium and slow rates of turnover. The decay of radioactivity (R) can therefore be expressed by the 3 exponential function $R = Ae^{-at} + Be^{-bt} + Ce^{-ct}$. a, b and c are the fractional turnover rates of the components. A,B and C are the initial radioactivities in the components and are proportional not only to the mass of component but also its rate of turnover (7). The relative masses of the components are therefore given by A/a, B/b and C/c. The values of these constants were determined from the experimental results (Fig. 1) by curve fitting with a computer. This enabled the mean fractional rate of turnover for the mixture to be calculated from the formula $k=(A+B+C)/(A/a+B/b+C/c)$. A value of $\bar{k}$ of 41.3 %/day (T1/2=1.7 days) was obtained.

The stochastic approach (8) makes no assumptions about the number of different proteins in the mixture. If the labelling by $^{14}CO_3^{=}$ constitues a genuine pulse, $\bar{k}$ is given by the peak height of the decay curve divided by the area under it between zero and infinite time. This was calculated by drawing the curve on linear graph paper and measuring the area. The area between the last experimental point and infinite time can be neglected since it is very small and in any case probably represents the small amount of recycling which is unavoidable. The value of $\bar{k}$ obtained by this method is 42.8 %/day.

The third method is derived from an empirical observation of a number of conceivable model mixtures. In each case the true value of $\bar{k}$ coincided with the apparent value obtained from the gradient of the straight line between the initial point and a point on the decay curve at a time between 3 and 4 times the mean half-life of the mixture, i.e. when the radioactivity had fallen to 10 % of its initial value (3, J.C. Waterlow, unpublished observations). No explanation can be found for this observation, but the value of $\bar{k}$ obtained by this method from the experimental decay curve in Fig. 1 (40 %/day) is very close to those obtained by more exact analysis. The advantage of this method is that it takes only a few days to complete. The previous two methods, although based on theoretical reasoning, are usable only when it is convenient and possible to construct a full decay curve over a large number of days.

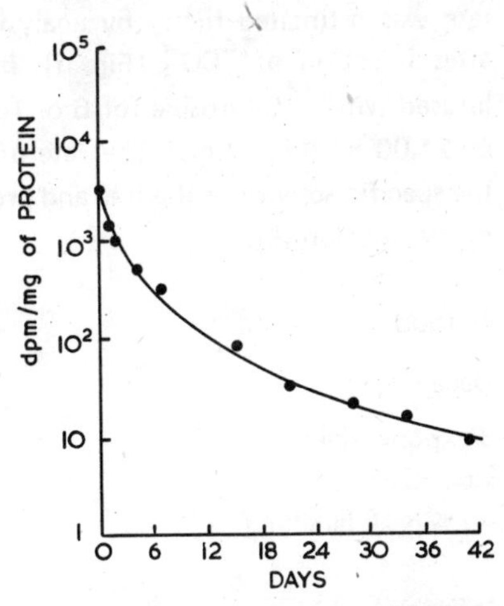

FIG. 1. The specific activity of liver protein in rats at different times after injection of $^{14}CO_3^{=}$. Female rats (240g) were given intraperitoneal injections of 1 mCi $NaH^{14}CO_3$ and groups of 4 killed at intervals of 1 hr to 41 days thereafter.

In order to compare the results obtained from the decay curve with those obtained by an entirely different technique, the rate of protein synthesis in liver was determined by measurement of the incorporation of label into protein during continuous intravenous infusion of ^{14}C-tyrosine (5). We believe that estimates of $\bar{k}$ from measurements of incorporation of label are less susceptible to problems caused by heterogeneity of turnover rates in protein mixtures than estimates from rates of decay (3,7). The figures shown in Table 1 are a little higher than those obtained from the decay curve. The difference may result from a small amount of recycling of label in the decay experiment and also from uncertainty regarding the specific activity of the precursor pool during incorporation. The results obtained from the decay curve also fall within the range of those obtained by other methods (9-11).

TABLE 1

Estimates of the Rate of Protein Turnover in Rat Liver Measured by Different Techniques

The mean fractional rate of protein turnover ($\bar{k}$) in liver of 240 g female rats was estimated firstly by analysis of the specific activity decay curve after injection of $^{14}CO_3^{=}$ (Fig. 1). In addition, rats treated similarly were infused with ^{14}C-tyrosine for 6 or 10 hr starting at either 9.00 hr (daytime) or 21.00 hr (night-time) The rate of protein turnover was calculated from the specific activity of the free and protein bound tyrosine in liver at the end of the infusion (5).

Method	$\bar{k}$(%/day)
Decay	
3 exponential	41.3
stochastic	42.8
10 % peak labelling	40.0
Infusion	
6 hr – daytime	59.1±2.9
6 hr – night-time	50.2±4.9
10 hr – daytime	40.7±2.7

REFERENCES

1. Miehe, M. in Intracellular Protein Catabolism (Hanson, H., Bohley, P., eds.) Leipzig, (1974), p. 142
2. Ip, M.M. and Swick, R.W., J.Biol. Chem. 249, 6836 (1974).
3. Garlick, P.J. and Millward, D.J., Proc. Nutr. Soc. 31, 249 (1972).
4. Millward, D.J., Clin. Sci. 39, 591 (1970).
5. Garlick, P.J., Millward, D.J. and James, W.P.T., Biochem. J. 136, 935 (1974).
6. Garlick, P.J. and Marshall, I., J. Neurochem. 19, 577 (1972).
7. Koch, A.L., J. Theoret. Biol. 3, 283 (1962).
8. Shipley, R.A. and Clark, R.E., Tracer Methods for In Vivo Kinetics, Academic Press, New York and London (1972).
9. Swick, R.W., J.Biol. Chem. 231, 751 (1958).
10. Haider, M. and Tarver, H., J. Nutr. 99, 433 (1969).
11. Fern, E.B. and Garlick, P.J., Biochem. J. 142, 413 (1974).

PRIMARY REACTION OF INTRACELLULAR PROTEIN CATABOLISM

Peter Bohley, Heidrun Kirschke, Jürgen Langner, Susanne Riemann, Bernd Wiederanders, Siegfried Ansorge and H. Hanson

Physiologisch-chemisches Institut der Martin-Luther-Universität, 402 Halle (Saale), Hollystr. 1 (GDR)

In 1968 we proposed a hypothetical mechanism for the selectivity of intracellular protein catabolism: different degrees of exposure of superficial hydrophobic areas of substrate proteins could cause different affinities for lysosomal membranes and/or proteolytic enzymes (Naturwissenschaften 55, 211 (1968).

Two years ago, we found significantly enhanced aggregations and precipitations of short-lived proteins from rat liver cytosol in vitro. Last year in Budapest we reported further results obtained after incubations under widely different conditions in vitro using leupeptin, pepstatin and other inhibitors of proteolytic enzymes. It was shown that proteolytic enzymes alone cannot be the primary cause for the selective aggregations of these short-lived proteins.

In the opinion of SCHIMKE and coworkers, the most important property of the selective catabolism of short-lived proteins is their high molecular weight, implying more susceptible bonds, more different conformational stages, less stable native conformations, and more errors in amino acid sequence. Confirming this result in 1971 (FEBS meeting, Varna), we found, however, also short-lived proteins with **lower** molecular weights (30-50 000 daltons) in rat liver cytosol.

Therefore, the size of substrate protein molecules can be only **one** factor influencing catabolic rates, and we performed experiments to confirm our hypothesis concerning the preferential degradation of apolar proteins.

Double-labeled proteins from rat liver cytosol were prepared as described before (Bohley and cow. 1972) using NaH $^{14}CO_3$ or ^{14}C-guanidinoarginine four days before, and ^{3} H-arginine thirty minutes before, the sacrifice of male rats. Rat liver cytosol (10^7 g . min) was ultrafiltered extensively, until the trichloroacetic acid soluble radioactivity did not exceed 1 % of total radioactivity (for ^{3}H as well as for ^{14}C). These proteins were shaken for 20 seconds in two-phase systems at 2° C, the mixtures cleared up by centrifugation for 2 minutes at 2°C and the amount of ^{3}H- and ^{14}C- radioactivity determined in the different phases. The results are

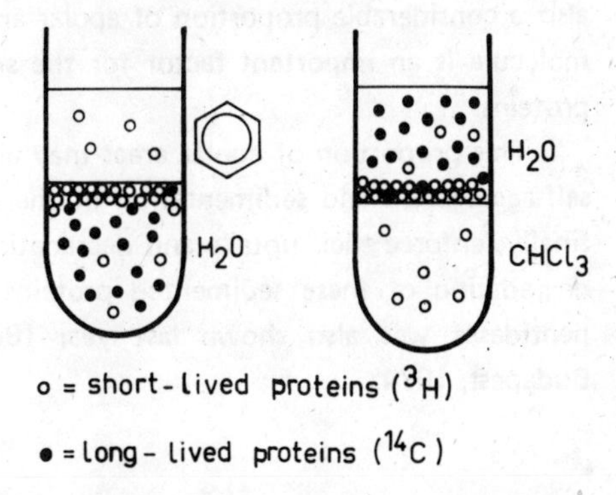

described in the scheme of Fig. 1: independently of the kind of organic solvent (in this case benzene or chloroform), there occurs a preferential enrichment of short-lived (^{3}H-labeled) proteins at the interphase.

It is possible to isolate these interphase-proteins carefully by sucking, which demands some manual skill. In all cases we found the same behaviour of short-lived proteins: their amount in the interphases is higher than the

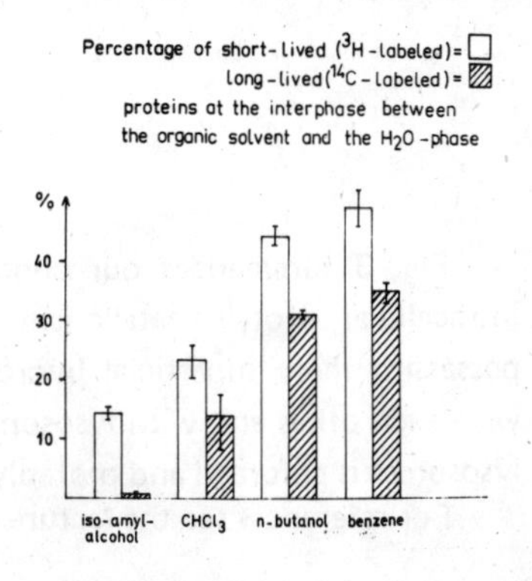

amount of long-lived (^{14}C-labeled proteins), a very small amount is found in the organic solvent, and the amount of short-lived proteins in the water-phase is diminished. A quantitative comparison is shown in Fig. 2, and the same result was obtained using CCl_4, nonane, heptane, hexane, cyclohexane, heptanol and iso-butanol.

Moreover, short-lived proteins are always enriched in the flotating, fat-rich layer of the cytosol after high-speed centrifugations at 10^6 and $>10^7$ g . min.

In conclusion, not only a high molecular weight (Schimke 1970), but also a considerable proportion of apolar areas at the surface of the protein molecule is an important factor for the selective catabolism of short-lived proteins.

This proportion of apolar areas may also be the reason for the selective self-aggregation and sedimentation of these short–lived proteins and may, finally, enforce their uptake and degradation in lysosomes. The preferential degradation of these sedimented proteins by solubilized lysosomal endopeptidases was also shown last year (Bohley et cow., FEBS meeting, Budapest, 1974).

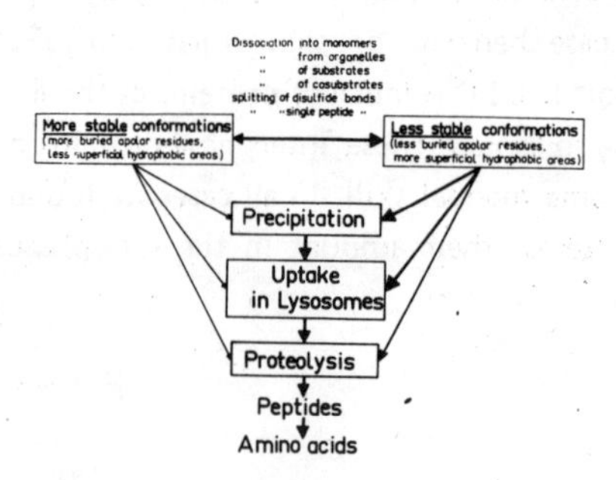

Fig. 3 summarizes our concept concerning the primary reactions of intracellular protein catabolism. Proteins with less stable conformations possessing more superficial hydrophobic areas tend especially to aggregate with each other and with lysosomal membranes. Therefore, their uptake in lysosomes is enforced and proteolysis can follow.

For references see the lecture of H. Hanson and coworkers.

DEGRADATION OF ABNORMAL PROTEINS IN ESCHERICHIA COLI: IN VITRO PROTEOLYSIS OF CYANOGEN BROMIDE PEPTIDES

Alan R. Hipkiss and John T. Kemshead

Department of Biochemistry, University of London King's College, Strand, London WC2R 2LS, England.

INTRODUCTION

Exposure of E. coli to the antibiotic puromycin hydrochloride results in the synthesis of shortened polypeptide, which is rapidly degraded unlike protein of normal chain length (1,2). However, the mechanism by which the cell differentiates between normal length and the puromycin-induced prematurely terminated polypeptide to degrade specifically the latter is not understood. In an attempt to gain some insight into this process, we have produced shortened polypeptide **in vitro** and then examined the degradability of the resultant polypepetide in **E. coli** cell-free extracts.

MATERIALS AND METHODS

5.6 l of **Escherichia coli** B148 were grown as previously described (2) in the presence of 20 μCi L-(1-^{14}C)-leucine (Radiochemical Centre, Amersham) (0.093 mCi/mmole) and then harvested by centrifugation while in logarithmic growth. The concentrated cells (4x10 ml in 0.05M sodium phosphate buffer pH 7.5) were broken by sonication and the 150,000 xg supernatant fraction obtained as described previously (2).

To produce shortened peptides, 34 ml of 90 % formic acid and 1.7g of cyanogen bromide (CNBr) (Eastman Kodak) were added to 14 ml of 150,000 xg supernate and left for 48 hours at room temperature. The product was freeze-dried and taken up in NaOH until dissolution was obtained. HCl was added to return the pH of the solution to approximately 7.5, and the slight precipitate which was formed was removed by centrifugation and filtration through a glass fibre disc. Final volume was 5 ml.

The products of the CNBr cleavage were fractionated according to size using a column of Sephadex (Pharmacia) G150 (40 cm x 3.2 cm diameter), and 0.05 M sodium phosphate pH 7.5 as eluent.

Proteolysis (conversion of acid-precipitable radioactivity into acid-soluble radioactivity) was determined at 37° in a shaking water bath, as previously described (2).

RESULTS AND DISCUSSION

Fig. 1(a) shows the effect of CNBr clevage of ^{14}C-leucine-labelled **E. coli** 150,000 xg supernatant fraction upon its degradability as determined by production of acid-soluble radioactivity. Compared to the native preparation (Fig. 1(c)), addition of crude extract produced a large stimulation in proteolysis (up to 30 % in two hours). Similarly, addition of crude extract to the formic acid denatured material (i.e. CNBr omitted from the cleavage procedure but otherwise treated identically) also gave only low levels of proteolysis (3-4 % in one hour (Fig. 1(b)).

Following fractionation of the CNBr-cleaved material by gel filtration using Sephadex G150, the degradability of the CNBr-peptides with respect to their apparent molecular weight (i.e. elution volume) in crude extract was determined. Fig. 2 shows that the peptides of low apparent molecular weight (i.e. large elution volumes) were far more degradable (up to 38 % proteolysis in 2 hours) than those of high apparent molecular weight (down to approximately 10 % proteolysis in 2 hours). By comparison with known standards, the mean molecular weights of the fractions used were estimated. Fig. 3, which correlates proteolysis with apparent molecular weight of the substrate CNBr-peptides, suggests that as the apparent molecular weight of the substrate decreases below 25,000, its degradability in the crude extract increases markedly.

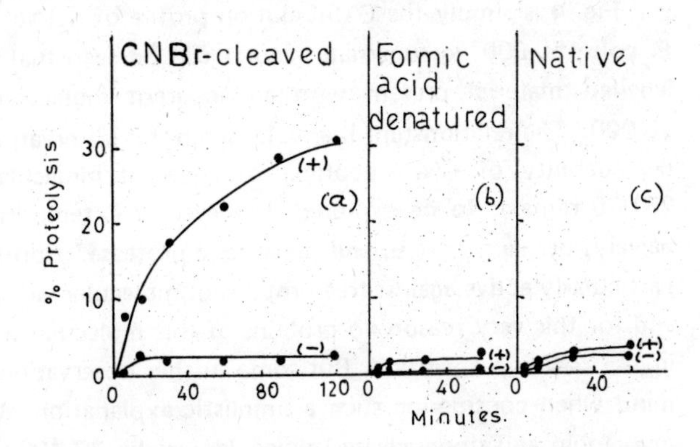

Fig. 1. Effect of E. coli cell extract on proteolysis of CNBr-treated, formic acid denatured, and native E. coli 150,000 x g supernate. (+) In presence of cell extract, (—) in absence of cell extract.

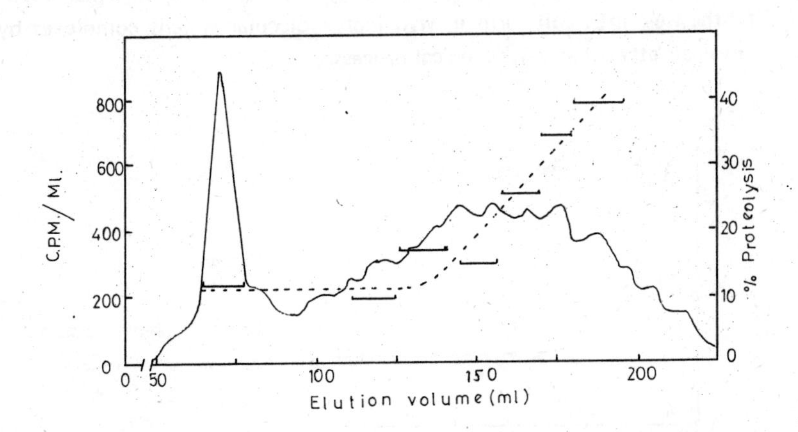

Fig. 2. Fractionation of CNBr-treated E. coli 150,000 x g supernate through Sephadex G150.(—) Elution profile (cpm/ml), (- -) % proteolysis (in 2 hours) in presence of cell extract.

Fig. 4 is simply the G150 elution profile of native ^{14}C-leucine-labelled **E. coli** 150,000 xg supernate, and it can be seen that there is little or no labelled material present with an apparent molecular weight less than 25,000. The relationship, if any, between this observation and the very high degradability of CNBr peptides of apparent molecular weight less than 25,000 remains to be explored. However, it is tempting to suggest, rather naively, perhaps, that **E. coli** possesses a protease/peptidase system which is particularly active against free proteins of molecular weight less than 25,000, and for this very reason no proteins of this molecular weight are detectable in the native preparation. But some further observations must be borne in mind when considering such a simplistic explanation. We have tested both myoglobin and ribonuclease (molecular weights 17,400 and 12,000 respectively) as potential substrates for the proteolytic activity present in the **E. coli** cell- free extract. However, we have no evidence for proteolysis of either protein. Furthermore, we have found by SDS-polyacrylamide gel electrophoresis that the maximum molecular weight of the peptide species present in the CNBr-treated preparation was 12,000. This means, therefore, that the fractions taken from the G150 column for assay for proteolytic susceptibility consist of peptide oligomers or complexes whose molecular weights are very much lower than the apparent molecular weights suggested by their G150 elution volumes. Such aggregation of the CNBr peptides provides support for Prouty and Goldberg's suggestion (1) that abnormal proteins synthesised in**E. coli** form **in vivo** high molecular weight complexes by a physical rather than physiological process.

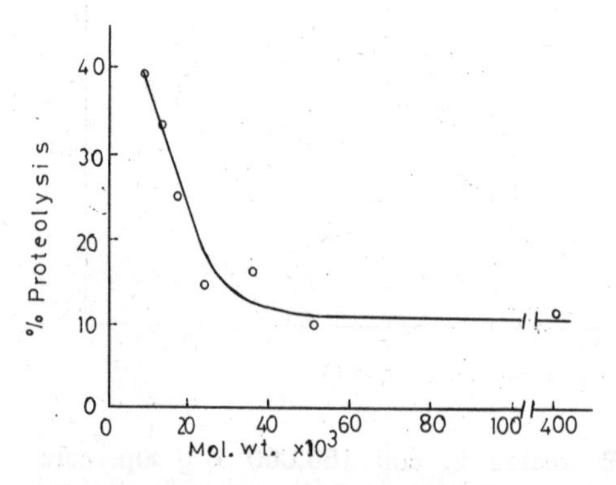

Fig. 3. Variation of % proteolysis (in 2 hours) of CNBr-peptides from 150,000 x g supernatant fraction with apparent mean molecular weight.

The present observations are consistent with our previous suggestion that **in vitro** degradation of **in vivo**-synthesised abnormal proteins is in some way related to the molecular weight and chain length (2), since the denatured protein of normal chain length (formic acid-only treated) was stable in crude extract, while at least some of the CNBr peptides were degradable. Recent experiments have indicated that we are observing proteolysis of the lower molecular weight CNBr peptides since SDS-polyacrylamide gel electrophoresis showed that as the proteolytic susceptibility of the G150 fractions increased, the relative proportion of the higher molecular weight peptides decreased.

The nature of the primary determinant of **in vitro** and **in vivo** proteolytic susceptibility appears therefore to be related in some way to substrate molecular weight (either oligomer or monomer). It remains to be seen if the reason for certain complexes being relatively unsusceptible to in **vitro** proteolysis is either due to the unsusceptibility of the individual peptide components of that complex, or due to complex formation which then reduces the degradability of the individually susceptible monomeric peptide species.

REFERENCES

1. Prouty, W.F. and Goldberg, A.L., Nature New Biol. (London) 240, 147-150(1972).
2. Kemshead, J.T. and Hipkiss, A.R., Eur. J. Biochem. 45, 535-540(1974).

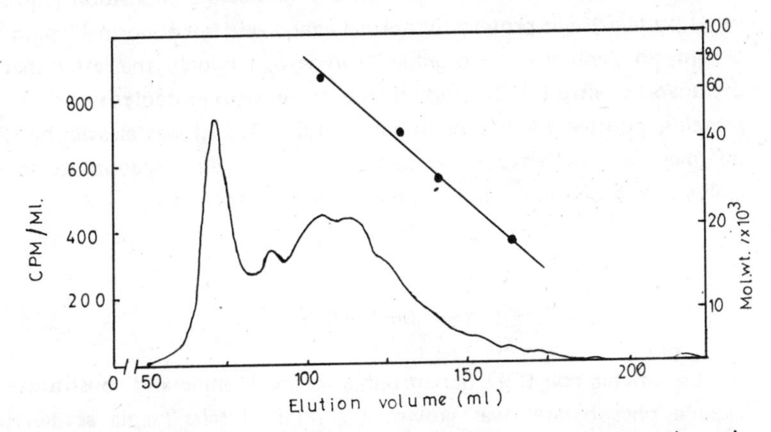

Fig. 4. Fractionation of native **E. coli** 150,000 xg supernate through Sephadex G15C (—) Elution profile (cpm/ml); (—) elution volume of known proteins.

DEGREDATION OF ABNORMAL PROTEINS IN ESCHERICHIA COLI– SUSCEPTIBILITY OF CNBr GENERATED PEPTIDES FROM ALKALINE PHOSPHATASE TO PROTEOLYSIS IN E. COLI CELL EXTRACTS

John T. Kemshead and Alan R. Hipkiss

Department of Biochemistry, King's College, Strand, London WC2R 2 LS, England.

INTROUDCTION

E. coli possesses a mechanism for the specific degradation of abnormal proteins (1,2). Attempts to elucidate the mechanism by which the cell can differentiate between normal and abnormal proteins have shown that whilst proteins of normal length but abnormal amino acid composition (e.g. canavanyl proteins) and proteins of normal amino acid but abnormal length (e.g. puromycin peptides) are degradable **in vivo,** it is only the latter that are degradable **in vitro** (3). To study further the **in vitro** proteolysis of shortened proteins, purified alkaline phosphatase (EC. 3.1.3.1.) was cleaved by CNBr treatment and the ensuing products tested for their susceptibility to proteolysis by incubation with **E. coli** crude cell extract.

METHODS

Escherichia coli C 97 (auxotrophic for methionine and constitutive for alkaline phosphatase) was grown in a minimal salts media as described previously (4). Alkaline phosphatase was purified from these cells by methods described in the above communication (4). Following purification, the enzyme was cleaved by CNBr treatment according to the method of Kelley (5). Proteolysis was determined by assaying the conversion of acid precipitable to acid soluble radioactivity (2). **E. coli** cell extracts used as a protease source were prepared as previously described and used at a final concentration equivalent to 40 ml of cells at OD 1.0 (2). Initial separations of CNBr peptides to peaks A and B were obtained using Sephadex G100 columns (67 cm x 2.6 cm diam.) with 0.1M NH_4HCO_3 as the eluent. Other gel filtration columns (Sephadex G100 and G50 67 cm x 2.6 cm diam) were used at 37° with 0.1M Tris pH 7.5 as the eluent.

RESULTS

Whilst native alkaline phosphatase was stable when incubated either with or without crude cell extract, CNBr cleavage of the enzyme generated species degradable in the presence of extract but completely stable in the absence of cell extract (Fig. 1). In order that the material susceptible to proteolysis could be identified, an attempt was made to separate the CNBr generated peptides by gel filtration. Elution of the peptides from a Sephadex G100 column at room temperature using 0.1M NH_4HCO_3 as the eluent resolved the material into 2 peaks A. M.Wt 36-42000 and B. 17-22000 (Fig. 2). However, previous studies on unfractionated CNBr cleaved alkaline phosphatase had shown that this material was mainly of a molecular weight less than 12,000. To determine that both peaks A. and B. contained peptides in the form of aggregates, these fractions were lyophilised and the products examined upon SDS polyacrylamide gel electrophoresis. Fig. 3 shows that Peak A contained material of 12.000 and 10,000 Mol. Wt. along with some minor less densely staining species, and Peak B contained no material of molecular weight greater than 7,000.

As it was possible to show peptides in Peak A not present in Peak B, this step was used as a primary separating procedure in determining the nature of the peptide species susceptible to proteolysis. Material from Peak A and Peak B was lyophilised, dissolved in 0.1M Tris pH 7.5 and then incubated either with or without a crude cell extract. The material susceptible to proteolysis in A and B was assayed, using 2 different methods – 1. conversion of acid precipitable to acid soluble radioactivity and 2. gel filtration of the peptides following incubation either with or without crude cell extract. Whilst the peptides from Peak A. were stable upon incubation

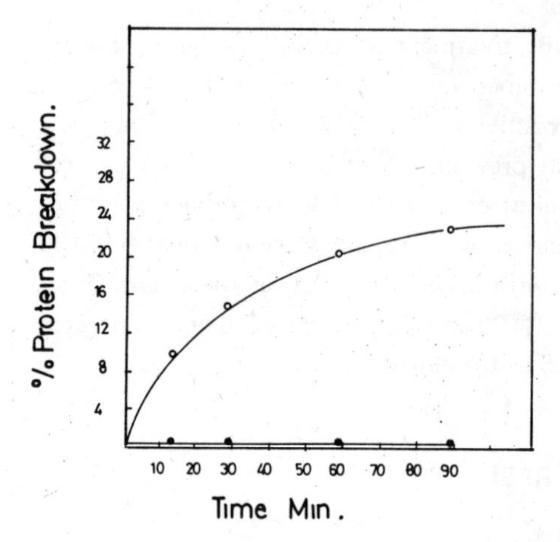

Fig. I Susceptibility of CNBr cleaved alkaline phosphatase (12,000 counts x min^{-1} x ml^{-1}) to proteolysis when incubated either with (O—O—O) or without (●—●—●) cell extract

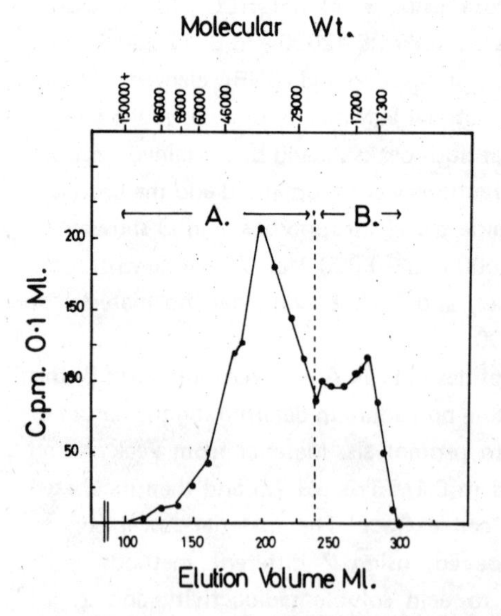

Fig. II Gel filtration of CNBr peptides from alkaline phosphatase (Sephadex G100 – 22^{0} – 0.1M NH_4HCO_3).

without cell extract, some 18 % of the radioactivity in the peptides was rendered acid soluble following incubation with cell extract. Gel filtration of Peak A. material at 37° after incubation either with or without cell extract showed that only the smaller molecular aggregates of less than 29,000 appeared susceptible to proteolysis (Region Q of Fig. 4). Analysis of Peak B material showed that this was again stable when incubated without cell extract but some 65 % of the radioactivity in the material was made acid soluble when incubated with cell extract. Comparison of the elution profile of Peak B material after incubation either with or without crude cell extract showed that material indentifiable throughout the whole of the elution profile was susceptible to proteolysis (Fig. 5).

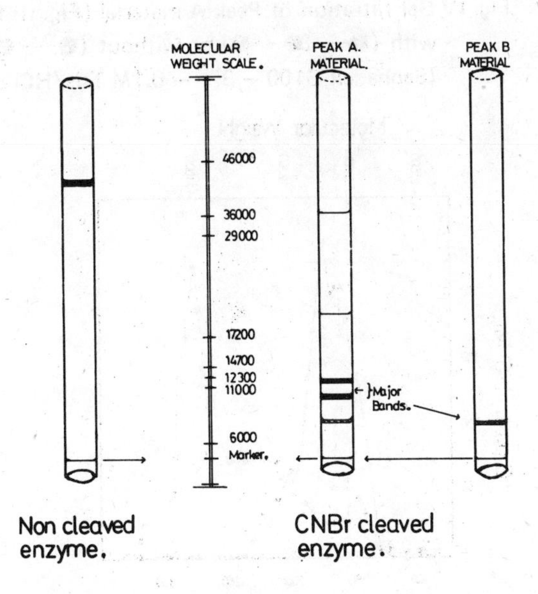

Fig. III SDS Polyacrlyamide gel electrophoresis of native enzyme (a) and partially fractionated CNBr peptides (b,c) from alkaline phosphatase (12.5 % gels run at 6mA/tube for 2 hrs-50 μgms of protein applied to each tube).

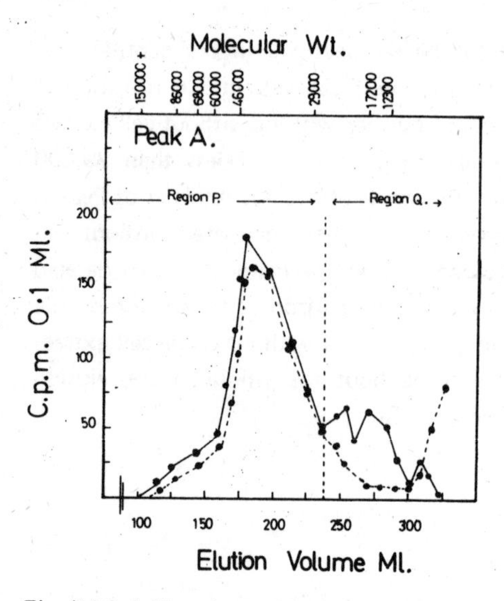

Fig. IV Gel filtration of Peak A material (Fig. II) following incubation either with (●-- ●--●) or without (●---●---●) E. coli cell extract (Sephadex G100 – 37° – 0.1M Tris/HCl pH 7.5).

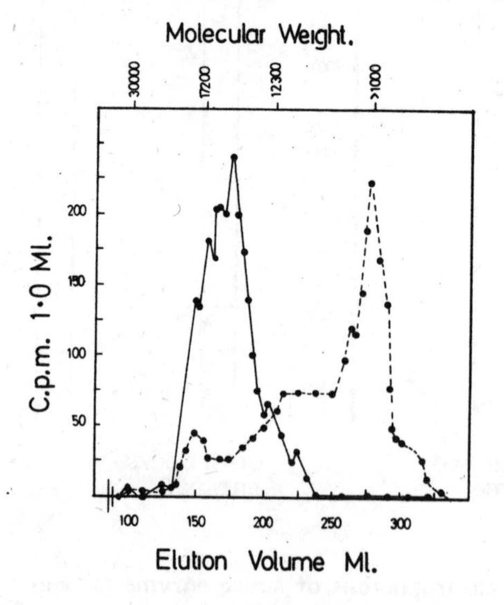

Fig. V Gel filtration of Peak B material (Fig II) following incubation either with (●---●-- ●) or without (●—●—●) E. coli cell extract (Sephadex G50–37° – 0.1M Tris/HCl pH 7.5).

DISCUSSION

By gel filtration of CNBr peptides from alkaline phosphatase, it was possible to show that only the peptide aggregates not susceptible to proteolysis are of a molecular weight in excess of 29,000 (Fig. 4, Region P). It is interesting to speculate on what is the pr imary determinant in rendering aggregates of peptides in region P, Fig. 4 not susceptible to proteolysis. It is posible that the peptides in region P are unique in terms of their inate non-susceptibility to **in vitro** proteolysis whether in the aggregated or non-aggregated form. Alternatively, it is also possible that the size of the peptide complexes in region P are simply too large to be recognised and degraded by the in **vitro** protease system. Work is presently in progress to distinguish between these two possibilities, and early experiments suggest that the latter possibility is the most feasible explanation for the peptides in region P not being susceptible to proteolysis.

REFERENCES

1. Goldberg, A.L., Proc. Nat. Acad. Sci. U.S.A. 69, 422-426(1972).
2. Prouty, W.F. Goldberg, A.L., J. Biol. Chem. 250, 1112-1123(1975).
3. Kemshead, J.T. Hipkiss, A.R.,Eur. J. Biochem. 45, 535-540(1974).
4. Kemshead, J.T. Hipkiss, A.R., In preparation.
5. Kelley, P.M., Neumann, P.A., Shriefer, K., Canceda, F., Schlesinger, M.J., and Bradshaw, R.A., Biochemistry 12, 3499-3503(1973).

CHANGES IN METABOLIC STABILITY OF STRUCTURAL AND ENZYME PROTEINS IN EXPERIMENTAL HEPATOMEGALY

John G. Nievel

Department of Medicine, King's College Hospital Medical School, Denmark Hill, London SE5 8 RX, England

A variety of molecules occurring in the natural environment as well as compounds used as anti-oxidants, food additives, pesticides and drugs of therapeutic potential influence at some stage the biochemical mechanism in the hepatocytes. Model compounds were selected to study biochemical/molecular changes produced by environmental chemicals. The effect of Actinomycin-D and the strongly hepatotoxic and carcinogenic diethylnitroseamine (DENA) was used to demonstrate reaction to acute toxic hazards. Food additives coumarin and butylatedhydroxytoluene (BHT) or the sedative phenobarbitone effect was studied to elucidate the underlying mechanism during chronic toxic injury and to assess the reserves of compensatory liver functions against chemical overload. Although acute toxic compounds may interfere with pre-formed molecules and enzymes, both highly toxic and low toxic molecules following chronic administration and at relatively low dose levels interfere with the dynamic processes of macromolecular synthesis responsible for the regular renewal of enzymic and structural components of the liver. The effect may take place at various levels through which the genetic information from the nuclei is carried. The interference with this process may be stimulation or inhibition of the rate of synthesis of macromolecules in the nuclei or cytoplasm. The increased rate of nucleic acid/protein synthesis may lead to hypertrophy of liver if the effect is disproportionate and does not simultaneously stimulate catabolic processes. Inhibition of the synthesis of nucleic acids and protein will manifest itself in a pathological change mostly when the pre-formed macromolecules are decayed at the end of their life-time.

In most cases an acute reaction to chemical injury of the environment is both inhibition of macromolecular synthesis and an enhanced activity of the catabolic enzymes, leading to a rapid intracellular depletion of nucleic acids and proteins. Acute and chronic doses of DENA and Actinomycin-D reduced the relative liver weight and the protein content of the endoplasmic reticulum in the liver. Both inhibition of RNA and protein synthesis as well as an increased breakdown of pre-formed RNA and proteins were detected in the liver. A similar effect was observed on the fate of drug metabolizing enzymes which metabolize DENA (1). In contrast to these acute effects, low doses of highly toxic compounds or relatively high doses of low toxic compounds such as coumarin, BHT and phenobarbitone induce apparently different types of changes, as the inhibitory component of the toxic compounds is relatively mild allowing complex changes to develop. However, at molecular level, all basic elements of the effect of foreign compounds can be recognized. Following initial depression, coumarin, BHT and phenobarbitone stimulated protein synthesis in the cytoplasm and enhanced the metabolism of the inducers and various other foreign compounds. These changes were dose related and within six days of chronic administration through a sequence of molecular events, resulted in hypertrophy of endoplasmic reticulum and hepatomegaly. A basically similar phenomenon was recognizable in hepatomegaly induced by both ‚toxic' and ‚non-toxic' compounds. The biochemical events leading to chemically induced hepatomegaly appear to be different from those occurring during compensatory hypertrophy or regenerating liver. The initial changes induced by the model compounds originated from the cytoplasm and radiokinetic studies on the incorporation of orotic acid into RNA, changes in the polysome/ribosome pattern, titration of the messenger receptor sites, effect of Actinomycin-D on the stimulation of ribosomal protein synthesis and various enzymes in the cytoplasm, or studies on the RNA catabolism and overall turnover, give no indication of a primary effect in the nuclei. Using multiple labelled radiokinetic experiments as a function of duration of treatment and dose of the model compounds used, a disproportionate synthesis of various enzymes and proteins was detected.

At the outset of the sequential events, the intracellular breakdown of various structural proteins and enzymes was reduced, which however, following repeated administration of inducers was reversed. This effect coincided with an excessive rate of protein synthesis, inducing liver growth. The proteins involved in the initial effect were structural components of the particles and membranes of the liver and possibly various unidentified enzymes, however, the frequently claimed reduction of phosphatases at this stage such as glucose-6-phosphatase, ATPase, RNAase, was unlikely to be due to reduced synthesis or breakdown of these enzymes, as the content of these proteins in the whole liver was not changed. It is more likely that the change in the former enzyme activities was due to the fact that protein having no such enzyme activity has been synthesized or increased in amount due to reduced breakdown at the location of these enzymes. Glucose-6-phosphatase may also be inhibited by the expanded pyrimidine nucleotide pool (vide infra). Biochemical and histochemical studies demonstrated an increase in both nucleoside diphosphatases and RNAase after the second day of administration of model compounds. During this period enhanced drug metabolizing enzyme activity for steroids and various foreign compounds was observed on the endoplasmic reticulum. However, in contrast to the cytosol, when the components of the endoplasmic reticulum were separated by molecular sieving and equilibrium density centrifugation no protein fraction of outstandingly high turnover rate was detected. Instead the enhanced metabolism of various hormones, enodogenous metabolites and foreign compounds represented a proportionate involvement of the whole endoplasmic reticulum of the liver. The increment in liver protein and drug metabolizing enzyme activity initially appeared to be a reversible process and the stimulated drug metabolizing activity and liver growth returned to normal with a half-life of 1-2 days following cessation of treatment. Repeated administration of high doses, however, depleted the compensatory reserves of the liver and the intracellular breakdown could not maintain a parallel rise with the enhanced rate of macromolecular synthesis. Only at this stage was an increased rate of RNA synthesis observed and ultrastructural and histochemical analysis indicated development of nodular hyperplasia. Histochemical analysis revealed a sudden rise in succinate dehydrogenase, isocitrate dehydrogenase, lactate dehydrogenase, glucose-6-phosphatase dehydrogenase and $NADH_2$-diaphorase near the wall of sinusoids and parallel with the hypertrophy of smooth endoplasmic reticulum an increase of glucose-6-phosphatase dehydrogenase was also detected in the hyperplastic foci. Coinciding with changes in the respiratory enzyme, increased acid phosphatase, adenosine triphosphatase and

5'-nucleotidase were also observed. α-glycerophosphate dehydrogenase and monoamine oxidase or the activity of alkaline phosphatase did not change. The toxicological/pathological significance of these hyperplastic nodules has not yet been fully clarified, however, features of molecular changes resemble in many respects those of tumors.

No specific indicators are as yet available which would delineate transition between various stages of increasing signs of toxicity and pathology in the growing liver. Even less information is available regarding the triggering factors which initiate these molecular changes. There are various indications however, which suggest that the initial interaction of the inducers with a protein fraction in the cytosol (2) closely linked with an early effect on the lipoprotein matrix of the endoplasmic reticulum may play an important role. The effect of the inducers on the turnover of the macromolecules is dose related (3) and their constant presence in the initial sites appears to be essential. Shortly following the primary effect the pyrimidine nucleotide pool expands and in the case of the relatively most toxic model compound used (coumarin) this effect remained even after 7 daily doses. In the effect of coumarin and possibly other inducers, corticosteroids played some role. Many similarities were clearly recognized between changes in the endoplasmic reticulum, ribosomal protein synthesis, enhanced enzyme activities, nucleotide metabolism, glucose-6-phosphatase, and fat deposition in the liver induced by coumarin or corticosteroids. These changes, however, are not due to stress imposed by the compounds on the animals, nor were they related to a central nervous system effect of the model compound.

It appears that an initial effect in the cytoplasm intimately related to some protein factors, plays some triggering role in the whole process. In spite of various compensatory efforts of the liver, the control of macromolecular synthesis in liver may ‚decompensate' in the face of chemical overload. The induced liver growth even during the reversible phase, before the ‚hepatic reserves' are depleted, is not exclusively beneficial for the body. However, chemically induced hepatomegaly has some ‚protective' characteristics once the animal is challenged with acute doses of highly toxic compounds. Under these circumstances we observed reduced toxicity both at cellular and whole body level in rats when they were exposed to acute doses of DENA or actinomycin as the result of an effect which was apparently not due to enhanced elimination of these compounds from the body.

REFERENCES

1. Magour, S., and Nievel, J.G., Biochem. J., 123, p8 (1971).
2. Nievel, J.G., 9 th Meet. Fed. Europ. Biochem. Soc., Budapest,p400 (1974).
3. Nievel, J.G., Fd. Cosmet. Toxicol., 7, 621 (1969).

SUBCELLULAR COMPARTMENTATION AND DEGRADATION OF INSULIN IN LIVER AT THE OUTSET OF HYPOGLYCAEMIA INDUCED BY ORAL ANTIDIABETIC SULPHANYLUREAS

John G. Nievel and John Anderson,

Department of Medicine, King's College Hospital Medical School, Denmark Hill, London SE5 8RX, England

It is now well documented that glibenclamide and possibly most of the other sulphanylureas reduce blood glucose content, mainly by stimulating the release of insulin in the pancreas (1). The primary receptor for glibenclamide may be in the pancreas, however, no significant interaction between Langerhans Islets and ^{3}H- or ^{14}C-glibenclamide was observed. The rapidly induced macromolecular synthesis resulting in an increase in the size of the nuclei of β-cells within a few hours following the administration of sulphanylureas (2) suggests that some extra-pancretic ‚trophic' influence on the β-cells may be operating. It appears that a specific direct inhibition of macromolecular synthesis in the β-cells by preformed insulin molecules is unlikely for following administration of sulphanylureas, in addition to release of insulin, both α-and β-cells hypertrophize (3). The pituitary gland has possibly no role in the effect of sulphanylureas (4). Furthermore, various effects on fat tissue or glucose utilization also indicate a more complex effect than would be explainable simply in terms of stimulated insulin secretion. Analysis of various **in vivo** effects of glibenclamide together with the observation that the liver accumulates the highest amount of glibenclamide indicate the presence of some receptors inside the liver cell in spite of the report that sulphanylureas are distributed in the extracellular space (5).

White Wistar male and female rats were used in these experiments. Intraperitoneally injected ^{14}C-glibenclamide produced a peak level of radioactivity in the whole blood and plasma 1/2 hr following administration. Within the first 3 hrs of glibenclamide treatment the radioactivity in the plasma space was the same as that of the whole blood indicating that unmetabolized glibenclamide does not penetrate into the blood cells. The radioactivity curve of both the plasma and the whole blood overlapped up to 23 hrs following injection of radioactive glibenclamide suggesting that not only glibenclamide, but also its metabolites, which are circulating in the blood in increasing amounts during the second part of the labelling period, are incapable of passing through the membrane of blood cells. Amongst various organs studied, liver accumulated the highest amount of radioactivity following injection of either ^{3}H- or ^{14}C-glibenclamide. The radioactivity of the liver increased when that in the blood began to fall and reached a peak of radioactivity 3-4 hrs following oral intubation. A relatively high amount of radioactivity was detectable in the liver even 23 hrs following treatment, and the labelling kinetics of liver was found to be different from that of other organs in the abdominal cavity. Following both oral and intraperitoneal administration, radioactivity was detectable in the pancreas, gut and diaphragm due to diffusion from the vicinity of the injection or to the anatomy of circulation. No such effect was observed following intravenous injection.

Rat liver subcellular particles were prepared at various times following glibenclamide injection. Highly purified nuclei, mitochondria, lysosomes, microsomes and cytosol were separated by sucrose density gradient ultracentrifugation. During isopycnic ultracentrifugation, further subfractions, ie nuclear membranes, outer and inner membranes of mitochondria, purified lysosomes, rough and smooth endoplasmic reticulum and plasma membranes were obtained. Extracts from the particles, membranes and cytosol were further fractionated by diafiltration, molecular sieving, ultracentrifugation and electrophoresis. Radioactive glibenclamide injected 30 min or 18 hrs before sacrifice was mainly detected in the cytosol of the liver and to a lesser extent in association with various subcellular particles. The relative proportion of radioactivity associated with these particles was dependent on the labelling time used. Interaction between glibenclamide and subcellular fractions was detected both **in vivo** and **in vitro.** In comparative experiments radioactive and immunoreactive insulin was also found to associate with liver subcellular fractions. The affinity of the particles was apparently higher for insulin than for glibenclamide. Insulin remained associated with subcellular particles, following sedimentation of these fractions on a sucrose density gradient by zonal or isopycnic centrifugation. Pretreatment of various liver particles with glibenclamide prior to association with insulin reduced the interaction of insulin with these particles. This effect was mainly observed in lysosomes/mitochondria, mitochondria and in microsomes. These preparations also contained plasma membranes in a similar decreasing order. The interacting sites for which insulin and glibenclamide may compete is possibly protein, as the receptor capacity of various subcellular fractions was reduced by trypsin treatment. However, competition between insulin and glibenclamide was only observed when administration of a relatively high amount of glibenclamide preceded addition of insulin. Even high doses of glibenclamide could not replace insulin, if the particles had already bound insulin to their surface receptors. It appears that glibenclamide reduces the attachment of newly formed (or released) insulin to the particles. Insulinase activity was mainly detectable in the cytosol and to a lesser extent amongst the particles, in a decreasing order of specific activity of: microsomos, mitochondria and lysosomes. Under the conditions used, the distribution of insulinase activity in the cell indicated a relatively low insulin hydrolysis in the fraction which contained plasma membranes, the main insulin binding site. Sulphanylureas have been reported to increase the life-time of circulating insulin in the blood (6). However, under the conditions used, glibenclamide **in vivo or in vitro** had no significant effect on the insulinase activity of any of the subcellular fractions. Glibenclamide reduces hepatic uptake of insulin, an effect which may play some role in the longer life-span of circulating insulin although subcellular compartmentation of glibenclamide and insulin may also participate in this. The interaction of glibenclamide and insulin with various subcellular fractions of liver is apparently a dynamic process. Intracellular transfer of glibenclamide between cell sap, nuclei and cytoplasmic membranes was observed, and a small peptide was released which partly remained associated with glibenclamide. A high affinity site was detected for glibenclamide amongst plasma proteins. Glibenclamide bound to plasma proteins is rapidly exchanged with liver proteins **in vivo.** Liver protein of high affinity for glibenclamide was used to assess glibenclamide concentration in a competitive radioactive assay system. Glibenclamide had 46 times higher efficiency than tolbutamide, (moles:moles) in ‚chasing off' radioactive glibenclamide from the receptor complex. Equimolar amounts of phenylbutazone, sodium salicylate, acetylsalicylic acid, warfarin and sulfadiazine or orotic acid had slight or no effect in contrast to sulphanylureas. Phenylbutazone competes with glibenclamide for plasma sites but not for liver receptor.

In order to characterize the turnover rate of liver fractions associating with glibenclamide and insulin multiple labelled kinetic experiments were used. The individual isotopes were assessed in a liquid scintillation spectrometer linked on-line with a computer. The computer was programmed to calculate relative stability of the protein fractions from the isotope ratios. The fractions containing plasma membranes and endoplasmic reticulum had a relatively higher turnover rate in contrast to other subcellular fractions. The peptide which has close relationship to the subcellular fate of the glibenclamide in the liver was also found to be synthesized and turned over at a rapid rate **in vivo.** From turnover studies it appears that this peptide possibly does not originate from the endoplasmic reticulum, but has similar stability to that of plasma membrane. However, the peptide released at the outset of the hypoglycaemic action and association of glibenclamide with the liver, cannot exclusively be identified with the plasma membrane as liver fractions with the lowest plasma membrane content appeared to have the highest receptor capacity for glibenclamide. The exact role of this peptide in the hypoglycaemic effect of glibenclamide and its relationship to insulin levels cannot yet be assessed until it is purified and identified.

REFERENCES

1. Breidahl, H.D., Ennis, G.C., Martin, F.I.R., Stawell, J.R., and Taft, P., Drugs, 3p, 79 (1972).
2. Pfeiffer, F., Steigerwald, H., Sandritter, W., Bander, A., Mager, A., Becker, U., and Retiene, K., Dtsch. med. Wschr., 82, 1568 (1957).
3. Bander, A., Pfaff, W., and Schesmer, G., Arzneimittel-Forsch., 8a, 1448 (1969).
4. Bander, A., and Scholtz, J.B., Dtsch. med. Wschr., 81, 889 (1956).
5. Wick, A.N., Britton, B., and Grabowski, R., Metabolism, 5, 739 (1956).
6. Mirsky, I.A., Perisutti, G., and Diengott, D., Metabolism, 5, 156 (1956).

DEGRADATION OF INSULIN AND GLUCAGON IN HOMOGENATES AND SUBCELLULAR FRACTIONS OF ISOLATED PANCREATIC ISLETS

S. Schmidt, K.-D. Kohnert, H.Jahr, H.Zühlke, B. Wilke, H.Kirschke, and H. Fiedler

Zentralinstitut für Diabetes „Gerhardt Katsch", Bereich experimentelle Forschung, DDR-2201 Karlsburg and Physiologisch-Chemisches Institut der Universität Halle, DDR-402 Halle/Saale, GDR.

SUMMARY

^{125}I–labelled insulin was degraded by islet homogenates in the presence of GSH by an initial reductive cleavage of its disulfide bonds and, subsequently, by proteolytic degradation of the single chains. The proteolytic degradation of insulin was measured directly using the sensitive fluram method. The proteolysis of both insulin and glucagon occurs mainly in the cytosol and to a lesser extent in the fraction of nuclei and cell debris. Evidence is given for the presence of lysosomes in pancreatic islets.

Studies on insulin secretion by RENOLD (1) have shown that intracellular insulin catabolism seems to play an important part in regulating the hormone content of pancreatic islets. In several tissues insulin is catabolized by a reductive cleavage of its disulfide bounds and, subsequently, by a proteolytic degradation (2,3). We investigated the insulin and glucagon degradation in pancreatic islets in vitro.

MATERIAL AND METHODS

Bovine insulin (crystallized twice) obtained from VEB Berlin-Chemie (GDR), crystallized glucagon was from Eli Lilly GmbH (FRG), fluram (fluorescamine) from F. Hoffmann-La Roche and Co. AG (Schweiz), BANA (B-α-benzoyl-D, L-arginine-2-naphthylamide hydrochloride) and dimethylcasein from Serva (FRG).

^{125}I-labelled insulin was prepared using the ICI method(4).

Radioiodinated S-sulfonated insulin A chain was prepared by sulfitolysis of ^{125}I-labelled insulin (5). All other chemicals were of reagent grade and were obtained commercially.

Islets of Langerhans were isolated from Wistar rats (body weight 180-200 g) using the collagenase method (6,7). The homogenate of about 500 pancreatic islets was centrifuged to sediment nuclei and cell debris (N, 600 xg for 10 min.), secretion granules and mitochondria (SG/M, 10 000 x g for 20 min.), and microsomes (R, 100 000 x g for 60 min.) respectively. The supernatant is the cytosol(C). The pellets were rehomogenized in 100 µl of 0.01 M phosphate buffer, pH 6.5. The insulin immuno reactivity (8), the protein content (9), and the marker enzymes of the individual cell organelles (glutamate dehydrogenase, GLDH, EC 1.4.1.3, 6-phosphogluconate dehydrogenase, 6-PGDH, EC 1.1.1.44, malate dehydrogenase, MDH, EC 1.1.1.37) were determined in the homogenate and the subcellular fractions after one freezing by a fluorimetric determination of generated pyridindinucleotides (10); acid hydrolases (acid galactosidase, A. Galact. ase, EC 3.2.1.23, acid phosphatase, A. Phosphatase EC 3.1.3.2; acid glucoronidase, A. Glucur. ase, EC 3.2.1.31) were determined by splitting the corresponding methylumbelliferon derivatives and fluorimetric measurement (10).

Antisera against highly purified cathepsin B1 and a proteinase fraction („L 20") prepared from rat liver lysosomes (11,12) were obtained by immunization of rabbits as described elsewhere (12). For details of degradation measurements see legends to the figures and Table 1.

RESULTS AND DISCUSSION

In order to study the manner of insulin degradation, insulin containing a tracer amount of ^{125}I-labelled insulin was incubated in the presence of glutathione and EDTA with islet homogenate for various times and chromatographed on Sephadex G-75 (Fig. 1). When the radioactivity disappeared in the insulin peak, a radioactive peak eluted at the position of the sulfonated A chain was formed. In the case of incubation in

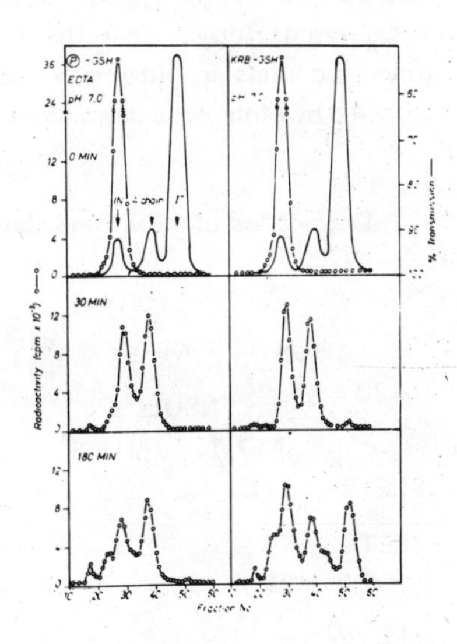

Fig. 1: Gel chromatography pattern of insulin (15 μg) containing a tracer amount of ^{125}I-labelled insulin that had been exposed to homogenate of isolated rat pancreatic islets (46 μg protein per incubation) either in 1 mM glutathione, 0.1 M potassium phosphate – 5 mM EDTA (pH 7.0) (left) or 1 mM glutathione, Krebs-Ringer- bicarbonate buffer (pH 7.0) (right) at 37 °C for the indicated time periods. Incubations were terminated by cooling the samples in ice and by subsequent addition of 20 μl of 0.1 M N-ethylmaleimide and 20 μl of glacial acetic acid. 100 μl of the entire mixture (120 μl) were applied to a Sephadex G-75 column (1 cm x 50 cm) equilibrated and eluted with 50 % acetic acid at a flow rate of 8.5 ml/h. 1 ml fractions were collected and counted for radioactivity. Radioactivity in cpm, o———o, % transmission at 280 nm, ——.

Vertical arrows denote from left to right, the insulin peak (IN), the peak of S-sulfonated A chain and the peak of low molecular weight components (I) consisting of salt and N-ethylmaleimide.

Krebs-Ringer-bicarbonate (without EDTA) after 180 min. a radioactive peak appeared at the position of low molecular weight components. When ^{125}I-labeled A chain was exposed to homogenate preparations of pancreatic islets and the products of incubation were also examined by gel chromatography, it could be shown that the insulin A chain is degraded to amino acids via smaller peptides. This could also be confirmed by paper electrophoresis of the degradation products which were formed (not shown). These results suggest that the first step of ^{125}I insulin degradation in pancreatic islets in vitro is a reductive cleavage of its disulfide bonds followed by proteolytic degradation.

TABLE 1

Degradation of insulin and glucagon by homogenates of pancreatic islets of rats.

$$\text{specific activity} = \frac{\text{pMol TCA-soluble NH2}}{\mu\text{g homogenate protein.min}}$$

	INSULIN		GLUCAGON	
	pH 7.0	pH 5.5	pH 7.0	pH 5.5
ISLETS	27.5	19.0	140.1	47.0
ISLETS + EDTA	24.8	–	64.8	–
ISLETS + 2 ME	85.3	19.5	142.5	38.8

2 ME = 2-Mercaptoethanol

10 μl homogenate (protein content 1-2 μg), 10 μl substrate (10 μg insulin or 6.7 μg glucagon in 50 mM phosphate buffer), and 10 μl 50 mM phosphate buffer with or without 2-mercaptoethanol (30 mM) or EDTA (15 mM) were incubated at 37° C for 20 min. After addition of 40 μl buffer (or 20 μl 15 mM $HgCl_2$ and 20μl 25 mM EDTA in buffer in presence of 2-mercaptoethanol) and 30 μl 20 % TCA the incubation samples were cooled for 10 min at 5°C and centrifuged. Two aliquots of the supernatant (30 μl) were neutralized with 20 μl NaOH. After addition of 50μl 0.2 M borate buffer, pH 9.0, and 50 μl fluram-reagent (150 μg/1ml Dioxan) the fluorescence intensity was measured in a Beckman spectrofluorimeter SF 1078 using a rectangular micro cuvette (quartz, 1x10 mm). Excitation 374 nm, emission 486 nm.

Little is known up to now on proteolytic enzymes in pancreatic islets (13). In order to obtain more information, we measured the proteolytic degradation of insulin, glucagon, and dimethylcasein using the fluram method and the BANA-splitting activity.

In Table I the specific activities of insulin and glucagon degradation are shown. At pH 7.0 the breakdown of glucagon is about five times higher than that of insulin. 2-Mercaptoethanol stimulates insulin degradation at pH 7.0 but not the proteolysis of glucagon. EDTA has no influence on insulin degradation whereas glucagon is split to a smaller extent. These results demonstrate the usefulness of the fluram method for measuring proteolytic activities in pancreatic islets. With this highly sensitive method a proteolytic degradation of insulin (14) in homogenates and subcellular fractions of islets is measurable after a short incubation time (10-20 min).

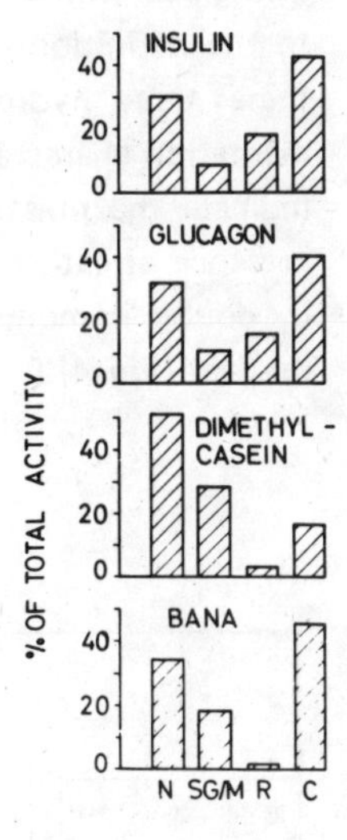

Fig. 2: Proteolytic activities in subcellular fractions of pancreatic islets (rat). The proteolytic degradation of insulin, glucagon, and dimethylcasein was measured using the fluram method (for details of the method see legend to Table I). BANA-splitting: 10 μl homogenate (protein content 1-5 μg) were incubated at 37°C in 120μl of BANA (4.4μg) in 50 mM potassium phosphate (pH 7.0). After 90 min the fluorescence intensity of the liberated β-naphthylamine was measured in a Beckman spectrofluorimeter SF 1078 (13). – All columns represent 8 measurements with 4 different homogenates. The measured total activity of N, SG/M, R, and C were set to 100 %.

A high percentage of insulin- and glucagon – degrading activity was found in the cytosol and in the fraction of nuclei and cell debris (Fig. 2). The similar subcellular distribution of degrading activities for insulin and glucagon in pancreatic islets suggests that there is no activity which degrades insulin specifically. We found two groups of BANA-splitting activities: 1. pH optimum at 7.0, and 2. pH optimum in the range of 5.5 – 6.0 in the presence of EDTA. The latter activity was mainly found in the fraction of secretion granules/mitochondria. We suggest that in this fraction lysosomes are present. Up to now only ORCI (15) could show with some electron-micrographs that lysosomes are present in pancreatic islets. There were no reports on biochemical investigations in the literature. Therefore we characterized our subcellular fractions using some enzymes and insulin as markers (Fig. 3.). Acid phosphatase, acid galactosidase, and acid glucoronidase as marker enzymes for lysosomes were checked according to the definition of lysosomes given by DE DUVE (16). The properties of these three hydrolases (acid pH optima, structure-linked latency, sedimentation characteristics in differential and density gradient centrifugation) indicate the presence of lysosomes in pancreatic islets. In addition, the presence of cathepsin-like proteinases in pancreatic islets was demonstrated by double immuno diffusion with antibodies to lysosomal proteinases from rat liver (Fig. 4).

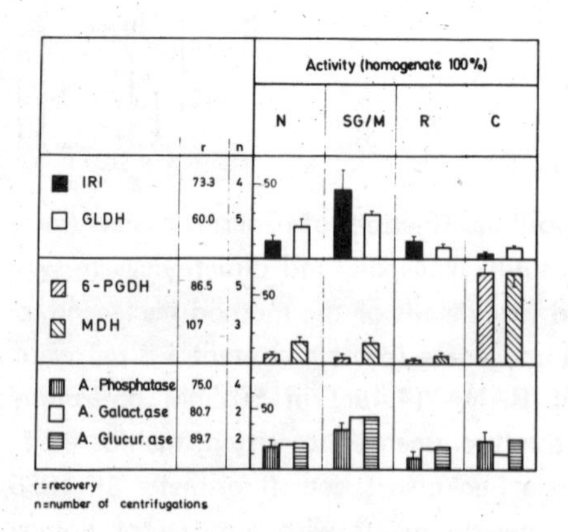

Fig. 3: Characteriaztion of subcellular fractions of rat pancreatic islets.

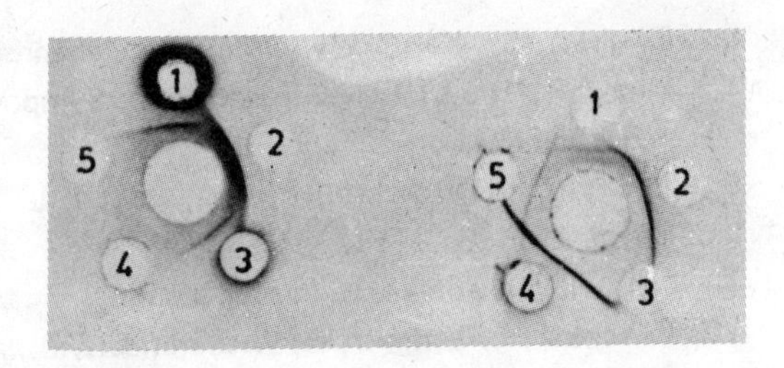

Fig. 4 (left): Double immunodiffusion with antiserum to a lysosomal proteinase fraction L 20 of rat liver. The central well received 20μl of L 20 antiserum. The peripheral wells received each 5μl of rat liver homogenate, 38 μg protein (1), proteinase fraction L 20 of rat liver, 40 μg protein (2), rat islet homogenate, 38μg protein (3), rat pancreas homogenate, 27 μg protein (4), and buffer (5).

Homogenates were prepared in 0.9 % sodium chloride; other proteins were in 0.01 M potassium phosphate buffer (pH 6.0). Diffusion was performed at pH 8.6 for 72 h.

Fig. 4 (right): Ouchterlony double immunodiffusion with anti-cathepsin B1 –γ– globulin fraction (central well) and in the peripheral wells proteinase fraction L 20, 20 μg protein (1), cathepsin B1 from rat liver, 9μg protein (2), buffer (3), rat islet homogenate, 75μg protein (4), and rat pancreas homogenate, 81μg protein (5).

REFERENCES

1. Renold, A.E., Diabetes 21, Suppl. 2, 619-631 (1972).
2. Varandani, P.T., Shroyer, L.A., and Nafz, M.A., Proc. Nat. Acad. Sci. USA 69, 1681-1684 (1972)
3. Varandani, P.T., Biochim Biophys. Acta 295, 630-636 (1973).
4. Ansorge, S. and Kaiser, B., Wiss. Z. Karl-Marx-Univ. Leipzig, Math. —Naturwiss. R. 18, 559—568 (1969).
5. Varandani, P.T., Biochim. Biophys. Acta 127, 246-249 (1966).
6. Lacy, P.E. and Kostianowsky, M., Diabetes 16, 35-42 (1967).
7. Hahn, H.-J., Lippmann, H.-G., and Schulz, D., Acta biol. med. germ. 25, 421-429 (1970).

8. Ziegler, M., Karg, U., Gens, J., Johannsen, B., Michael, R., and Michaelis, D., VIII. Nuklearmedizin. Symposium, Reinhardsbrunn, pp. 141-145 (1971).
9. Lowry, O.H., Rosebrough, N.J., Farr, A.L., and Randall, R.J., J. Biol. Chem. 193, 265-275 (1951).
10. Jahr, H. and Zühlke, H. in preparation
11. Kirschke, H., Langner, J., Wiederanders, B., Ansorge, S., and Bohley, P. Acta biol. med. germ. 28, 305-322 (1972).
12. Kirschke, H., Langner, J., Wiederanders, B., Ansorge, S., Bohley, P., and Hanson, H. in: Intracellular Protein Catabolism (Hanson, H. and Bohley, P. eds.), Verlag J. Ambrosius Barth, Leipzig, p.210 (1974)
13. Zühlke, H., Jahr, H., Schmidt, S., Gottschling, D., and Wilke, B., Acta biol. med. germ. 33, 407-418 (1974).
14. Duckworth, W.C., Heinemann, M., and Kitabchi, A. E., Biochim. Biophys. Acta 377, 421-430 (1975).
15. Orci, L., Stauffacher, W., Rufener, C., Lambert, A.E., Roullier, C. and Renold, A.E., Diabetes 20,385 – 388 (1971).
16. De Duve, C., J. Cell. Biol. 50, 20D-56D (1971).

HETEROGENEITY OF RAT LIVER LYSOSOMES WITH RESPECT TO PROTEOLYTIC ACTIVITY AT pH 6.0

J. Langner, P. Hoffmann, S. Riemann, B. Wiederanders, S. Ansorge, P. Bohley, H. Kirschke and H. Hanson

Physiologisch-chemisches Institut der Martin-Luther-Universität, 402 Halle (Saale), GDR

Experiments revealing lysosomal heterogeneity have been published during the past years from various laboratories (1).

The results obtained so far agree with respect to the existence of heterogeneities of distribution patterns of lysosomal marker enzymes, but the physiological significance of these findigs seems to be in no case quite clear, or even no more than speculative.

Our present contribution adds to the earlier knowledge the distribution pattern of proteolytic enzymes with activity at pH 6. The starting point of our experiments was the observation in large-scale zonal centrifugations of the separation of the peaks for β-glycerophosphatase and proteinase. We found it worthwhile to investigate this further by means of analytical gradients, aiming at high resolution but not so much at an increase in specific activity.

The **experimental procedure** was briefly as follows. A rat liver particle sediment was collected between 35,000 and 340,000 g x min and it was washed twice with 0.25 M sucrose. This so-called L-fraction contained 30-42 % of lysosomal enzymic activities and was on average 6-7fold enriched over the homogenate.

0.5 or 1 ml L-fraction (with 20 to 50 mg of protein) was centrifuged isopycnically on a 9 ml sucrose gradient (30-55 %, w/v, 1.5 ml cushion of

60 % sucrose) for 90 min at 40.000 rpm (corresponding to 1.28–2.56 . 10^7 g x min) or under rate zonal conditions on a Ficoll gradient of the same volume (0 – 25 % Ficoll in 0.25 M sucrose, w/v, 1.5 ml 60 % sucrose as cushion) for 15 to 18 min at 15.000 rpm corresponding to 3 – 6. 10^5 g x min). Thereafter the gradents were fractionated by puncturing the tubes and collecting 25 to 40 fractions of 6 or 3 drops. The fractions were analyzed for protein, β-glycerophosphatase, β-galactosidase; proteinase activity with azocasein as substrate at pH 6.0 with addition of 5 . 10^{-3} M GSH and EDTA; RNase I, and glutamate dehydrogenase. The concentration of gradient media was determined refractometrically.

RESULTS AND DISCUSSION

Figures 1 and 2 show typical examples of distribution patterns of the various parameters determined. The numbering of fractions is from bottom to top of the gradient. The lysosomal enzyme activities peak near 48 % sucrose, corresponding to a specific gravity (D_{20}) of about 1.22 (Fig. 1). As is clearly visible, the peaks of β-galactosidase and β-glycerophosphatase coincide, whereas the peak of proteolytic activity at pH 6 comes out one fraction earlier. This is still more pronounced in the Ficoll gradient (Fig. 2) than with sucrose. This separation of the lysosomes has constantly been obtained in 18 centrifugations performed. The two groups of lysosomes that have been partially separated we denominate as group A, slower sedimenting, and containing more phosphatase and galactosidase, and group B, faster sedimenting, with more proteinase containing particles.

The mitochondrial marker glutamate dehydrogenase parallels the protein curve (with the exception of the last protein peak lacking glutamate dehydrogenase) and has therefore been omitted. The heterogeneity of mitochondria revealed in Fig. 1 (Protein curve) has not been further investigated; on the other hand, we were not able to separate in these experiments mitochondria from lysosomes as has been achieved by Burge et al. (1) with the M.S.E. zonal rotor type HS.

Rahman et al. (2) found two groups of lysosomes in their zonal centrifugations of rat liver homogenate the first with acid phosphatase and cathepsin C (slower sedimenting), the second with acid RNase and cathepsin D (faster sedimenting). In our experiments, the RNase distribution pattern

(Fig. 3) gives reproducibly two peaks, which do not clearly coincide with the two others, A and B. Therefore, at the present stage it is not quite possible to correlate with certainity our subgroups with those of Rahman's experiments, but the lysosomal heterogeneity might be still more complicated than thought of as yet.

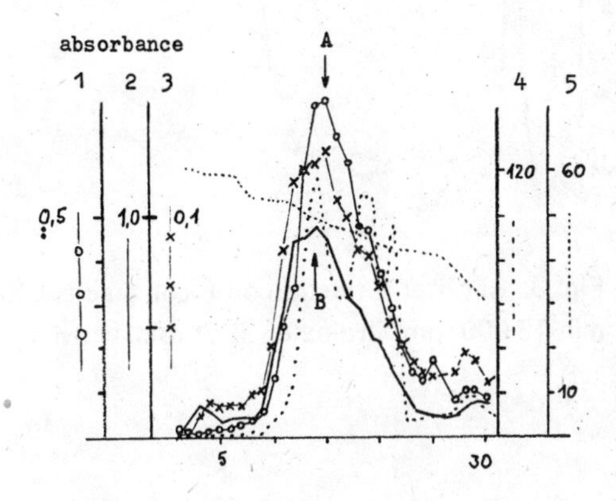

Fig. 1: Rat liver L-fraction on sucrosse gradient (30-55 %), 90 min 40 000 rpm. The ordinates designate: 1. β-glycerophosphatase, 2 proteolytic activity (pH 6,0; azocasein), 3 β-galactosidase, 4 protein μg/20 μl, 5 sucrose %
Abscissa: Fraction number

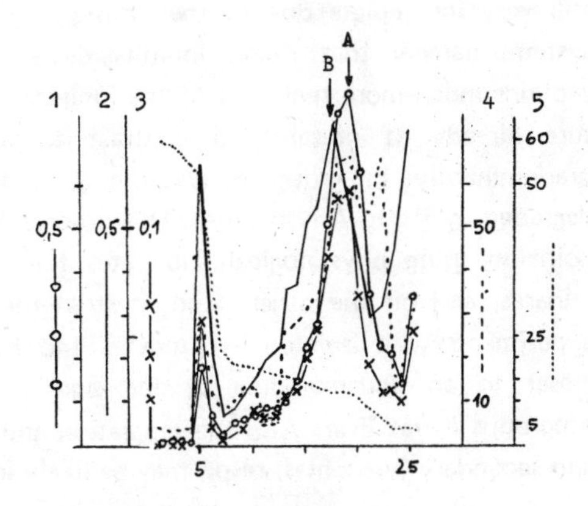

Fig. 2: Rat liver L-fraction on Ficoll gradient (0-25 % in 0,25 M sucrose), 15 min 15 000 rpm. Ordinates as in Fig. 1

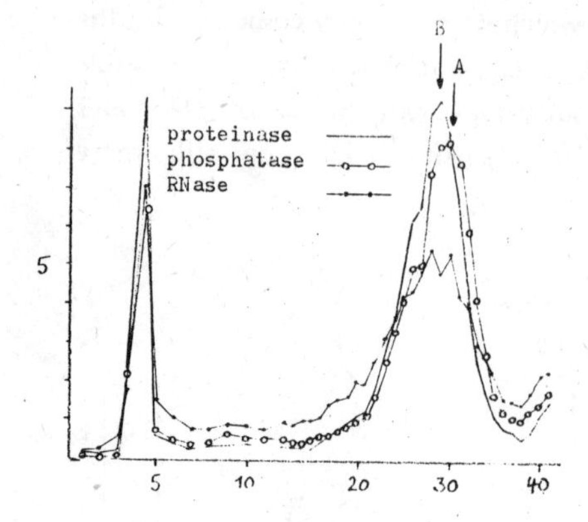

Fig. 3: Rat liver L-fraction on Ficoll gradient (0-25 % in 0,25 M sucrose), 18 min 15 000 rpm. Ordinate: % of total activity, Abscissa: Fraction number

To get an insight into the possible physiological significance of these two lysosomal subgroups A and B, we have incubated both with a double-labelled (short and long-lived) rat liver cytosol protein fraction according to Bohley et al. (3). The result of this experiment unfortunately is still very incomplete due to the unforeseen effect of Ficoll in this test system, namely that Ficoll inhibits degradation of short-lived cytosol proteins much more than that of the long-lived ones. One point however is sure already at present: in contrast to azocasein that is degraded predominantly in group **B** lysosomes, cytosol proteins, however, are degraded by group **A** lysosomes. This observation once more illustrates our point in using physiological substrates for studies on intracellular proteinases, and on the other hand, it strengthens the necessity for further experiments with isolated hepatocytes and Kupffer cells, resp., to come closer to an understanding of the significance of different lysosomal subgroups in rat liver. Also the separation and characterization of primary and secondary lysosomes, resp., may be likely involved in such experiments.

REFERENCES

1. Burge, M.L.E., and R.H. Hinton., in Separation with zonal rotors. (E. Reid, ed.) Guildford, Wolfson Bioanalytical Centre, University of Surrey (1971).
2. Rahman, Y.E.,J.F. Howe, S.L. Nance and J.F. Thomson., Biochim. Biophys. Acta 146, 484 (1967).
3. Bohley. P., J. Langner, S. Ansorge, M. Miehe, C. Miehe and H. Hanson: Communication at the 7th FEBS-Meeting, Varna 1971, Abstr. No. 360, p. 162.

THE AGE DEPENDENCE OF INTRACELLULAR PROTEOLYSIS IN THE RAT LIVER

B. Wiederanders, S. Ansorge, P. Bohley, H. Kirschke, J. Langner and H. Hanson

Physiologisch-chemisches Institut der Universität, Halle, DDR-402 Halle (Saale), Hollystrasse 1, GDR

If ageing is a problem of diminished adaptation, then studies of ageing changes of the intracellular protein catabolism might elucidate its more common regulatory phenomena. Therefore, last year we started studies on intracellular proteolysis in the aged organism.

In this specialized field published results are scarce and often contradictory. More recent findings support the opinion that a decrease of protein turnover occurs in aged organisms (1-3).

A decreased degradation rate of intracellular proteins has to be proposed, because the synthesis rate of them is known to be retarded (4). The reasons for a decreased protein breakdown may be:

1) Changes in the amount or in the properties of degradative enzymes or their effectors.

2) Reduced digestibility of the substrate proteins.

3) The primary reactions of protein breakdown within the cell are increasingly disturbed in the aged organism. In this report we can summarize some findings concerning points 1 and 2.

The total proteolytic activity in the rat liver increases during the whole life cycle. There exists a linear correlation between the DNA content of the rat liver and its total proteolytic activity at pH 3.0 and pH 6.0 towards the artificial substrates azo-Hb and azocasein.

The correlation between these values and body weight is linear up to the second month, and later it is hyperbolic (5). The expected loss of proteolytic capacity in the livers of old animals was not found.

Next we determined the specific proteolytic activity of rat liver in relation to age. We found the highest values during the weaning period, but this fact will not be discussed here. More important is the finding of a small but significant ($p<0.005$) loss of specific proteolytic activity in the livers of animals older than 20 months, compared with adult younger ones. The substrate was azocasein, pH of incubation was 6.0, 5 mM glutathione and EDTA resp. were present. The same result can be achieved at pH 3.0 with azo-Hb as substrate (5).

To see which cell fraction is responsible for this shift, we fractionated the liver homogenates of old and young rats and determined the specific proteolytic activity at pH 3.0 and 6.0 in each of the cell fractions. The reduced specific proteolytic activity at both pH values could be demonstrated in old animals more or less in all cellular fractions with the exception of the cytosol and the microsomal fractions. The most remarkable loss of specific proteolytic activity was seen in the lysosomal fraction of old rats (48 % and 30 % at pH 3.0 and 6.0 resp.).

Let us consider now another aspect given at the begining, the changes in effector-concentration. As Kirschke and Bohley found, the lysosomal proteases obviously responsible for important steps in intracellular protein breakdown are SH-enzymes. They are dependent on the presence of SH-compounds. Therefore, we determined the concentration of total, free and proteinbound SH-groups in relation to age. The results we got were not as informative as we had hoped. The concentration of free SH-groups shows a weak increase during the aging process. This fact is known from the literature (6) and we could confirm it also. But on the other hand, the concentration of free SH-groups depends strongly on the dietary situation of the organism. Thus the free SH-concentration falls to half of the normal value in rat livers after fasting for 24 hours. i.e., the observed weak rise of free SH-concentration in the livers of old rats can be compensed for very easily and very rapidly by dietary events. Therefore we conclude that this result has to be interpreted with care in respect to our problem.

Let us turn now to the last kind of experiment concerning the substrate properties of proteins to be degraded. We isolated the cytosol of rat livers by a one step high speed centrifugation of the homogenate. This cytosol was concentrated and equilibrated to pH 6.0 under pressure by using a Diaflo membrane PM 10. The cytosol of rats of about 4 months of age and of old rats, 18 and 21 months resp., served as substrates for a number of different proteolytic enzymes. All the incubations with the extracellular proteinases papain, trypsin, pepsin and pronase were done under so called optimum conditions. Cathepsin D was allowed to act at pH 3.4, the soluble lysosomal enzyme mixture (assigned as LS) and L 20 at pH 5.0 with 5 mM glutathione and EDTA resp. L 20 represents the lysosomal enzyme fraction containing the cathepsins B1, B3, and L (7). The lysosomal enzymes came from rats of about 8-12 months of age. Proteolysis was measured by determining the increase of TCA soluble amino nitrogen (8). The Fig. shows the result. The left hand column shows the amount of cytosol from the young animals degraded within one hour by the enzyme noted below, taken as 100 %. The right hand columns show the relative amount of old animal cytosol degraded under nearly identical conditions. Despite the large standard deviations and the small number of experiments, the tendency is clearly visible: lysosomal proteases acting at neutral pH values degrade the young cytosol proteins faster than the cytosol of the old rats. Extracellular proteases did not show this preference. Cathepsin D also does not show clearly the preferred digestion of young cytosol.

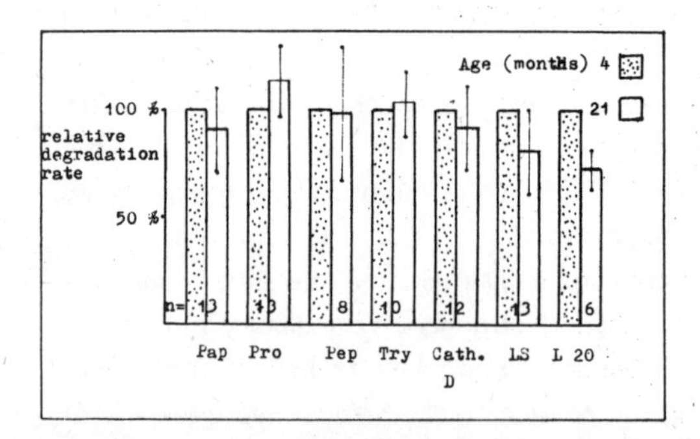

As a conclusion we might note that at least two reasons might lead in the older animals to a reduced protein degradation; these are a fall in the specific proteolytic activity particularly of the lysosomal proteases, and a reduced digestibility of substrate proteins. Whether the results we got from in vitro experiments with the rat liver are true also for other organs and whether they represent the in vivo situation remains open however.

REFERENCES

1. Ove, P., Obenrader, M., and Lansing, A.I., Biochim. Biophys. Acta 277, 211 – 221 (1972).
2. Comolli, R., Feriolo, M.E., and Azzola, S., Exp. Gerontol. 7, 369-376 (1972).
3. Millward, D.J. (1975) this symposium.
4. Mainwaring, W. I. P., Biochem. J. 113, 869-878 (1969).
5. Wiederanders, B., Ansorge, S., Bohley, P., Kirschke, H., Langner, J., and Hanson, H., Z. Alternsforsch., in press.
6. Harisch, G., and Schole, J., Z. Naturforsch. 29c, 261-266 (1974).
7. Kirschke, H., Langner, J., Wiederanders, B., Ansorge, S., Bohley, P., and Hanson, H., this symposium.
8. Langner, J., Ansorge, S., Bohley, P., Kirschke, H., and Hanson, H., Acta biol. med. germ. 26, 935-951 (1971).

THE BREAKDOWN OF A TETRAPEPTIDE BY YEAST AMINOPEPTIDASE I: KINETIC ANALYSIS AND COMPUTER SIMULATION

G. Metz, F. Göbber and K.-H. Röhm

Institut fur Physiologische Chemie der Universität, D-355 Marburg (Lahn), GFR, Lehrstuhl für Teoretische Chemie der Universität, D-74 Tübingen, GFR.

SUMMARY

The removal of the NH_2-terminal amino acids from Val-Ala-Thr-Ala, Ala-Thr-Ala and Thr-Ala catalyzed by yeast aminopeptidase I (a metal-dependent enzyme belonging to the leucine aminopeptidases) was studied. The kinetic parameters derived from these data were used to simulate the time course of the complete breakdown of the tetrapeptide with the aid of a digital computer program.

Close agreement between the calculated values and experimental data obtained by amino acid analyzer runs suggests the applicability of the underlying kinetic model (independent binding of all intermediates to the enzyme and mutual competitive inhibition). The kinetics of the reaction as compared with the hydrolysis of related peptides strongly indicates the presence of a subsite on the enzyme with high affinity towards Thr and Val.

INTRODUCTION

It is well known that many proteinases exhibit a marked second site specificity, that is, the rate of hydrolysis of a given peptide bond depends on the amino acid sequence adjacent to the point of cleavage. This was convincingly demonstrated with papain by Schechter and Berger (1). A particularily interesting situation arises in the case of exopeptidases which split numerous bonds successively. It is obvious that the kinetics of such a system (which may be of practical importance in protein sequence work) will provide information on the enzyme's subsite specificity which is likely to impose quite different kinetic characteristics on each step, depending on the neighbourhood of the respective bond. First of all, however, one has to ascertain whether the degradation of the chain is in fact a linear sequential reaction of the type $A \xrightarrow{E} B \xrightarrow{E} C$ ....,. the intermediates of which are freely competing for the active site of the enzyme, or if the degradation process includes only one initial binding step followed by the successive splitting off of amino acids without the release of the residual chain.

In the present paper evidence is given that the breakdown of the tetrapeptide Val-Ala-Thr-Ala by yeast aminopeptidase follows the linear sequential mechanism. The role of nonproductive binding and the influence of the amino acid sequence on the kinetics of the system are discussed.

METHODS

Aminopeptidase I, an enzyme described by Johnson (2) for the first time, was purified 600-fold from autolysates of brewer's yeast by ammonium sulfate fractionation, heat precipitation and chromatography on Sephadex G 150, DEAE-cellulose and Sepharose 6B (3). Preparations with specific activities from 500 to 900 U/mg (towards Leu-Gly-Gly, for conditions see below) were used in the kinetic runs. The peptides containing threonine were prepared in our laboratory by Dr.E. Schaich and Dr. A. Raschig. Most of the other substrates were purchased from Fluka, Buchs or Serva, Heidelberg.

The kinetic experiments described in this paper were uniformly performed at 40° C in 0.1 M phosphate-borax buffer, pH=7.8, containing 0.1 M Cl^{-} and 0.05 mM Zn^{++}. The hydrolysis of dipeptide substrates and of those tripeptides yielding dipeptides hardly attacked by the enzyme was followed spectrophotometrically at 235 nm (4). In other cases samples were removed from the reaction mixtures and after quenching with acid run on a Beckman Unichrom amino acid analyzer. The column dimensions and buffers varied with the substrate under study. The intermediates and products of Val-Ala-Thr-Ala hydrolysis were separated on Beckman M-82 resin(0.9 x 60 cm) using the standard 0.2 N citrate buffer, pH=3.28. Elution times (related to Glu=1) were: Val-Ala-Thr-Ala = 1.25, Ala = 1.47, Ala-Thr-Ala = 1.63, Val = 1.93. Following a buffer change to pH 4.25, Thr-Ala was eluted at t = 2.2.

The simulations were carried out on a Hewlett-Packard 9820 A calculator with plotter. The program used (5) integrates a system of ordinary first order differential equations according to the fourth-order Runge-Kutta-Merson formula (6). Since it independently selects the maximal step lengths in time without exceeding a present single-step error the computing time is considerably shortened. In our case the simulation was finished within several minutes with each set of parmeters at 100 – 1000 operations/second.

RESULTS AND DISCUSSION

To judge from its substrate specificity, the enzyme belongs to the leucine aminopeptidases (see Table 1). Unexpected, however, are the low activities toward substrates with amino terminal Thr or Val and the high rates measured in Thr occupies the second position. A more detailed kinetic analysis (see Fig. 1) shows that the enhanced activities in the latter case are due to considerably lowered K_M values (Ala-Gly:40mM-Ala-Thr:6mM; Ala-Gly-Gly:12mM-Ala-Thr-Ala:3.2mM) Amino terminal Val or Thr, on the other hand lead to very low maximal velocities. An explanation for these anomalies is provided by the assumption that there exists a distinct subsite (S_2) with high affinity toward Thr and Val located adjacent to the catalytic site (S_C), thus interacting with the side chain of the substrate's second amino acid so far as productive complexes are concerned. If Val or Thr are situated in other positions, nonproductive binding (see Fig. 1) will prevail, hindering the turnover of the respective compound. This is particularily obvious if one compares Ala-Thr with Thr-Ala, the latter being barely attacked but readily bound by the enzyme (K_I =1.6 mM).

The low K_M of Val-Ala-Thr-Ala (which should easily form two nonproductive complexes) is in agreement with this idea, since nonproductive binding decreases **both** K_M and V by the same factor.

		V	K[mM]
Ala-Gly	p	1.5	40
Ala-Gly-Gly	p	100	12
Ala-Thr	p	1.6	6
Ala-Thr-Ala	p	190	3.2
Thr-Ala	np	0	1.6
Val-Ala	np	0	5
Val-Ala-Thr-Ala	np		
Val-Ala-Thr-Ala	np		
Val-Ala-Thr-Ala	p	7	4

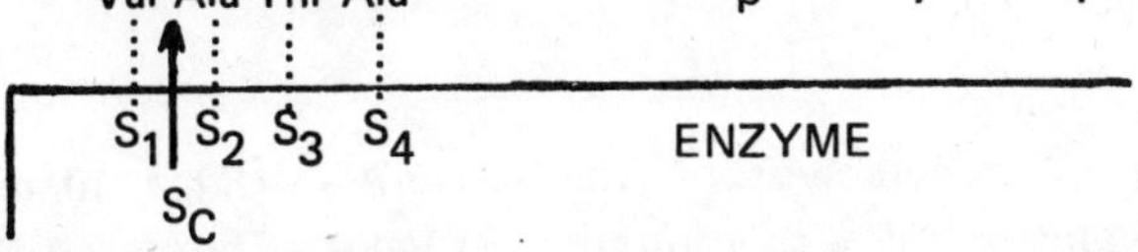

S_1 S_2 S_3 S_4 ENZYME

S_C

Fig. 1 Proposed binding of substrates to yeast aminopeptidase. Productive complexes are marked with p, others with np. In the latter cases the K are inhibition constants (measured with Ala-Thr-Ala as substrate) S_1 to S_4 are bunding sites, S_C the catalytic site.

TABLE 1

Substrate specificity of yeast aminopeptidase I;

Intial velocites at [S] = 10 mM were determined. The rate of Leu-Gly-Gly cleavage was set as 100 %.

Substrate	Activity	Substrate	Activity
Leu-Gly-Gly	100	Gly-Gly	0.09
Ala-Gly-Gly	91	Gly-Ala	0.52
Phe-Gly-Gly	89	Val-Ala	0.001
Val-Gly-Gly	3	Thr-Ala	0.001
Gly-Gly-Gly	7	Ala-Gly	1.9
Ala-Thr-Ala	510	Ala-Thr	1.6
Val-Ala-Thr-Ala	9		
BOC-Val-Ala-Thr-Ala	0	Leu-p-NHNP	1.3

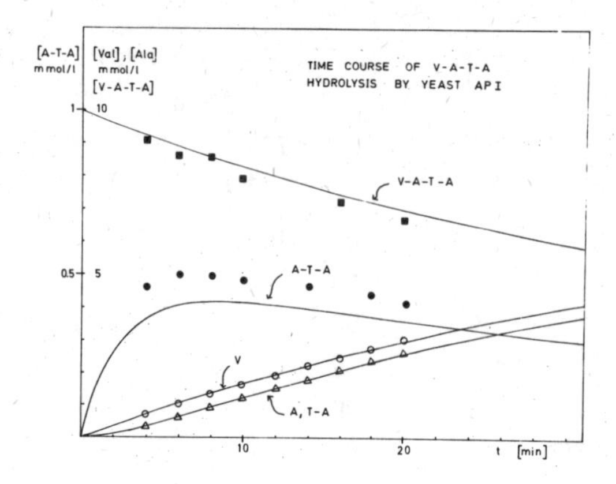

Fig.2 The hydrolysis of Val-Ala-Thr-Ala ($[S]_o$= 10 mM) by yeast aminopeptidase I: The concentrations of Val-Ala-Thr-Ala (■) Ala-Thr—Ala (●), Val (○) and Ala (△) are plotted vs. the incubation time. For experimental details see'methods'. The values of Thr-Ala are identical with Ala and have been omitted. Note the different scale for Ala-Thr-Ala. -The curves were calculated by the simulation procedure from the reaction scheme.

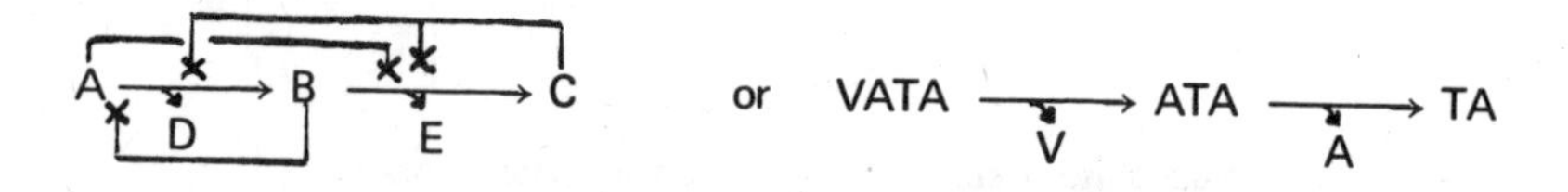

which is described by the following set of differential equations

$$-d[A]/dt = d[D]/dt = V_A[A]/(K_1 + [A]);\ K_1 = K_A\left(1 + \frac{[B]}{K_B} + \frac{[C]}{K_C}\right)$$

$$d[B]/dt = d[A]/dt - V_B[B]/(K_2 + [B])$$

$$D[C]/dt = d[E]/dt = d[A]/dt - d[B]/dt\ ;\ K_2 = K_B\left(1 + \frac{[A]}{K_A} + \frac{[C]}{K_C}\right)$$

The values of the constants and the initial concentrations were

V_A = 0.2 U/ml $\quad$ K_A = 3.0 mM; K_B = 3.2 mM ; K_C = 1.6 mM

V_B = 5.4 U/ml $\quad$ $[A]_o$ = 10 mM, $[B]_o = [C]_o = [D]_o = [E]_o$ = 0 mM

The considerable dissimilarity of the individual steps in the breakdown of Val-Ala-Thr-Ala (Ala-Thr-Ala is a much better substrate than the tetrapeptide, Thr-Ala is an inhibitor) due to the second site specificity of the enzyme thus creates a rather characteristic kinetic situation which should allow a discrimination between the two mechanisms mentioned above.

In Fig. 2 experimental data are compared with a simulation based on the linear sequential model (see the legend to Fig. 2) and carried out with numerical values of the constants corresponding to those given in Fig. 1. The close coincidence between both sets of data (especially in the case of Ala-Thr-Ala the maximal concentration of which as well as the time of this maximum are of high diagnostic value) confirms the applicability of the underlying model.

REFERENCES

(1) Schecther, I. and Berger, A., Biochem. Biophys, Res. Comm. 32, 898-902 (1968).

(2) Johnson, M.J., J. Biol. Chem. 137, 575-586 (1941).

(3) Metz, G. and Röhm, K.-H., in preparation.

(4) Röhm, K.-H., Z. Physiol. Chem. 355, 675-686 (1974).

(5) Göbber, F., Initial Value Program with Variable Step Length, Program No. 0082, Model 20/21 Calculator Users Club, Hewlett-Packard GmbH, D-703 Böblingen, Postfach 250.

(6) Jentsch, W., Digitale Simulation kontinuierlicher Systeme, Oldenbourg Verlag, München (1969).

CROTON OIL INDUCED COLLAGENOLITIC ACTIVITY IN THE MOUSE SKIN

G. Wirl

Institut für Molekularbiologie der österreichischen Akademie der Wissenschaften, Abt. Biologie, 5020 Salzburg, Österreich

Since J.C. Houck's work on some inflammation models in 1964, it is known that croton oil, injected intradermally into the skin of rats, produces local inflammation which is characterized by a remarkable loss of dermal insoluble collagen (1). The same effect in the skin was observed after injection of sodium hydroxide and histamine hydrochloride or after topical burning of a thin cotton disk, and it was suggested that, independent of the nature of the irritant, the general inflammatory process involves primarly a loss of insoluble collagen from the wound.

In the same year, Goldstein described collagenolytic activity in the isotonic saline extract from both intact and necrotic skin of the rat that was activated by trypsin at pH 5,5 and degraded collagen at the same pH. But neither the enzyme nor its precursor has been well characterized (2).

This paper reports that 0,2 ml of a 5 % solution of croton oil in acetone, dropped on the mouse skin, results in a rapid increase of collagenolytic activity in this region within 24 hours.

Using a method that was elaborated on the uterus post partum of the mouse (3) it was possible to extract an enzyme from the 6000 x g sediment of the skin homogenate that degrades native collagen at neutral pH. For the extraction step a Tris buffer, containing 1,0 M NaCl or 5 M urea was generally used and the extraction was performed at about 6°C . Fig. 1 shows the scheme as employed for the extraction: Shortly, the 6000 x g sediment

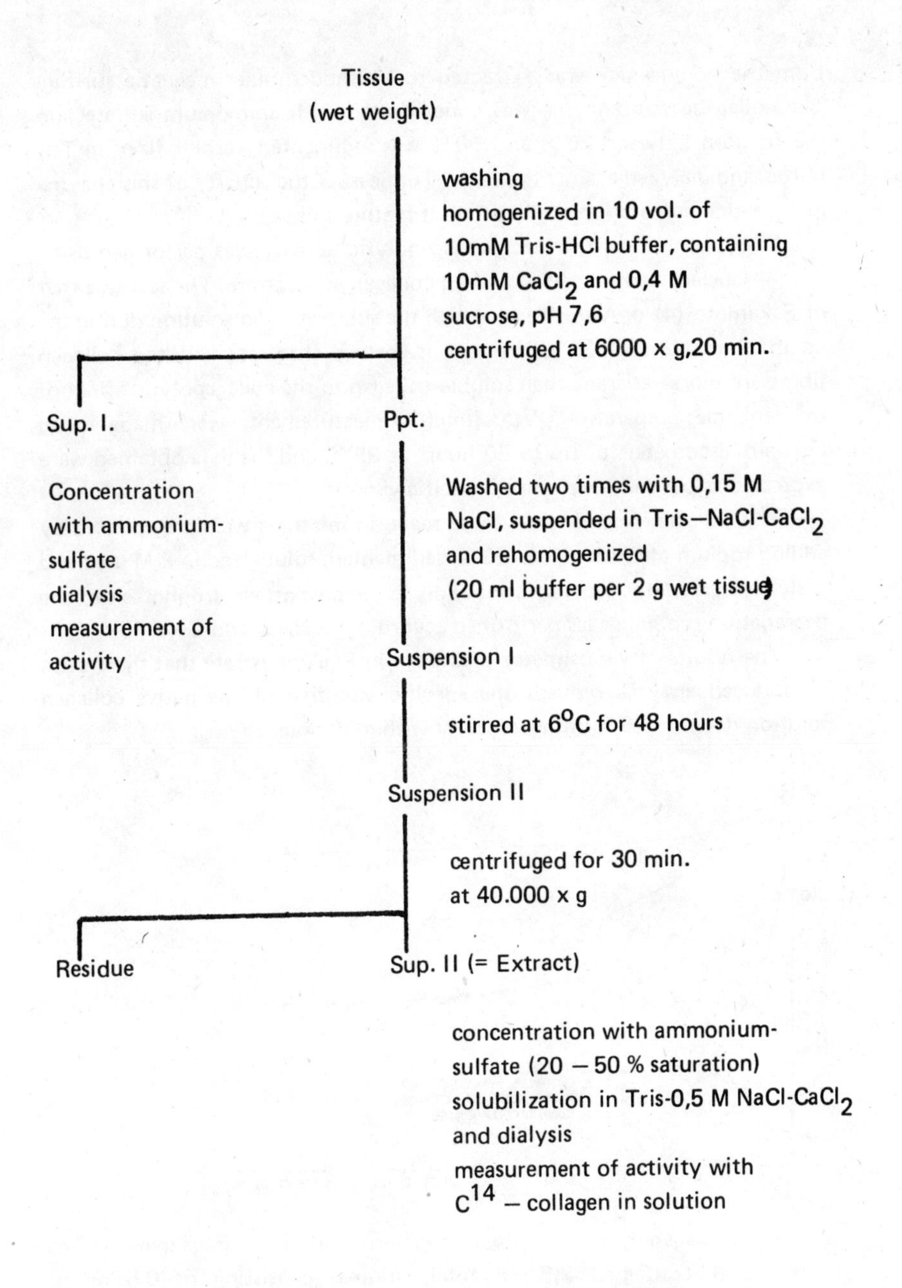

Fig. 1. Scheme employed in extraction of collagenase from the sedimentable tissue fraction (6000 x g sediment)

from the homogenate was extracted for 48 hours under magnetic stirring. The collagenolytic enzyme was concentrated with ammonium sulfate and the fraction between 20 % and 50 % was sedimented, resolubilized in Tris buffer and dialyzed against the same. To measure the activity of this enzyme in the crude preparation three different methods were used:

The quantitative assay for collagenolytic activity was performed using C^{14} – labeled, neutral salt extracted collagen as substrate. The test was that of Sakamoto (4) or Vaes (5) in which the substrate is in solution during the incubation period at 25°C, because it seemed that reconstituted collagen fibres are more resistant than soluble collagen to the collagenolytic effect of this enzyme preparation. Viscosimetric measurements were made in an Ostwald viscometer for up to 30 hours at 25°C and the data obtained were expressed as per cent decrease of specific viscosity.

The collagen fragments in these reaction mixtures were precipitated by adding sodium chloride up to 20 %, sedimented, solubilized in 8 M urea and dialyzed against the same solution. Polyacrylamide gel electrophoresis of the degradation products was performed according to Nagai and Gross (6).

The results of viscosimetry presented in Fig. 2 illustrate that the croton oil induced enzyme reduces the specific viscosity of the native collagen solution down to 25 % initial viscosity within 30 hours.

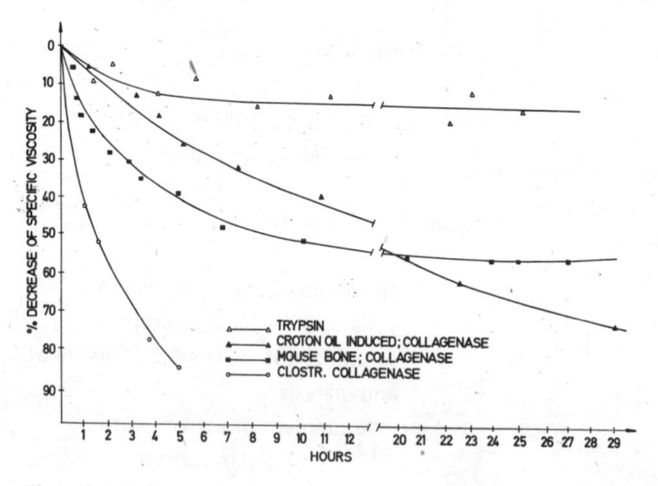

Fig. 2 Collagenolytic activity as determined by viscosimetry. The viscosimeter contained 2,5 ml total volume, consisting of 0,5 ml of approximately 0,5 % guinea pig collagen, 0,5 ml buffer (pH 7.6) and 1,5 ml enzyme solution in the same buffter. Incubation temperature: 25°C

When the viscosity reaction mixtures after variable periods were tested for collagen fragmentation by gel electrophoresis, the resultant patterns were typical for mammalian collagenases.

In all instances collagen cleavage products of αA and βA were obtained with complete absence of any additional finer bands, as you can see in Fig. 3 on the left gel column that represents 20 hours of incubation. The right gel demonstrates the pattern of collagen fragmentation after incubating together with mouse bone collagenase. The middle one represents a control incubation with EDTA.

To give a further characterization of the collagenolytic enzyme, the effect of some inhibitors of mammalian collagenases was examined. EDTA, cystein, penicillamin and cysteamin were prepared at appropriate concentrations in Tris buffer and added to the test sample. The serum was preincubated for 1 hour at $0^{o}C$ together with the enzyme solution. From the results shown in Table I it is clear that EDTA, cystein and penicillamine are potent inhibitors of collagenolytic activity down to a final concentration of 10^{-3}M. Cysteamine is capable of partially inhibiting the enzyme activity at a concentration of 10^{-2}M. Human serum is an effective inhibitor at 1:100 dilution and was without effect at 1:200 dilution. As croton oil induces a

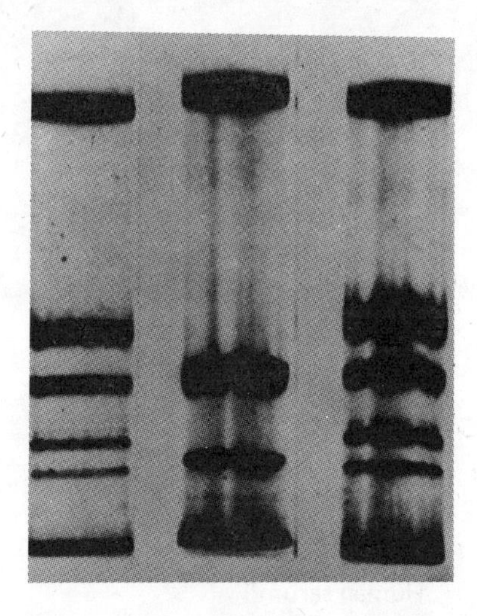

Fig. 3. Gel electrophoresis showing collagen fragmentation: after 20 hrs of incubation (left gel); control incubated with EDTA (middle gel); collagen incubated with mouse bone collagenase (right gel).

severe, local vascular reaction in the skin that is responded by hyperemia and edema formation, it is generally agreed that croton oil has an inflammatory capacity (7). Twenty four hours after the application of croton oil histological by a massive infiltration of blood cells can be observed, and it is known from the investigations of Lazarus (8) and Wahl (9) that polymorphonuclear human leucocyte granules and mononuclear macrophages contain a neutral collagenase.

TABLE 1

Inhibition of croton oil induced collagenase

Inhibitor	concentration (mM)	Inhibition %
Na_2EDTA	10^{-2}	98
	10^{-3}	85
Penicillamin	10^{-2}	92
	10^{-3}	75
	10^{-4}	35
Cysteine	10^{-1}	96
	10^{-2}	60
	10^{-3}	43
	10^{-4}	26
Cysteamin	2.10^{-2}	86
	10^{-2}	78
	10^{-3}	13
Human serum	1 : 10	94
	1 : 20	95
	1 : 100	43
	1 : 200	8

Stimulated by these results, we attempted to test some anti-inflammatory drugs for their inhibiting capacity on the collagenolytic activity present in the skin 24 hours after application of the irritant. The results presented in Table II suggest now that Indomethacin, Dexamethason and Phenylbutazone are potent inhibitors while Acetysalicylic acid has only a small effect.

Croton oil not only has inflammatory properties but also acts as a co-carcinogen, that is, as a tumour promoting agent. There appears to be an association between epidermal hyperplasia, inflammation and mouse skin tumour promotion, and although neither process is well understood, both the role of the hyperplasia and inflammation have been continuously debated.

TABLE 2

Effect of antiinflammatory drugs on collagenase activity.

Drug	Concentration (mM)	Inhibition %
Acetylsalicylic acid	10	10
	5	11
	1	0
Phenylbutazone	20	31
	10	38
	5	5
	1	0
Indomethacin	20	89
	10	75
	5	25
	1	0
Dexamethason	2	33
	1	48
	0,5	12

However, the question seemed to me to be of considerable interest whether other co-carcinogens, carcinogens or only irritant substances are able to produce collagenolytic activity in the mouse skin. The effect of the different substances so far tested in my laboratory is shown in Table III. In this table collagenolytic activity is expressed in cpm per gram original wet tissue. Only values above the trypsin blank, which means 8 % degradation, were registered as enzyme activity. As can be seen, there is a high activity in the mouse skin 24 hours after application of the croton oil. The other

TABLE 3

Effect of different irritant substances on extractable collagenase

Extraction was performed 24 hours after dropping the substances. Results expressed as the mean of 10 animals, analyzed in triplicate. Incubation time was 12 hours at 25°C.

Control:			
Aceton			
Carcinogenic:			
9,10 DMBA 200 ug		1600[1]	+[2]
——,,——— 10 ug		0	0
Methylcholanthrene	1 mg	0	++
3,4 Benzpyrene	2 mg	2500	+++
Co – carcinogenic:			
Croton oil	5 %	15800	+++++
Tween 60	100 %	2850	+
Irritant only:			
Cantharidin	50 ug	4400	++
Acetic acid	30 mg	11200	++
Terpentine	50 %	11700	++
Benzene		0	+

(1) cpm per gram wet tissue

(2) 0 – +++++ representing the spectrum of inflamed subcutaneous tissue

co-carcinogen Tween 60 induces only a small activity at this time. A relatively high activity was induced by the skin irritants cantharidin, terpentine and acetic acid. On the other hand the carcinogens induce either no or only a small activity when tested after 24 hours.

To summarize, the data presented in this paper strongly suggest the presence of collagenase activity in the mouse skin after application of the co-carcinogen croton oil, as the degradation of native collagen at neutral pH, the degradation pattern and the inhibition of activity by EDTA, cystein and serum are characteristic for mammalian collagenase. The results of viscosity measurements are untypical for this enzyme shown by comparison with activated mouse bone collagenase, but this difference is probably due to the further cleavage of the primary breakdown products by other proteases present in the crude enzyme preparation e.g, collagen peptidases.

Furthermore, these results suggest that croton oil produced wounds are a reasonable and reliable model system for the study of the chemical pathology of inflammation, and therefore this study does not result in information unique to this irritant alone as is demonstrated by the topical application of acetic acid and terpentine.

The question of the purification of the enzyme remains open and the question of the molecular weight of the collagenase and thus the type of cells which produce the enzyme.

It is highly probable that there is more than one active enzyme in this system. As mentioned before, croton oil involves both inflammation and hyperplasia in the mouse skin. Consequently polymorphonuclear leucocytes and macrophages may play an active role in production of collagenase in this system. On the other hand, it can be proposed that proliferative, pseudo-embryonic epithelial cells, of follicular as well as interfollicular origin with some kind of dedifferentiation may also contribute to the measurable pool of collagenase activity. Such results would be in agreement with investigations done on wound healing models by Eisen, Grillo and Gross (10).

REFERENCES

(1) J.C. Houck and R.A.Jacob, Proc. Soc. Exp. Biol. & Med. 116: 1041 (1964).
(2) E.R. Goldstein, Y.M.Patel and J.C. Houck, Science 146: 942 (1964).
(3) G.Wirl, Hoppe-Seyler's Z. Physiol. Chem. 356: 1289 (1975).
(4) S. Sakamoto, P. Goldhaber, M.J. Glimcher, Proc. Soc. Exp. Biol. and Med. 139: 1057, (1972).
(5) G.Vaes, Biochem. J. 126 : 275 (1972).
(6) Y.Nagai. J.Gross and K.A.Piez, Ann. N.Y. Acad. Sci. U.S. 121: 494 (1964).
(7) A. Janoff. A. Klassen and W. Troll, Cancer Res. 30: 2568 (1970).
(8) G.S. Lazarus, J.R. Daniels, J. Lian and M.C. Burleigh, Americ. J. Pathol. 68: 565 (1972).
(9) L.M. Wahl, S. M. Wahl, S.E. Mergenhagen and G. R. Martin, Proc. Ntl. Acad. Sci. U.S. 71: 3598 (1974).
10) H.C. Grillo and J. Gross, Develop. Biol. 15: 300 (1967).

PROTEOLYTIC DEGRADATION OF INSULIN AND GLUCAGON BY ENZYMES OF RAT LIVER

S. Ansorge, H. Kirschke, J. Langner, B. Wiederanders, P. Bohley and H. Hanson

Physiologisch-chemisches Institut der Universität Halle, DDR-402 Halle (Saale), Hollystrasse 1, GDR

Beside the system of thiol-proteindisulfide oxidoreductases (TPO) described in Reinhardsbrunn which catalyzes the cleavage of disulfide groups of insulin, the liver disposes of a spectrum of proteolytic enzymes capable of degrading insulin and also glucagon at a wide range of hormone concentrations. This communication presents new results on the breakdown of ^{125}I-insulin and -glucagon by the so-called insulin-specific proteinase (cytosol) and the lysosomal cathepsins B1, B3, and L.

The so-called insulin-specific proteinase (ISP)

In 1971, BRUSH (1) described a proteolytic enzyme system in rat skeletal muscle which is thought to inactivate specifically insulin. It has been named insulin-specific proteinase because of its comparable low K_m and its marked specificity for insulin as compared with a limited number of other substrate proteins. The ISP of the rat liver was partially purified by BURGHEN et al. (2). This enzyme was shown to be a thiol proteinase with a mol. wt. of 80 000. Recently we also started to study this enzyme. Using DEAE–Sephadex (pH 7.8) it could be separated from the 2 TPO_s of the cytosol and partially purified. The pH optimum of the breakdown of insulin by the ISP was determined to be pH 8. In contrast to the TPO_s, the ISP was found to be a metallo-enzyme (3). As shown in Table 1 it can be inhibited by chelating agents, e.g. o-phenanthroline and EDTA. Removal of the protein-bound metal ions by treatment with o-phenanthroline followed by dialysis leads to irreversible and total loss of enzymatic activity. However, the o-phenanthroline-inhibited enzyme can be partially reactivated particularly by zinc, cobalt or manganese ions. Other ions tested are without effect (Mg, Ca) or inhibit the enzyme (Cu).

TABLE 1

Effect of metal ions on o-phenanthroline–inhibited ISP of rat liver

$\frac{\text{Metal}^{++} \text{(mM)}}{\text{o-Phenanthr. (mM)}}$	ISP activity (percent of non-inhibited enzyme)					
	Zn^{++}	Co^{++}	Mn^{++}	Mg^{++}	Ca^{++}	Cu^{++}
$\frac{0}{1}$	0	0	0	0	0	0
$\frac{0.3}{1}$	82	53	36	2	1	3
$\frac{1}{1}$	10	58	41	1	1	1
$\frac{1}{0}$	27	80	84	94	99	2

In contrast to the postulated specificity, the action of the so-called insulin-specific protease is not restricted to insulin (3). As shown in Table 2, glucagon is degraded with comparable rates, even at physiological hormone concentrations. The pH optimum of glucagon breakdown was found to be the same as with insulin (pH 8). Although the K_m for insulin is about 26 times lower than the K_m for glucagon, the physiological efficiencies (k_{cat}/K_m) differ only slightly from each other. Moreover, this calculation is well supported by the breakdown rates determined directly by measurements at physiological hormone concentrations. From these values we may conclude that glucagon is degraded even better than insulin by the ISP.

It should be mentioned that very recently KITABCHI's group, who created the name insulin-specific protease, also reported that the ISP is capable of degrading glucagon at physiological concentrations (4). On the basis of these findings one might assume that this enzyme is involved in both the physiological breakdown of insulin and glucagon.

The cathepsins L, B1, B3

Some years ago we demonstrated by cell fractionation studies that the lysosomal fraction of rat liver, especially the soluble part of lysosomes, is capable of degrading insulin as well as glucagon at pharmacological hormone concentrations (5,6). Recently, we extended these studies to purified lysosomal proteinases and hormone concentrations in the physiological range. The preparations of purified lysosomal proteinases used in these experiments were the same as described in Dr. KIRSCHKE's communication.

The pH optimum of breakdown of ^{125}I-glucagon was determined to be 4,5 – 5 in the case of cathepsins B1 and L, and 6-6.5 in the case of cathepsin B3.

TABLE 2

Kinetic data and physiological efficiencies (k_{cat}/K_m) of insulin and glucagon degrading enzymes of rat liver

Enzyme	K_m (nM)	V ($nM \cdot min^{-1}$)	k_{cat} (min^{-1})	$\frac{k_{cat}}{K_m}$ ($min^{-1} \cdot /uM^{-1}$)
		INSULIN	BREAKDOWN	
TPO	31.0	74.0	12.33	0.40
Cathepsin L	12.5	247.2	2.19	0.18
ISP	0.33	2.9	0.085	0.26
		GLUCAGON	BREAKDOWN	
ISP	8.6	42.2	2.53	0.29

$$v = \frac{k_{cat}}{K_m} \cdot [S] \cdot [E]; \quad ([S] \ll K_m).$$

All cathepsins tested at pH 6.0 were found to be capable of degrading insulin as well as glucagon. Glucagon is much more rapidly degraded than insulin by each enzyme independently if high or low concentrations were used. The highest specific activity was measured with cathepsin L. This is in agreement with the finding of our laboratory that cathepsin L is the most active protein degrading endopeptidase of the lysosomes acting near neutral pH.

Some kinetic parameters of cathepsin L determined with insulin as substrate are compared in Table 2 with those obtained for the so-called insulin-specific proteinase (ISP) and the thiol-proteindisulfide oxidoreductase (TPO), resp. Although the K_m values and the maximal velocities (V) differ extremely, the calculated physiological efficiencies (k_{cat}/K_m) of all three enzymes are approximately in the same range. These findings show that cathepsin L exhibits about the same capability of degrading insulin at physiological concentrations as the TPO and the so—called ISP of rat liver.

It should be pointed out, however, that we do not know exactly the route of extracellular insulin at or within the liver cell and if these enzymes are generally accessible for these hormones. Therefore, we cannot yet decide which of these enzyme systems plays the main role in the process of the inactivation of insulin and glucagon in vivo.

REFERENCES

1. Brush, J.S., Diabetes 20, 140-145 (1971).
2. Burghen, G.A., Kitabchi, A.E., and Brush, J. S., Endocrinology 91, 633-642 (1972).
3. Ansorge, S., Bohley, P., Kirschke, H., Langner, J., Wiederanders, B., and Hanson, H., in 9. Karlsburger Symp. (Bibergeil, H., Fiedler, H., and Poser, K., eds.), (1974) in press.
4. Duckworth, W.C., Heinemann, M., and Kitabchi, A.E., Biochim. Biophys. Acta 377, 421-430 (1975).
5. Ansorge, S., Bohley, P., Kirschke, H., Langner, J., and Hanson, H., Europ. J. Biochem. 19, 283-288 (1971).
6. Ansorge, S., Kaiser, B., Senger, U., Bohley, P., Langner, J., and Kirschke, H., Acta Biol. Med. Germ. 28, 761-777 (1972).

INTRACELLULAR PROTEIN CATABOLISM IN MUSCLE

John W.C. Bird and William N. Schwartz

Rutgers University New Brunswick, N.J. U.S.A.

We feel that our main function at this symposium is to act as devil's advocates. Several of the speakers this week have discussed protein degradation in many different tissues. However, I would like to gently remind the audience that about half of the total weight of the body, or 40 % of the total protein of the body is muscle, and of this protein, 60 % is represented by contractile protein (1). The average turnover time of the contractile proteins has been estimated to be about 6 days, varying from 4 days for actin to about 11 days for myosin (2). It is obvious that the protein reservoir is not only large, but highly mobile in the musculature of the body.

You are all aware of the recent book entitled „Tissue Proteinases", edited by our colleagues Dr. Barrett and Dr. Dingle (3), containing the papers from an earlier symposium on tissue proteinases. It is interesting that this magnificent collection of papers does not contain the word **muscle** in its index! We feel that this ommission is an appropriate indication of the status of our knowledge concerning intracellular protein catabolism in muscle.

In our lecture today we would like to first briefly review the lysosomal apparatus of muscle as a possible source of the degradative machinery in muscle; secondly, present some new data which lends support to this hypothesis; and finally, discuss a couple models which may be used as working hypotheses for continued work in this area.

Muscle lysosomal apparatus

About 10 years ago (4–9) it was established that the acid hydrolases in muscle tissue fulfilled the criteria set forth by de Duve (10) to be classified as lysosomal enzymes. However, at that time there was a great deal of controversy as to whether the lysosomes in muscle tissue originated from muscle cells. Many thought that the enzyme activity was exclusively derived from phagocytic cells in the muscle tissue. This hypothesis was supported by electron microscopic studies at that time which were unable to demonstrate the presence of the usual type of lysosomes found in secretory tissues (11).

Our first clue to the presence of lysosomes, **per se,** in muscle cells was by the use of fluorescence microscopy (12). We injected our animals with the fluorescent metachromatic dye acridine orange, which has been shown to concentrate within lysosomes causing them to appear as bright orange granules in fluorescence microscopy (13–15). In gastrocnemius muscles from normal animals bright orange granules were seen in macrophages, mast cells, leucocytes, and endothelial cells surrounding muscle fibers. The orange granules were rarely seen within the muscle fibers of normal animals. However, after subjecting rats to 6 days of starvation prior to dye injection, large numbers of orange granules were also seen within the muscle fiber, predominantly in the perinuclear region, suggesting that the lysosomal apparatus had been activated for the process of autophagy. After rate–zonal fractionation of post–nuclear homogenates, the distribution of the acridine orange as determined by fluorescence spectroscopy was the same as that of particle–bound acid hydrolases. Dingle and Barrett (16) have also demonstrated that the treatment of subcellular fractions with acridine orange, after differential centrifugation, was an easy and rapid method for the examination of fractions for lysosomes. All of this evidence strongly supports the view that the orange fluorescent granules seen in tissue sections of muscle are lysosomes, and that some of these lysosomes are indigenous to muscle fibers.

Skeletal muscle cells are capable of very minimal amounts of pinocytosis under normal conditions, thus the technique of loading lysosomes with Triton WR–1339, or Dextran–500 by the method of Thines–Sempoux (cited in 17) is not feasible for the isolation of muscle cell lysosomes (18). However, because muscle tissue is quite heterogenous in its cellular composition, a distinct advantage can be taken of the principle of loading lysosomes. Lysosomes of non–muscle origin in muscle tissue will take up the injected substances and thereby change their equilibrium densities in sucrose gradients. With this method, lysosomes of muscle cells can be separated from lysosomes of other cell types in muscle tissue. By using this technique in conjunction with lysozyme (muramidase) as an enzyme marker for leucocytes and macrophages (19), we have been able to estimate the proportion of lysosomes from phagocytic cells in our muscle preparations. Approximately 85 % of the acid hydrolase activity is associated with lysosomes indigenous to muscle tissue.

Because starvation was known to activate the lysosomal apparatus in skeletal muscle (12,18), and since it presented a good non–pathological stress, this model was pursued in greater detail in collaboration with Professor de Duve's laboratory in Louvain (20, cited in 21). The main objective of the experiments was to obtain electron microscopic evidence for lysosomes in skeletal muscle. Animals were starved for up to eight days, and the following changes in the gastrocnemius muscles were observed: (1) a linear decrease in the wet–weight of the muscles, shown to be proportional when compared to total body weight; (2) no significant changes in protein concentration of the muscle tissue until after 4 days of starvation, when the protein concentration decreased; (3) a transient increase in percent dry weight of the muscles just prior to the decrease in protein concentration; (4) a linear increase in the specific activities of cathepsin D and p–nitrophenyl-phosphatase for 6 days of starvation; and (5) a significant increase in the free activity of the acid hydrolases by the fifth day of starvation. These measurements indicated that autophagy was initiated in the muscle tissue around the 4th to 5th day of starvation of the animals.

In another series of experiments we measured the capacity of the tissues for protein synthesis during starvation by measuring the uptake of ^{14}C–leucine (22). There was a significant decrease in incorporation of the label in total homogenates by the third day of starvation, which fell to 40 % of the controls by the sixth day of starvation. Further analysis of the subcellular fractions indicated that the first component to show a significant decrease in ^{14}C–leucine incorporation was the myofibrillar proteins (2 days) followed by the soluble proteins at 5 days of starvation. Allison et al. (23) reported a similar pattern for total muscle proteins using ^{35}S–methionine, which they correlated with a reduction in RNA content and protein synthesis.

To summarize the sequence of events that we have measured in skeletal muscle during the first 6 days of starvation, they appear in the following order: a decrease in protein synthesis by the second day with a concomitant increase in specific activity of lysosomal proteinase; formation of autophagic vacuoles by the fifth day followed by significant decreases in protein concentration.

Electron microsocopy of the gastrocnemius muscles after 5 days of starvation demonstrated dense bodies with the morphological appearance of lysosomes in the region of the Z–line and beneath the sarcolemma (21). There were large dilations of the lateral cisternae, usually most prominent from the longitudinal tubules which are found in the region of the I–band. Also prominent were multivesicular bodies which appeared to be either contiguous or confluent with longitudinal tubules. It was not possible to determine the exact membranous origin of these bodies on a strictly morphological basis. Other general observations at the EM level after starvation were autophagic vacuoles containing mitochondria and unidentified membrane, fiber splitting at the periphery, apparent pinocytotic activity of the sarcolemma, and occasional vacuoles loaded with glycogen.

There has been considerable conjecture concerning the origin of muscle lysosomes (see 21 for review). A working hypothesis in our laboratory has involved the possible role of the sarcoplasmic reticulum in forming autophagic vacuoles and/or lysosomes, much the same as does endoplasmic reticulum in other cell types (24–26). Electron microscopy has demonstrated lysosome–like dense bodies which appear to be budding off from the region of the lateral cisternae of the sarcoplasmic reticulum in skeletal muscle (21). Several years ago, Pearce (27) noted acid phosphatase, and Fishman (28) observed β–glucuronidase localized on the lateral cisternae and longitudinal tubules. More recently, Hoffstein et al. (29) have shown acid phosphatase and aryl sulfatase in dog myocardium localized in small circular or oblong profiles between bundles of myofilaments, with curvilinear profiles in close apposition to transverse tubules, thus corresponding to the placement of elements of the sarcoplasmic reticulum.

Biochemical studies (18, 30–32) have also indicated a relationship of the lysosomal acid hydrolases to preparations of fragmented sarcoplasmic reticulum. Three lysosomal acid hydrolases and calcium–activated ATPase, an enzyme marker for sarcoplasmic reticulum (33,34), were found to have the same modal equilibrium density and frequency distribution in a sucrose gradient (31). When lysosome–rich gradient fractions were incubated with Ca^{2+} and ATP in the proper environment, a portion of the acid hydrolase activity and calcium–activated ATPase activity shifted to the same new equilibrium density. The changes in calcium–activated ATPase were in agreement with those reported by Greaser et al. (34) for fragmented sarcoplasmic reticulum. Upon examination of electron micrographs of our lysosome fractions, isolated on zonal equilibrium density sucrose gradients, we were unable to differentiate between the images of our fractions (21) and those of fragmented sarcoplasmic reticulum published by other laboratories (35,36). Therefore, there appears to be reasonable biochemical and morphological evidence to support the hypothesis that some of the lysosomes in muscle cells originate from the sarcoplasmic reticulum.

Another source of lysosomes in muscle fibers could be from the Golgi apparatus, as has been reported in other cell types (see review by de Duve and Wattiaux, 37). The very limited Golgi apparatus in muscle is located in the perinuclear region, which is close to the sarcolemma in striated muscles. Schiaffino and Hanzlikova (38) have reported that the Golgi apparatus appears to be involved in the formation of both primary lysosomes and digestive vacuoles in denervated muscles of developing rats. A Golgi origin for lysosomes in muscle is appealing, and even though Golgi is relatively sparse in muscle fibers, it might be sufficient to account for the low acid hydrolase activity measured biochemically in skeletal muscle. However, considering the compartmentalization of the muscle fiber, from a logistical point of view it would appear somewhat difficult for an organelle to move from the region of the sarcolemma down through the myofibrills.

Even though the origin of lysosomes in muscle cells remains open to question, the presence of lysosomes indigenous to these cells is now firmly established. A major question to be answered concerns the role of these organelles in intracellular protein catabolism, and more specifically the role, if any, in the degradation of myofibillar proteins.

(See ref. 21 for supportive data and electronmicrographs of above discussion).

Our studies of protein degradation in skeletal muscle have centered on the roles of the lysosomal endopeptidases since they can initiate protein degradation by hydrolyzing bonds in intact molecules of a protein substrate. Cathepsin D has been the subject of many studies, but unfortunately has not been shown to hydrolize myofibillar proteins. Few reports dealing with cathepsin B1 in skeletal muscle can be found in the literature (39,40). Considering the possible importance of cathepsin B1 in protein degradation, we have begun further investigations of this enzyme in skeletal muscle. The hydrolysis of BANA determined colorimetrically by the method of Barrett (41) was used as a measure of cathepsin B1 activity.

Preliminary experiments demonstrated cathepsin B1 to be particulate, latent, and to have a typical lysosomal distribution after differential centrifugation. However, the amount of activity recoverable in all fractions was extremely low, raising the question of possible inhibition in the muscle homogenates. This suspicion was confirmed, using high speed supernatants (i.e. after 100,000g x 1 hr.) from well homogenized skeletal muscle as a source of inhibitor.

Initially, liver was used as an enzyme source due to its relatively high active titer of cathepsin B1. Rat liver was subjected to the initial steps of the purification procedure of Barrett (42). After removal of particulate debris by high speed centrifugation (100,000g x 1 hr.), the extract was allowed to autolyze overnight at pH 4.5 and was then subjected to acetone

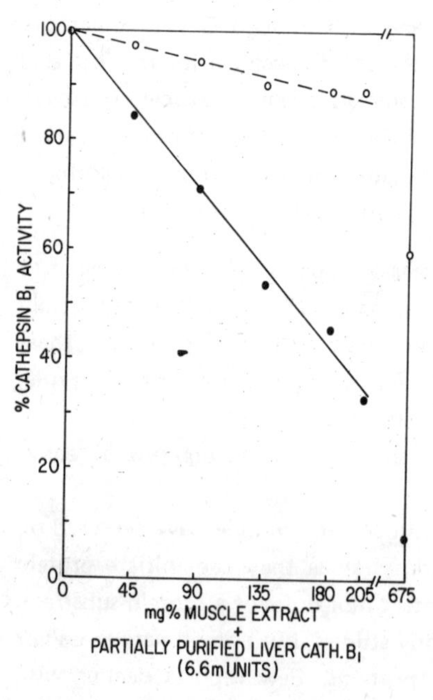

Figure 1 – Inhibition of liver cathepsin B1 activity by high speed muscle supernatant ●——●, bovine serum albumin substituted for muscle extract o- - -o.

fractionation. The 47 %–64 % acetone precipitate was resuspended in 5mM Na_2EDTA, dialyzed against H_2O, and subsequently used as an enzyme source for the inhibition studies. The only deviation from the established procedure was the finding that autolysis at 25°C, rather than at 40°C, gave better results for rat liver. Muscle extract (i.e. high speed supernatant) was mixed in varying proportions with the liver enzyme preparation and BANA-hydrolyzing activity measured. A linear relationship was demonstrated between the final concentration of muscle extract in the assay mixture and the degree of inhibition of the liver enzyme (Figure 1).

Hind leg muscles, except soleus, from Sprague–Dawley rats were homogenized (20 % (w/v) in.15 M KCl, 1mM Na_2EDTA; final pH 6.5) and then subjected to a purification regimen similar to that for liver, except that the final acetone precipitate was resuspended in, and dialyzed against, 50 mM acetate buffer, pH 5.5, with 1 mM Na_2EDTA. The BANA-hydrolyzing activity of the muscle extract increased five-to six-fold during autolysis, suggesting the removal of inhibition from the system (Table 1). The partially purified muscle enzyme preparation was then added to the liver enzyme and no inhibition was observed (Fig. 2), thus confirming that the inhibition could be eliminated by the purification procedures described above. The addition of muscle extract to the partially purified muscle enzyme resulted

TABLE I

Summary of partial purification of Cathepsin B1 from liver and skeletal muscle of the rat.

	Liver Relative Specific Activity	Yield (%)	Muscle Relative Specific Activity	Yield (%)
Homogenate	1.0	100	1.0	100
100,000g x 1 hr. supernatant	1.6	65	2.2	54
Autolysis	2.5	94	32.8	253
Acetone	65.5	55	63.3	56

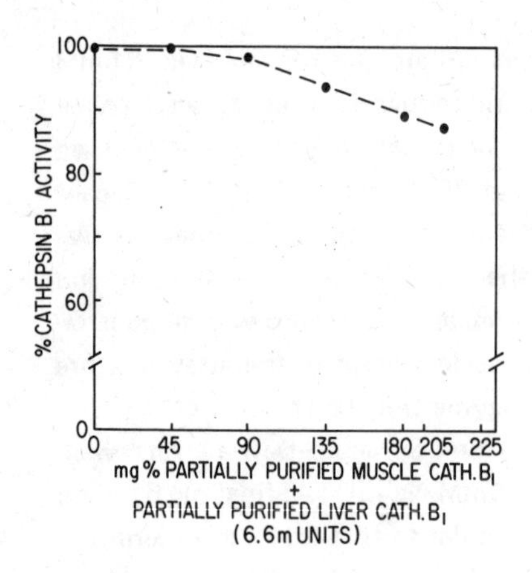

Figure 2 – Effect of partially purified muscle enzyme preparation on liver cathepsin B1 activity ●- - - -●.

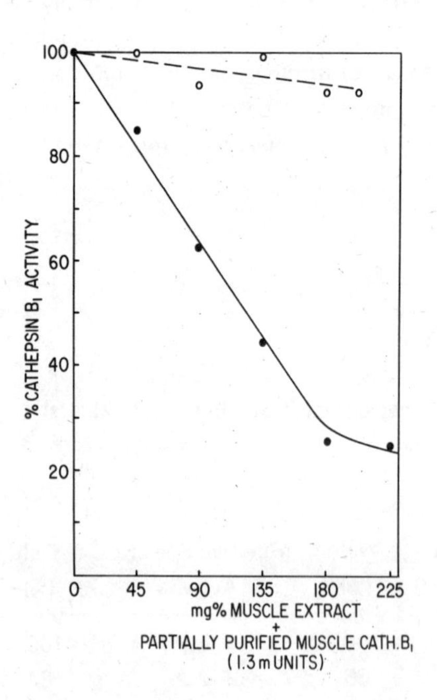

Figure 3 – Inhibition of muscle cathepsin B1 activity by high speed muscle supernatant ●——● . Bovine serum albumin substituted for muscle extract ○- - - -○.

in a pattern of inhibition similar to that obtained with the liver enzyme (Fig. 3). Thus, we were able to release the muscle enzyme from inhibition and restore the inhibition by adding back the original muscle extract.

We have begun preliminary characterization of skeletal muscle cathepsin B1, using the partially purified preparation described above. Gel filtration on Sephadex G-75 (superfine) revealed one peak of BANA-hydrolyzing activity corresponding to a molecular weight of 27,000 daltons. Optimum activity on BANA was found at pH 5.6–6.0. The effects of potential inhibitors were investigated essentially as described by Barrett (42) (Table 2). Iodoacetic

TABLE 2

Effect of potential inhibitors on crude cathepsin B1 from muscle.

COMPOUND	FINAL CONC. (mM)	INHIBITION (%)
IODOACETIC ACID	1.0	97
$HgCl_2$	0.1	92
	0.01	17
$HgCl_2$ + EDTA	0.1 + 1.0	97
	0.01 + 1.0	27
$ZnSO_4$	10.0	73
	1.0	0
$ZnSO_4$ + EDTA	10.0 + 1.0	87
	1.0. + 1.0	22
$CaCl_2$	10.0	58
	1.0	7
$CaCl_2$ + EDTA	10.0 + 1.0	67
	1.0 + 1.0	0
$MgCl_2$	10.0	36
	1.0	10
$MgCl_2$ + EDTA	10.0	11
	1.0	0
MYOGLOBIN	500 μg/ml	47
(Horse)	50	10

acid and mercuric chloride resulted in the characteristic inhibitions expected for a thiol–type enzyme. The effects of the major divalent cations found in skeletal muscle (Zn^{2+}, Ca^{2+}, and Mg^{2+}) were studied, but significant inhibition was noted only at relatively high, non–physiological concentrations (i.e. 10 mM) and was not reversible by EDTA. A significant degree of enzyme inhibition was measured in the presence of physiological concentrations of myoglobin.

The data that we have accumulated thus far indicates that the BANA-hydrolyzing activity of the partially purified enzyme preparation is due to one enzyme and that this enzyme is very similar to cathepsin B1 from other sources (i.e. molecular weight, pH optimum, effects of sulfhydryl reagents). The nature of the inhibitor in the muscle extract is not yet clear. The inhibitor is nondialyzable, functionally removed during autolysis, and significantly decreased by heating for 10–30 minutes in a boiling water bath. The inhibitor appears to be different than the proteinase inhibitor serum α_2–macro-globulin described by Barrett and Starkey (43,44), as the inhibition of BANA hydrolysis is greater than would be expected due to $\alpha 2$–macroglobulin and inhibition of other proteinases (i.e. cathepsin D) does not appear to be involved. Further investigation is underway to clarify the indentity of the inhibitor(s), the type of inhibition, and the possibility of a competitive substrate. It seems quite clear, however, that the amount of BANA–hydrolyzing activity in rat skeletal muscle **homogenates** is not a valid indicator of the amount of enzyme actually present.

Measurement of higher levels of cathepsin B1 activity than those previously reported for skeletal muscle has renewed our interest in the possible involvement of cathepsin B1 in myofibrillar degradation. Preliminary experiments measuring proteolytic activity of the partially purified enzyme on myosin and actomyosin (purified by the method of Perry (57)), suggest that there is some hydrolysis by a thiol–dependent enzyme. These studies will be reported in a further communication.

MODELS

Much has been said during this Symposium, both in seriousness and jest, concerning new models. We would also like to propose a working model, which is based on known facts and some speculation, of a possible mechanism of controlled or compartmentalized degradation of the intracellular proteins in muscle cells. Reiner (45) has suggested that the term „model" is a useful bridge between „law" and „theory", and defines the term model as „a mental image (or an artefact based on such a mental image) of the imagined machinery of a phenomenon." The following discussion should be considered with this definition in mind.

We envisage two general mechanisms: one, operating in normal physiology and concerned with the day to day turnover of all the different intracellular proteins (State I, fig. 4); and another mechanism, a modification of State I, which would principally be involved with pathophysiology (State II, fig. 4).

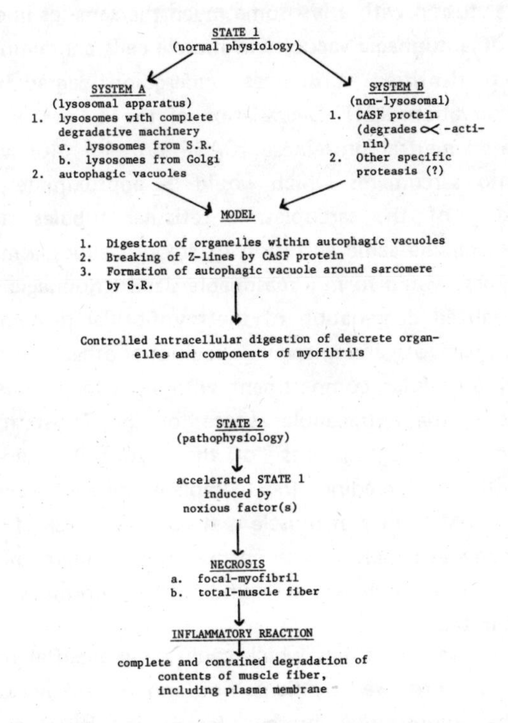

The normal physiology of State I could have two or more systems involved. One system, which we will call System A, involves the **lysosomal apparatus.** The other system, System B, may involve a **non–lysosomal** method of degradation. In the earlier part of this lecture we have shown that there is substantial evidence for a lysosomal apparatus in skeletal muscle cells with the complete degradative machinery. The lysosomes appear to be derived from the sarcoplasmic reticulum, and to a lesser extent from the somewhat sparce Golgi apparatus. The origin of membrane for the formation of autophagic vacuoles remains somewhat unclear, but could conceivably come from longitudinal or transverse tubules as well as the „traditional" smooth endoplasmic reticulum.

The only convincing evidence of a non–lysosomal intracellular muscle proteinase of which we are aware is the Ca^{2+}–activated neutral proteinase described by Busch et al. (46), which selectively degrades the α–actinin of muscle Z lines. Other specific proteinases could conceivably be present.

In our model, intracellular organelles would be degraded in autophagic vacuoles after fusion with a lysosome much the same as in other cell types. EM images of autophagic vacuoles in muscle cells containing mitochondria and other unidentified substances undergoing degradation have been common observations (21). Degradation of the Z lines of a myofibril by Ca^{2+}–activated neutral proteinase (CASF protein) (46) would break the myofibril into sarcomeres which would be approximately 2 microns in length. Fusion of the sarcoplasmic reticular tubules surrounding the sarcomere, much the same as fusion of smooth endoplasmic reticulum in other cell types, would form a reasonable sized autophagic vacuole for the compartmentalized degradation of the myofibrillar proteins. (The S.R. is homologous to smooth endoplasmic reticulum of other cells and seems to be a separate intracellular compartment with no connections to the sarcoplasmic matrix, the extracellular space, or the T-system (47). By this manner, one or more segments of the myofibril could be selectively degraded without degrading the complete muscle fiber. A common observation in EM studies of muscle is the apparent lack of structure in one or more sarcomeres in series, with normal appearing sarcomeres on all sides. However, a definite membrane surrounding the abnormal sarcomere has not been demonstrated.

An alternative hypothesis which could be injected at this point of the discussion is one proposed by Schiaffino and Hanzlikova (38). They have suggested that myofibrillar protein degradation is **initiated** by specific intracellular extralysosomal proteinases (much like the degradation of α–actinin above), and that enhanced accumulation of degradation products of myofilaments within the cell could trigger the formation of new lysosomes and induce autophagy, much the same as heterophagy may stimulate lysosome formation in other cell types (48). Unfortunately this appealing hypothesis would require specific intracellular proteinases which remain to be demonstrated.

Our second postulated mechanism (State II), involved in pathophysiology, could be an accelerated State I triggered by one or more noxious factors such as: a variety of physical agents, chemical products, diet disturbances, infections, ischemia, or immunologic reactions (see ref. (49) for review). All of these possibilities could result in necrosis; from focal necrosis of a portion of the myofibril to total necrosis of the muscle fiber due to solubilization of the lysosomal acid hydrolases. A good example of State II is ischemia in muscle tissue. Ischemia would result in hypoxia . The primary effect of hypoxia on the cytoplasm is thought to be a circumscribed (focal) injury which becomes walled off by intracellular membrane, or an autophagic vacuole (52). Since cellular autophagy appears to be an energy–requiring process (53) prolonged anoxia would prevent the formation of vacuoles and diffuse cytoplasmic damage would ensue (54,55). Engel (56) has demonstrated complete necrosis of muscle fibers, singlely or in groups, by occlusion of small arterioles leading to the fibers. It would appear that anoxia has solubilized the lysosomal acid hydrolases and the entire muscle fiber has become an autophagic vacuole. Hoffstein et al. (29) have also demonstrated this effect on myocardium. After 24 hours of ischemia lysosomes appeared to be disrupted with general cellular degeneration and erosion of I-band material (actin).

Many authors believe that there is a common pathway of pathologic changes after the initiation of necrosis regardless of the course (50,51,49). The inflamatory reaction starts a few hours after injury, and 12 to 24 hours later it is marked by the presence of large numbers of various mononucleated cells which increase in number for three to four days, moving from the periphery toward the interior of the necrotic tissue (49). These cell types contain large numbers of lysosomes and are capable of complete degradation of the contents of the myofiber, with the exception of the extracellular basement membrane. (It is interesting that the capability of macrophage lysosomes to degrade myofibrillar protein is accepted, whereas the ability of muscle cell lysosomes to degrade myofibrillar protein remains controversial).

CONCLUSION

It is quite evident that the mechanism(s) of protein turnover or degradation in either physiological or pathological conditions in muscle is not understood. And yet, as has been emphasized by many authors, it remains a central and indeed crucial question. We feel that the final answer to the mechanisms involved will require antibodies to specific proteases, that can be tagged and visualized ultrastructurally.

ACKNOWLEDGEMENT

The authors wish to thank Ms. Laura Wood for her expert technical assistance. The experiments discussed were supported by N.I.H. Grant NB–07180, the Charles and Johanna Busch Memorial Fund, the Muscular Dystrophy Association, Inc., a N.I.H. Career Development Award to J.W.C.B., and the Rutgers Research Council.

REFERENCES

1. Young, V.R. In: H.M. Monro (Ed.), Mammalian Protein Metabolism, Vol. IV, Academic Press, N.Y., pp 586–674 (1970)
2. Millward, D.J.,Clin. Sci. 39:591 (1970)
3. Barrett, A.J. and J.T. Dingle,Tissue Proteinases, North–Holland Pub. Co., Amsterdam, pp. 353 (1971)
4. Weinstock, I.M. and W. Lukacs,Enzymol. Biol. Clin. 5:89 (1965)
5. Bond, J. and J.W.C. Bird,Fed. Proc. 25:242 (1966)
6. Syrovy, I., I. Hajek and E. Gutmann,Physiol. Bohernoslov 15:7 (1966)
7. Park, D.C., and R.J. Pennington,Enzymol. Biol. Clin. 8:149 (1967)
8. Parrish, F.C., Jr., and M.E. Bailey,J. Agr. Food Chem. 15:88 (1967)
9. Bird, J.W.C., T. Berg and J. Leathem, Proc. Soc. Expt. Biol. Med. 127:182 (1968)
10. de Duve, C.,The lysosome concept, In: de Reuck, A.V.S. and M. P. Cameron (Eds.) Ciba Found. Symp. Lysosomes, pp 1–31 (1963)
11. Pellegrino, C. and C. Franzini,J.Cell Biol. 17:327 (1963)
12. Canonico, P.G. and J.W.C. Bird,J. Cell Biol. 43:367 (1969)
13. Koenig, H.,J. Cell Biol. 19:87A (1963)
14. Allison, A.C. and M.R. Young,Life Sci. 3:1407 (1964)
15. Zelenin, A.V.,Nature 212:425 (1966)
16. Dingle, J.T. and A.J. Barrett,Proc. Roy. Soc. B 173:85 (1969)

17. Beaufay, H., Methods for the isolation of lysosomes. In: J.T. Dingle (Ed.), Lysosomes A laboratory handbook, North–Holland Pub. Co., Amsterdam, pp. 1–45 (1972)
18. Canonico, P.G. and J.W.C. Bird, J. Cell Biol. 45:321 (1970)
19. Baggiolini, M., J.G. Hirsch and C. de Duve, J. Cell Biol. 40:529 (1969)
20. Bird, J.W.C., W.T. Stauber and C. de Duve, unpublished data (1970)
21. Bird, J.W.C., Skeletal muscle lysosomes. In: J. Dingle and R. Dean (Eds.) Lysosomes in Biology and Pathology, Vol. IV, North–Holland Pub. Co., Amsterdam, (In Press) (1975)
22. Gross, N. and J.W.C. Bird, unpublished data (1972)
23. Allison, J.B., R.W. Wannemacher and W.L. Banks, Fed. Proc. 22:1126 (1963)
24. Holtzman, E. and A.B. Novikoff, J. Cell Biol. 27:651 (1965)
25. Holtzman, E., A.B. Novikoff and H. Villaverde, J. Cell Biol. 33:419 (1967)
26. Novikoff, P.M. and A.B. Novikoff, J. Cell Biol. 53:532 (1972)
27. Pearce, G.W., The sarcolemma and sarcotubular systems in normal and dystrophic muscle. In: Res. Muscular Dystr. Proc. 3rd Symp., pp. 146 (1965)
28. Fishman, W.H., J. Histo. Cyto. 12:306 (1964)
29. Hoffstein, S., D.E. Zennaro, G. Weismann, J. Hirsch, F. Streuli and A.C. Fox, Am. J. Path. 79: 193 (1975)
30. Owens, K., W.B. Weglicki, R.C. Ruth, M.G. Gottwik, D.B. McNamara and E.H. Sonnenblick, Structural and functional characteristics of membrane fractions from cardiomyopathic and dystrophic muscle. In: Fleckenstein, A. (Ed.) Myocardiology: recent advances in studies on cardiac structure and metabolism (In Press) (1975)
31. Stauber, W.T. and J.W.C. Bird, Biochim. Biophys. Acta 338:234 (1974)
32. Smith, A. and J.W.C. Bird, J. Mol. Cell. Cardiol. 7:39 (1975)
33. Hasselbach, W. and M. Markinose, Biochem. Z. 333:518 (1961)
34. Greaser, M.L., R.G. Cassens, W.G. Hoekstra and E. J. Briskey, J. Cell Physiol. 74:37 (1969)
35. Hasselbach, W. and L.G. Elfvin, J. Ultrastruct. Res. 17:598 (1967)
36. Heuson–Stiennon, J. J. Wanson and P. Drochmans, J. Cell Biol. 55:471 (1972)
37. de Duve, C. and R. Wattiaux, Ann. Rev. Physiol. 28:435 (1966)
38. Schiaffino, S. and V. Hanzlikova, J. Ultra. Res. 39:1 (1972)
39. Distelmaier, P., H. Hubner, and K. Otto, Enzymologia 42:363 (1972)

40. Montgomery, A., D.C. Park, and R.J. Pennington, Comp. Biochem. Physiol. 49B:387 (1974)
41. Barrett, A.J., Anal. Biochem. 47:280 (1972)
42. Barrett, A.J., Biochem. J. 131:809 (1973)
43. Starkey, P.M. and A.J. Barrett, Biochem. J. 131:823 (1973)
44. Barrett, A.J. and P.M. Starkey, Biochem. J. 133:709 (1973)
45. Reiner, J.M., Behavior of Enzyme Systems, 2nd Ed., Van Nostrand Reinhold Co., N.Y., pp. 345 (1969)
46. Busch, W.A., M.H. Stromer, D.E. Goll and A. Suzuki, J. Cell Biol. 52:367 (1972)
47. Price, H.M., Ultrastructural pathologic characteristics of the skeletal muscle fiber. In: C.M. Pearson (Ed.). The striated Muscle, Williams and Wilkins Co., Baltimore, pp. 144–184 (1973)
48. Cohn, Z.A. and M.E. Fedorko, The formation and fate of lysosomes. In: J.T. Dingle and H.B. Fell (Eds.), Lysosomes in Biology and Pathology, Vol. I, North–Holland, Amsterdam, pp. 43 (1969)
49. Reznik, M., Current concepts of skeletal muscle regeneration. In: C.M. Pearson (Ed.). The Striated Muscle, Williams and Wilkins Co., Baltimore, pp. 185–225 (1973)
50. Adams, R.D., M.D. Denny–Brown, and C.M. Pearson, Diseases of Muscle. A Study in Pathology. 2nd ed. N.Y. Harper and Row (1962)
51. Veress, B., T. Kerenyi, I. Huttner and H. Jellinek, J. Pathol. Bacteriol. 92:511 (1966)
52. Ericsson, J.L.E., Mechanism of cellular autophagy. In. J.T. Dingle and H.B. Fell (Eds.), Lysosomes in Biology and Pathology Vol. II, North–Holland, Amsterdam, pp. 345–394 (1969)
53. Trump, B.F. and R.E. Bulger, Fed. Proc. 24:616 (1965)
54. Ericsson, J.L.E., P. Biberfeld, and R. Seljelid, Acta. Pathol. Microbiol. Scand. 70:215 (1967)
55. Trump, B.F., P J. Goldblatt and R.E. Stowell, Lab. Invest. 11:986 (1962)
56. Engel. W.K. Duchenne Muscular Dystrophy: A histologically based ischemia hypothesis and comparison with experimental ischemic myopathy. In: C.M. Pearson (Ed.), The Striated Muscle, Willams and Wilkins Co., pp. 453–472 (1973)
57. Perry, S.V., In: S.P. Ċolowick and N.O. Kaplan (Eds.) Methods in Enzymology Vol. II, Academic Press, N.Y. pp. 582–588 (1955)

PROTEIN TURNOVER OF RAT SKELETAL MUSCLE AFTER DENERVATION

D.F. Goldspink

The Physiology Department, Queen's University, Belfast, Northern Ireland.

SUMMARY

Protein synthesis and protein breakdown of the soleus and extensor digitorum longus muscles were studied in vitro during the early stages of denervation atrophy. 24 hr after nerve section significant increases in rates of protein breakdown were found in both types of muscle. These changes in degradative rates were maintained throughout the 10 days of denervation studied and correlated with the net loss of protein from the atrophying muscles. During the first 2 days of denervation significant decreases in rates of protein synthesis also correlated with the muscle wasting. However, these early effects of denervation upon rates of protein synthesis were not maintained at later time points, and were in fact reversed i.e. increased at 7 to 10 days. Both muscles responded in a parallel fashion, although the onset of atrophy occurred more rapidly and to a greater extent in the soleus. Myofibrillar and soluble proteins prepared from normal and denervated muscles were equally susceptible to in vitro degradation by pronase.

INTRODUCTION

Skeletal muscle is known to grow (i.e. protein accumulation) or atrophy (i.e. protein depletion) in response to a variety of factors, e.g. hormones

(1,2), increased work (3,4), disuse (5,6) or the animal's state of nutrition (4,7). Although the capacity of this tissue to adapt in size offers great selective advantages to the animal, we know little of the physiological and biochemical events responsible for these changes in tissue size.

Denervation has for several years been used as a model system for investigating muscle atrophy (5). In order to explain the phenomenon of muscle wasting we have investigated the effects of denervation on average rates of protein synthesis and average rates of protein breakdown of muscle. Either a decrease in protein synthesis or an increase in protein catabolism, or both, might be expected to result in a net loss of protein and hence tissue atrophy.

MATERIALS AND METHODS

All experiments involved the use of young growing male rats (approximately 50-70 g). The muscles of one hind limb were denervated by sectioning the sciatic nerve. The contralateral limb was sham-operated and its muscles served as internal controls to the denervated tissues. Two muscles were studied in these investigations; the soleus (a predominantly 'red' muscle) and the extensor digitorum longus (EDL, a predominantly 'white' muscle).

Average rates of protein synthesis and protein breakdown were measured in intact, isolated muscles according to the method of Fulks et al.(8). Basically protein synthesis was determined by measuring the incorporation of tyrosine into muscle protein during incubation in Krebs Ringer bicarbonate (KRB) containing glucose (10 mM), 0.05 μCi ^{14}C-L-tyrosine (483 m Ci/m mol) and L-tyrosine (0.1 mM). Average rates of protein breakdown were determined independently of protein synthesis by measuring the release of tyrosine into intracellular amino acid pools and into the surrounding medium. Cycloheximide (0.5 mM) was added to the KRB in this case to block protein synthesis, thus preventing reutilization of the tyrosine released by degradation of the muscle proteins.

Myofibrillar and soluble proteins were prepared (9) from pooled denervated, or pooled control, muscles obtained from 12 rats. 0.5 mg protein preparation was incubated with 20 μg pronase in tris buffer (100 mM), pH 8.0 for 60 min at 37°C. Pronase activity was measured as the increase in absorbance at 224 nm of perchloric acid-soluble material liberated during incubation (10).

RESULTS AND DISCUSSION

Fig. I shows the percentage weight losses of the soleus and EDL muscles as a function of time after denervation. In general, both muscles showed a rapid initial loss of weight, this being followed by a slower rate of atrophy at approximately 5 to 7 days onwards. Atrophy in the soleus muscle was found to occur more rapidly and to a grater extent than in the EDL. In keeping with the losses in tissue weight were net losses of protein from the denervated muscles. Since the amount of tissue protein at any one time is governed by rates of protein synthesis and rates of protein breakdown, we measured both of these parameters to determine the contributory role of each in the muscle wasting process.

In the EDL (Fig. 2) protein breakdown, whether expressed per mg muscle or per whole muscle, was elevated after denervation, with the earliest significant changes occurring after 24 hr (60 %, $p < 0.001$) and remaining elevated throughout the ten days investigated. Such elevated rates of protein breakdown clearly correlate with the muscle atrophy. Over the first two days of denervation rates of protein synthesis, expressed either per mg muscle or per whole muscle, were found to be significantly decreased. These findings again correlate with the net loss of muscle protein. However, after about 4 days onwards this trend was reversed with rates of protein synthesis becoming significantly higher than those found in the control muscles.

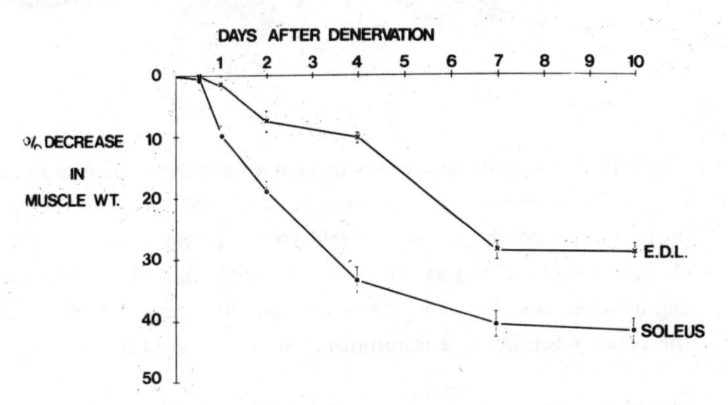

FIGURE 1. The percentage decrease in muscle weight following denervation is calculated as the mean of the individual differences in weight of the denervated and internal control muscles. Each value is the mean ± S.E.M. of at least six denervated and six control muscles.

A similar sequence of events was found for the soleus muscle (Fig. 3). Although protein catabolism per whole muscle was generally lower after nerve section, rates of breakdown per mg muscle were significantly elevated at all time points, commencing at 24 hr (17 %, $p < 0,025$). Decreases in protein synthesis (25 %, $p < 0.001$) were found as early as 12 hr after denervation. This more rapid onset of changes of protein synthesis in the soleus (compare EDL, Fig. 2) may possibly explain the more rapid onset of atrophy in the soleus muscle (Fig. 1). Protein synthesis in the whole muscle remained lower following denervation, but protein synthesis per mg muscle increased in a manner similar to that of the EDL (Fig. 2). Where changes in synthesis and breakdown per whole muscle move in opposite directions to those expressed per mg muscle (Fig. 3), the explanation appear to lie in muscle weight changes exceeding the respective changes in rates of protein synthesis or protein breakdown.

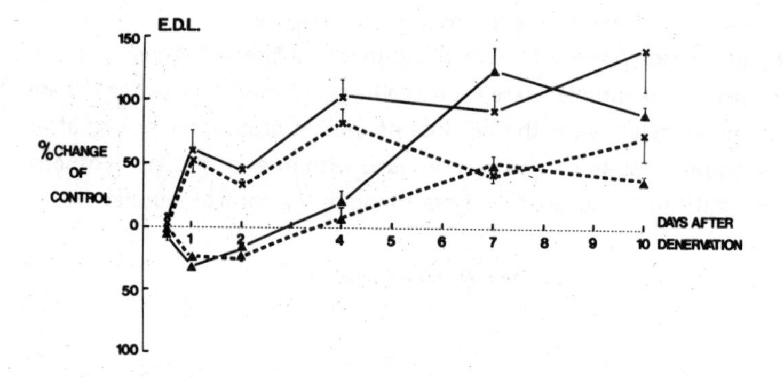

FIGURE 2. Percentage changes in protein synthesis are expressed either per mg muscle (X——X) or per muscle (X- - -X), and percentage changes in protein breakdown per mg muscle (▲—▲) or per muscle (▲-▲). Percentage changes are calculated as the mean of individual differences of synthetic, or degradative, rates between the denervated and control muscles. Each value is the mean ± S.E.M. of a minimum of six denervated and six control muscles.

Protein synthesis is measured as the in vitro incorporation of tyrosine from the medium into muscle proteins during a 2 hr in vitro incubation. Protein breakdown is measured as the release of tyrosine into intracellular amino acid pools and the surrounding medium during a 2 hr in vitro incubation.

Three points emerge from these data. Firstly, increased rates of protein breakdown are found in skeletal muscle following denervation. These in vitro data are supported by in vivo studies (Goldspink, D.F. – unpublished) showing a greater loss of radioactivity from prelabelled muscle proteins following denervation. This indicates the importance of degradative pathways in regulating amounts of tissue protein and, therefore, in influencing muscle size. Secondly, initial decreases in rates of protein synthesis favour the net loss of protein during atrophy, but then at the later time points the interesting increases in synthetic rates appear to oppose the wasting process. This reversal of trends of protein synthesis may, in part, explain the slowing down of the muscle atrophy at approximately 5 to 7 days onwards (Fig. 1). We are currently pursuing these studies to elucidate the physiological significance of these interesting changes in protein synthesis. Thirdly, the actual rates of protein synthesis and breakdown (n moles tyrosine incorporated or released, respectively) are higher in the 'red' soleus than in the 'white' EDL muscle (Goldspink D.F. – unpublished) i.e, protein turnover is higher in the soleus.

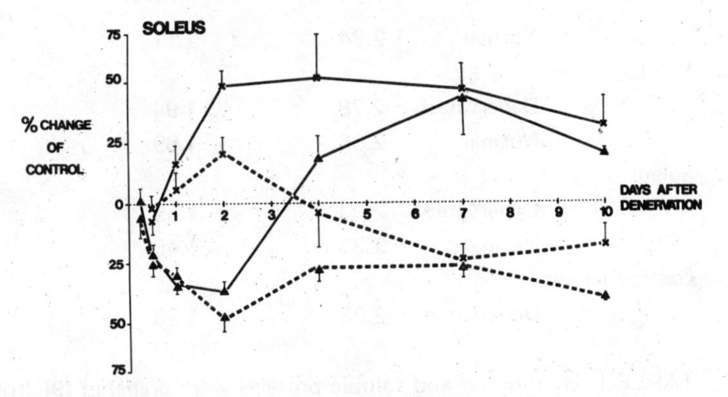

FIGURE 3. Percentage changes in protein synthesis are expressed either per mg muscle (X——X) or per muscle (X---X), and percentage changes in protein breakdown per mg muscle (▲—▲) or per muscle (▲--▲). All measurements and calculations were made in an indentical manner to that of Fig. 2.

Appreciable evidence is now accumulating which suggests that the halflife value of a protein is largely determined by the conformation of that protein (11,12). It has, of course, been known for several years that denaturation of a protein usually increases that protein's susceptibility to cleavage by proteolytic enzymes. With this in mind we examined the possibility that the elevated rates of protein breakdown caused by denervation migth arise from conformational changes in the proteins of denervated muscle.

Fractions of myofibrillar and soluble proteins were prepared from normal and denervated muscles 5 days after nerve section. Standard amounts of these protein preparations were incubated with a nonspecific protease, pronase. Three muscles were investigated, the EDL, soleus and gastrocnemius (Table 1). These data suggest that there is no preferential degradation of the proteins from the denervated muscles. In a few cases the differences in pronase activity suggested a slightly greater breakdown of proteins from

MUSCLE		PRONASE ACTIVITY (ΔA_{224})	
		Myofibrilar Protein	Soluble Protein
E.D.L.	Normal	2.84	2.11
	Denervated	2.78	1.94
Soleus	Normal	2.86	1.69
	Denervated	2.00	1.55
Gastrocnemius	Normal	2.33	1.46
	Denervated	2.38	1.23

TABLE I. Myofibrillar and soluble proteins were prepared (9) from pooled denervated, or pooled control, muscles obtained from 12 male rats (70-80 g). Each value is the mean of triplicate measurements of pronase activity against 0.5 mg protein preparations. Pronase activity is expressed as the increase in absorbance at 224 nm of PCA-soluble material liberated during a 60 min incubation at 37°C.

control muscles. If conformational changes do occur in vivo, leading to increased rates of protein breakdown, they could not be detected by this experimental approach. However, the isolation procedures used here may cause appreciable changes in the physical state of the muscle proteins possibly masking more subtle changes occurring in vivo. Therefore, it does not necessarily follow from these negative findings that such a mechanism does not operate for protein degradation in vivo.

ACKNOWLEDGEMENTS

The author acknowledges the excellent technical assistance of Mr. Paul Anderson and Miss Patricia Kilpatrick in these studies.

These investigations were supported by the Muscular Dystrophy Group of Great Britain and the British Heart Foundation.

REFERENCES

1. Goldspink, D.F. and Goldberg, A.L., Am. J. Physiol, 228, 302-309 (1975)
2. Manchester, K.L., in Mammalian protein metabolism IV, (ed. H.N. Munro). Acad. Press, New York and London (1970), PP. 229–297.
3. Goldberg, A.L., in Cardiac hypertrophy (ed. N.R. Alpert). Acad. Press, New York (1971), PP. 39-53.
4. Goldspink, G. in Differentiation and growth of cells in vertebrate tissues, (ed. Goldspink G.), Chapman & Hall, London (1974) PP. 88-93.
5. Gutmann, E., in The Denervated Muscle, Czech. Acad. Sci. Prague. Czechoslovakia (1962).
6. Williams, P.E. and Goldspink, G., J. Cell Sci., 9, 751-767 (1971).
7. Goldberg, A.L. and Goldspink, D.F., Am. J. Physiol., 228, 310-317 (1975).
8. Fulks, R.M., Li, J.B. and Goldberg, A.L., J. Biol. Chem., 250, 290-298, (1975).
9. Goldspink, D.F., Holmes, D. and Pennington R.J., Biochem. J., 125, 865-868 (1971).

10. Goldspink, D.F., Harris, J.B., Park, D.C., Parsons, M.E. and Pennington, R.J., Inter.J.Biochem., 2. 427-433 (1971).
11. Dice, J.F., Dehlinger, P.J. and Schimke, R.T., J.Biol.Chem., 248, 4220-4228(1973).
12. Goldberg, A.L. and Dice, J.F., Ann. Rev. Biochem. 43, 835-869 (1974).

ON THE LYSOSOMAL ROLE IN SKELETAL MUSCLE CATABOLISM FOLLOWING DENERVATION

W.T. Stauber *, A. M. Hedge, and B.A. Schottelius

Department of Physiology and Biophysics, University of Iowa, Iowa City, Iowa, 52242, USA

Increases in specific activities of several different lysosomal enzymes in muscle disease states have implicated the lysosome in muscle atrophy (12). However, lysosomal degradation of myofibrillar proteins has not been demonstrated. We have re-evaluated the lysosome and other degradative systems in denervation atrophy in order to better understand this important problem.

The chicken anterior latissimus dorsi (ALD) and posterior latissimus dorsi (PLD) skeletal muscles were chosen because they exhibit different responses to denervation. The ALD, a slow, oxidative muscle, hypertrophies following denervation while the PLD, a fast, glycolytic muscle, atrophies (4). Thus, they represent a unique comparative model for muscle wasting.

Specific activities of three lysosomal enzymes, cathepsin D, N-acetyl-β-glucosaminidase, and α-D-glucosidase, and the mitochondrial enzyme, cytochrome oxidase, are given in Table 1 for the ALD and PLD muscles, both normal and seven days after denervation. The increase in cathepsin D activity in the PLD muscle suggested that the lysosome might be involved in muscle degeneration by autophagic digestion of muscle components. When

---

* **W.T. Stauber is a MDAA Postdoctoral Fellow. This work was supported in part by USPHS grant NINDS 08550, and a grant from MDAA, Inc.**

autophagy increases in liver tissue, lysosomal enzymes increase in specific activity (3). However, there was a decrease in specific activity of α-D-glucosidase in the PLD muscle, and the increase in N-acetyl-β-glucosaminidase activity was similar in both the atrophying (PLD) and hypertrophying (ALD) muscle. Besides these unusual changes, the increase in cathepsin D activity of the PLD was also accompanied by an increase in cytochrome oxidase activity in an otherwise atrophying cell suggesting that autophagy might not be increased.

To help eliminate the possibility of false specific activities due to the loss of only one class of proteins (5), these lysosomal enzyme specific activities were calculated as ratios relative to cytochrome oxidase activity and are presented in Table 2. There is an apparent correlation between cathepsin D and cytochrome oxidase activities in all cases. N-acetyl-β-gluco-

TABLE 1

Enzyme specific activities

	Enzyme	Normal		Denervated	
	Cytochrome oxidase	699 ± 64	(11)	626 ± 44	(6)
ALD	Cathepsin D	134 ± 9	(13)	141 ± 12	(6)
	N-acetyl-β-glucosaminidase	95 ± 5	(7)	132 ± 6*	(5)
	α-D-glucosidase	30 ± 4	(5)	29 ± 3	(5)
	Cytochrome oxidase	403 ± 32	(11)	744 ± 31*	(6)
PLD	Cathepsin D	84 ± 9	(13)	150 ± 12*	(6)
	N-acetyl-β-glucosaminidase	55 ± 3	(7)	133 ± 13*	(5)
	α-D-glucosidase	57 ± 7	(5)	38 ± 1†	(5)

Enzyme Specific Activities. Specific activities in arbitrary units (absorbance/mgm protein/time) are presented as means ± standard errors of the number of experiments given in parentheses for each enzyme. *$P < .001$ †$P < .05$, indicate statistical significance of the value relative to the normal muscle. Reproduced from Stauber and Schottelius (9) by kind permission of Academic Press, Inc.

TABLE 2

Acid Hydrolase activity to: cytochrome oxidase activity

Acid Hydrolase	Muscle	Normal	Denervated
Cathepsin D	ALD	.213 ± .024	.230 ± .024
	PLD	.223 ± .022	.202 ± .014
N-acetyl-β-glucosaminidase	ALD	.143 ± .010	.212 ± .013
	PLD	.150 ± .010	.174 ± .012
α-D-glucosidase	ALD	.050 ± .009	.046 ± .005
	PLD	.152 ± .018	.051 ± .002

Comparative Lysosomal Enzyme Activity Ratios. Acid hydrolase specific activity as a function of the cytochrome oxidase activity. Values are means ± standard errors of six experiments. Reproduced from Stauber and Schottelius (9) by kind permission of Academic Press, Inc.

TABLE 3

Autolytic activity
(per cent protein degraded per day)

pH	ALD	PLD
3.8	4.25 + .38 *	2.96 + .17*
7.0	.87 ± .14 *	1.37 ± .20*
8.5	.69 ± .14	.55 ± .10

Autolytic Activity (per cent total protein degraded per day). Values represent means ± standard error of at least 10 experiments. *P<.05, indicates statistical significance of the values between ALD and PLD for each pH value.

saminidase was similar for the normal muscles, but differed following denervation. α-D-glucosidase ratios gave values that resembled the specific activities of this enzyme. The correlation between cathepsin D and cytochrome oxidase activities supports a lysosomal role in mitochondrial degradation, but gives no insight into the myofibrillar degradation system. Therefore, we looked at other proteolytic systems.

Autolytic activity was measured at the three pH values for which peak muscle protease activities have been described, acid, neutral and alkaline (7); the results are presented in Table 3. The pH 3.8 values are consistent with the cathepsin D data in that more activity was present in the ALD muscle. But at neutral pH (7.0), activity was found to be approximately two times greater for the PLD muscle. This neutral pH activity correlates well with the myosin-ATPase activity (8) or structural protein content (11) in Table 4, both of which are representative of the contractile protein content; such correlation suggests that this neutral protease might be responsible for the degradation of these skeletal muscle components.

Recently, contractile proteins have been shown to be attacked by a neutral pH, calcium-stimulated protease apparently unique to skeletal muscle (1). The neutral pH autolytic activity we observed was judged as possibly representative of this enzyme's activity. Consequently, the effect of 1 mM $CaCl_2$ (activator) and 1 mM EGTA (inhibitor) were tested (Table 5).

TABLE 4

Ratios of
pH 7.0 autolytic activity to:

Muscle	Myosin-ATPase Activity	Structural Protein Content
ALD	3.3	1.7
PLD	3.6	1.6

Ratios of pH 7.0 Autolytic Activity to Myofibrillar Content calculated from myosin-ATPase activity (8) and structural protein content (11).

TABLE 5

Autolytic activity
(per cent protein degraded per day)

pH	N (ALD)	EGTA (ALD)	Ca++ (ALD)	N (PLD)	EGTA (PLD)	Ca++ (PLD)
3.8	4.25 ± .38	5.51 ± .67	9.24 ± .36	2.96 ± .17	3.51 ± .35	4.91 ±* .40
7.0	.87 ± .14	.33 ± .14	2.37 ± .44	1.37 ± .20	.05 ±* .01	3.05 ±* .40
8.5	69 ± .14	.28 ± .14	2.76 ± .38	.55 ± .10	.16 ±* .10	2.69 ±* .41

Autolytic Activity (per cent of total protein degraded per day). EGTA, indicates 1 mM EGTA and Ca^{++}, indicates 1 mM $CaCl_2$ added to incubation mixture. Values represent means ± standard errors for at least 8 determinations. $P < .05$, indicates statistical significance of values relative to the normal respective muscle.

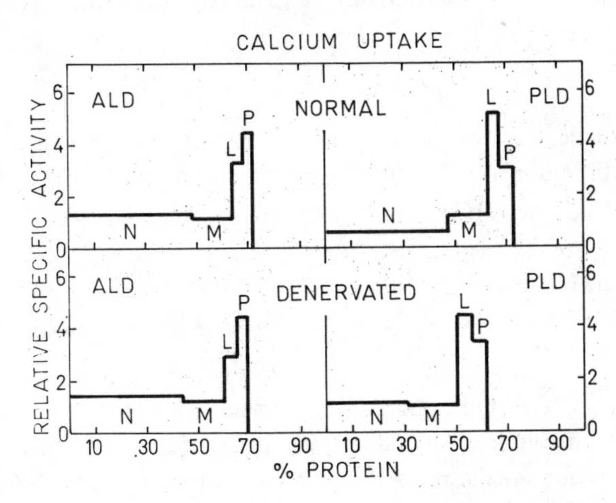

FIGURE 1. Distribution patterns of calcium uptake in normal and denervated muscles after differential centrifugation. Relative specific activity and fractions described elsewhere (2). Reproduced from Stauber and Schotelius (10) by kind permission of Academic Press, Inc.

Autolytic activities were elevated by calcium at all pH values but only at neutral and alkaline pH were the activities inhibited by EGTA. The most pronounced effect was on the autolytic activity at pH 7.0 for the PLD muscle.

The physiological significance of this calcium-stimulated proteolytic activity was not evident, so we measured the calcium uptake ability of ALD and PLD subcellular fractions obtained by differential centrifugation (Figure 1) as an indication of the muscles' ability to inhibit (by removal of cytoplasmic calcium) this stimulatory effect seen at each pH. There is little change in distribution pattern of calcium uptake for the ALD muscle, but the PLD has decreases in the heavy (M) and light (L) mitochondrial fractions and increases in the nuclear-myofibrillar (N) and microsomal (P) fractions following denervation. The specific calcium uptake ability of each fraction, with the exception of the light mitochondrial fraction (L) of the denervated ALD is unaltered (Table 6). The denervated PLD muscle has a decreased

TABLE 6

Calcium uptake (μM/mgm protein) Tissue Fractions

Muscle		N	M	L	P	T
ALD Normal	(8)	.31±.03	.26±.03	.75±.02	1.02±.04	.24±.03
ALD Denervated	(8)	.30±.03	.28±.04	.62±.04†	.92±.03	.25±.02
PLD Normal	(8)	.36±.02	.63±.14	2.62±.25	1.56±.16	.44±.02
PLD Denervated	(8)	.46±.04*	.43±.04	1.78±.16†	1.32±.06	.40±.03

Calcium uptake in tissue fractions following differential centrifugation. All values are in μ moles calcium/mgm protein/10 min. Total uptake represents the sum of the uptake in all fractions divided by the total protein in all fractions. Numbers in parentheses indicate the number of experiments. *$P < .05$ †$P < .02$, indicates statistical significance. Reproduced from Stauber and Schottelius (10) by kind permission of Academic Press, Inc.

uptake ability in the L fraction and increased in the N fraction. This latter increase may result from a decrease in myofibrillar content in the N fraction. However, the total uptake values did not change.

The total calcium uptake ability calculated as a ratio to the cytochrome oxidase activity is presented in Table 7 to achieve consistency with the acid hydrolase data and to evaluate if the total calcium uptake was reduced, but masked by a loss in myofibrillar protein content. There is no change for the ALD, but a 50 % decrease in the PLD. Obviously, the atrophying PLD muscle has a neutral proteolytic enzyme which can be activated by calcium and which exhibits greater activity than in the ALD muscle. Further, there is a calcium uptake change in the PLD following denervation. These two results suggest that a non-lysosomal system is responsible for the intial loss of myofibrillar material seen after denervation.

The comparative differences in the ability of these muscles to reduce cytoplasmic calcium levels may help explain the differences in protein turnover rates between normal red and white muscles (6). Red skeletal muscles have a decreased ability to compartmentalize calcium, and, therefore, some portion of this ion content may be available to activate proteolysis. On the other hand, white skeletal muscles are noted for their exceptional ability to reduce calcium levels and this removal of proteolytic activator may help account for their slower protein turnover rate. However,

TABLE 7

Ratio of calcium uptake to cytochrome oxidase activity

Muscle	Normal	Denervated
ALD (8)	.39±.05	.37±.02
PLD (8)	1.15±.05	.51±.05†

Ratio of calcium uptake ability (μ moles calcium/mgm protein/min.) to cytochrome oxidase activity (absorbance/mgm protein/min.). Numbers in parentheses indicate the number of experiments. P $< .001$, indicates the statistical significance. Reproduced from Stauber and Schottelius (10) by kind permission of Academic Press, Inc.

the physiological concentrations of calcium necessary to stimulate proteolysis and the intracellular localization of calcium during proteolysis are incompletely understood. Therefore, this intriguing hypothesis must await more detailed biochemical and physiological studies.

REFERENCES

1. Busch, W.A., M.H. Stromer, D.E. Goll, and Suzuki, A., J. Cell. Biol. 52, 367-381 (1972).
2. De Duve, C., Pressman, C., Gianetto, R., Wattiaux, R., and Appelmans, F., Biochem. J. 60, 604-617 (1955).
3. Deter, R.L. and de Duve, C., J. Cell Biol. 33, 437-449 (1967).
4. Feng, T.P., Jung, M.W., and Wu, W.Y., Acta Physiol. Sin. Abst. 25, 304-311 (1962).
5. Goldberg, A.L., J. Biol. Chem. 244, 3223-3229 (1969).
6. Goldberg, A.L. and Odessey, R., in Exploratory concepts in Muscular Dystrophy II. (A. L. Milhorat, ed.) Excerpta Medica, Amsterdam, (1974), PP. 187-199.
7. Pennington, R.J., in Muscle Diseases. (J.N. Walton, N. Canal, and G. Scorlato, eds.), Excerpta Medica Amsterdam, (1970), PP. 252-258.
8. Reasons, R.H. and Hikida, S., Exp. Neurol. 38, 27-32 (1973).
9. Stauber, W.T. and Schottelius, B.A., Exp. Neurol. (in press).
10. Stauber, W.T. and Schottelius, B.A., Exp. Neurol. (in press).
11. Syrovy, I., Hajek, I., and Gutmann, E., Physiol. bohemoslov. 14, 17-22 (1965).
12. Weinstock, J.M. and Iodice, A.A., in Lysosomes in Biology and Pathology. (J.T. Dingle and H.B. Fell, eds.) John Wiley, New York. (1969), pp. 450-466.

THE IMPORTANCE OF PROTEIN BREAKDOWN IN MUSCLE AS A REGULATORY PROCESS

D.J. Millward and P.J. Garlick

Department of Human Nutrition, London School of Hygiene and Tropical Medicine, Keppel Street, London WC1E 7HT.

INTRODUCTION

Protein breakdown has been shown to be a regulatory process in liver (1), heart (2), and skeletal muscle (3-5). The question considered here is the relative importance of the process of protein synthesis and breakdown in the regulation of skeletal muscle protein content.

The direct **in vivo** measurement of the rate of muscle protein breakdown is technically very difficult. The rate of protein synthesis can be measured accurately however with the constant infusion method (6) and if the rate of change of muscle protein is measured, the rate of protein breakdown can be calculated. We have therefore examined the changes in the rate of protein synthesis and breakdown **in vivo** in skeletal muscle of rats receiving a range of inadequate diets and in diabetic and hypophysectomised animals. We have also examined the rates in well-fed rats at different stages of development on a good diet and during rehabilitation from malnutrition. These results suggest that the regulation of muscle protein content is achieved primarily through changes in the rate of protein synthesis rather than protein breakdown.

MATERIALS AND METHODS

In all of these experiments $U^{-14}C$ tyrosine (12.9 μCi/ μmol) in 0.9 % NaCl (4μCi/ml) was infused into the tail vein at a rate of 2 μCi/h for 6 h. The rats were killed and the gastrocnemius and quadriceps muscles rapidly excised and homogenised in 10 % cold TCA. Protein content was determined by standard procedures and the specific radioactivity of the free (Si) and protein bound tyrosine (Sb) determined (6). The rate of protein synthesis (Ks) was then calculated from the equation

$$\frac{Sb}{Si} = \frac{R}{R-1} \cdot \frac{1-e^{-Kst}}{1-e^{-RKst}} - \frac{1}{R-1}$$

where the R=ratio of pool sizes of protein bound and free tyrosine in muscle, taken as 400 (6) and Ks=the fractional synhesis rate (FSR). The fractional breakdown was calculated as breakdown rate=synthesis rate – growth rate. The growth rate was determined by killing animals at daily intervals throughout the various procedures and calculating the fractional rate of change at the time of the infusion.

RESULTS AND DISCUSSION

All the treatments described in Table 1 resulted in a reduction in the rate of protein synthesis with a variable change in the breakdown rate. However, the results fall naturally into three groups. In the first group (1 day protein free diet, 10 days low energy intake) the treatment abolished growth by suppressing synthesis. Protein breakdown was only marginally reduced. No protein was lost from muscle however.

The characteristic feature of the second group of treatments was a more extensive suppression of protein synthesis and breakdown. In most cases no protein was lost from the muscles, but in the rats fed the protein free diet for thirty days net catabolism was observed at the rate of 3 % per day. This loss resulted from a greater fall in the synthesis rate than in the breakdown rate.

TABLE 1

Fractional rates of protein synthesis (FSR) and breakdown (FBR) in rat skeletal muscle

Rates of protein synthesis were measured by the constant infusion into the tail vein of U ^{14}C tyrosine for 6 h (6). Measurements were made in the control groups at the start of the treatments and the rate of change of protein content measured in the gastrocnemius and quadriceps muscles throughout the time course of the treatments in separate animals in addition to those infused. The breakdown rate was calculated as FBR = FSR – fractional growth rate. The diabetic rats received streptozotocin 6 days before the infusion, and the hypophysectomised rats were infused 10 days after the operation. In each group the body weight and muscle protein content was stable. The low energy rats were pair fed the food intake of the hypophysectomised rats. The muscle protein content in these rats was also stable.

Treatment		FSR % per day	FSR % control	FBR % per day	FBR % control
1. Control		18.7(2.0)		11.0	
Protein free diet	(1 day)	9.8(1.1)	52	9.8	89
Control		12.9(1.6)		8.7	
Low energy intake	(10 days)	7.1(0.9)	55	7.1	82
2. Control		12.9(1.6)		8.7	
Diabetes		4.8(0.8)	37	4.8	55
Hypophysectomy		5.8(1.3)	48	5.8	67
Control		18.7(2.0)		11.0	
Protein free diet	(9 days)	5.1(1.5)	27	5.1	46
	(30 days)	2.7(0.9)	14	5.7	52
Control		16.5(1.7)		8.9	
Starved	(2 days)	6.9(0.9)	42	6.9	77
3. Control (100 g)		16.5(1.7)		8.9	
Starved	(4 days)	4.7(0.5)	28	18.7	210
Control (400 g)		4.5(0.6)		3.7	
Starved	(4 days)	2.6(0.3)	57	6.4	172

The third group includes both young (100 g) and adult (400 g) rats which were starved for four days. The young rats were rapidly losing protein at the rate of 14 % per day and this was achieved by a marked depression of synthesis coupled to an increase in the breakdown rate. In the adult rats the response was qualitatively the same but the absolute magnitude of the rates was less.

These results illustrate the sensitivity of protein synthesis in muscle and the extent to which changes can occur. The fall in protein synthesis involved losses of ribosomes and reductions in their activity (i.e. the rate of synthesis per unit RNA). The lowest activity was observed in the diabetic rats which was half the control value.

Such sensitivity of protein synthesis does not result in rapid losses in muscle protein however, since as shown here in most cases falls in the synthesis rate are accompanied by falls in the breakdown rate. Even when net catabolism occurred in the chronically malnourished rats the breakdown rate was lower than in the control rats. Similar results are indicated in man during starvation when 3 methyl histidine excretion falls indicating reduced muscle protein breakdown (7), and during the catabolic period after surgery in which both synthesis and breakdown falls (8). The increase in the breakdown rate observed in the 4-day starved rats must therefore be seen as an emergency response. Certainly losses of muscle protein at such a rate could not be tolerated for more than a short time period.

It is not immediately apparent why the breakdown rate should fall with the synthesis rate, although there will be an obvious saving of energy as a result of the reduced turnover. This fact leads us to question however, the purpose of the high rates of protein turnover in the well fed rats.

One possible explanation is that growth of muscle necessitates high rates of turnover (9). Table 2 shows that the turnover rate in well fed rats is highest in the youngest rats which are growing more rapidly than the older animals. Furthermore when rats are rehabilitated after a period of malnutrition they grow rapidly and it is apparent from Table 2 that breakdown is increased, particularly during the second week of rehabilitation. Thus very high rates of protein synthesis are observed at this time.

We have suggested that this relationship between growth and protein breakdown may reflect myofibrilar proliferation perhaps resulting from splitting of the contractile filaments (9). If this is the case then any interruption of growth would be expected to reduce the breakdown rate as we observe. Certainly in the adult rat muscle protein turnover is slow.

The reduced rate of protein breakdown in the diabetic, 2-day starved and protein deficient rats might be thought surprising in view of the evidence that **in vitro** in the absence of insulin, the branched chain amino acids and glucose, the rate of protein breakdown is elevated (5). **In vivo,** however, it is unlikely that muscles are ever exposed to a complete absence of substrates, nutrients and hormones, as used to provide baseline measurements **in vitro.** Even in the diabetic rats the plasma insulin level was not insignificant (8μunits/ml). Thus unsupplemented **in vitro** incubations are more analogous to the 4-day starved rats with an abnormally high rate of protein breakdown. Certainly any addition of substrates or the appropriate hormones would be expected to reduce this breakdown rate.

We can conclude then that in skeletal muscle regulation of protein content is achieved through alteration of the synthesis rate; increases in the breakdown rate only occurring in response to exceptional situations.

TABLE 2

Fractional rates of protein synthesis (FSR) and breakdown (FBR) in muscle of growing rats

Measurements were made as described in Table 1. Refed rats were fed a stock diet after 30 days feeding on a protein free diet.

		FSR % per day	FBR % per day
Weanling	(23 days)	28.6(2.6)	22.3
Young adult	(65 days)	11.5(2.0)	9.8
Adult	(330 days)	4.9(0.6)	4.6
3 days refed		10.4(2.0)	5.5
8 days refed		19.8(2.8)	13.0
14 days refed		17.1(1.7)	11.0
Weight control to 8 day rats		13.8(2.6)	8.0

REFERENCES

1. Mortimore, G.E. and Mondon, C.E., J. Biol. Chem. 245, 2375-2382 (1970).
2. Rannels, D.E., Kao, R. and Morgan, H.E., J. Biol. Chem. 250, 1694-1701 (1975).
3. Jefferson, L.S., Rannels, D.E., Munger, B.L. and Morgan, H.E., Fed. Proc. 33, 1098-1104 (1974).
4. Goldberg, A.L, Howell, E.M., Li, J.B., Martel, S. B. and Prouty, W.F., Fed. Proc.33, 1112-1120 (1974).
5. Fulks, R.M., Li, J.B. and Goldberg, A.L., J. Biol. Chem.250, 290-298 (1975).
6. Garlick, P.J., Millward, D.J. and James, W.P.T., Biochem. J.136, 935-945 (1974).
7. Young, V.R., Haverberg, L.N., Bilmazes, C. and Munro, H.N., Metabolism 22, 1429-1436 (1973).
8. O'Keefe, S.J.D., Sender, P.M. and James, W.P.T., Lancet, Nov. 2, 1035 (1974).
9. Millward, D.J., Garlick, P.J., Stewart, R.J.C., Nnanyelugo, D.O. and Waterlow, J.C., Biochem. J. in press (1975).

PROTEIN DEGRADATION IN SPONTANEOUSLY DEGENERATING MUSCLES

Richard A. Lockshin

U. of Rochester Sch. of Med., Rochester, N.Y. U.S.A.

The programmed death of the intersegmental muscles of moths provides an ideal situation in which to study protein catabolism. In this tissue, 50 to 100 mg (fresh weight) of muscle is resorbed during the first two days of adult existence. The intersegmental muscles are fully contractile at the moment of emergence, but begin to degenerate within three hours. Thus the situation provides us with a sufficiently large amount of relatively pure tissue, which degenerates rapidly and with good synchrony. Under such circumstances we can readily study the control of protein catabolism as well as its enzymology.

Many of the workers present at this meeting are more sophisticated biochemists than I, and I regret that at this time I can tell you more about what does not happen than about what does. Nevertheless, our findings are sufficiently clear at present to raise questions for other workers. Briefly, they can be described as follows:

First, proteolysis begins while the muscle is fully contractile and functional. Loss of total protein begins within approximately three hours of emergence while most fibers maintain resting potentials within the normal range for at least 12 hours and, indeed, display normal appearing active membrane responses and even contract (1). A previously-reported drop in

---

(new address: Dept. of Biology, St. John's University, Jamaica, N.Y., U.S.A.)

Sponsored by the National Sciene Foundation. In collaboration with K. Srokose, J. Bidlack, and R. Schlichtig.

capacitance appears to be an experimental artifact (2, Lockshin and Beaulaton, in preparation). These studies were originally based on statistical correlations, but have now been confirmed by electron microscopic observations of individual fibers from which recordings had been made (Lockshin and Beaulaton, in preparation). Thus our evidence indicates that at least the ionic milieu of the fiber does not evolve during the early phases of involution.

A second observation is that dissolution of the myofilaments takes place in the cytoplasm, external to enveloping membranes of the lysosomal system. Although lysosomes are quite prominent in the tissue, increasing as much as 30-fold as degeneration progresses, the myofilament structure disappears in the sarcoplasm, and myofilaments are never seen within the autophagic vacuoles. These latter at various times are seen to contain glycogen particles, mitochondria, and ribosomes, but never myofilaments. Individual myofilaments evidently disappear quite rapidly, as transitional stages are rare (3-5, Beaulaton and Lockshin, submitted). Thus, even if myofilament proteins are ultimately degraded within lysosomes, they must be sufficiently altered so as to dissociate within the sarcoplasm. An incidental observation further supports this argument of pre-lysosomal alteration of proteins: when chloroquine is injected into the insects, the myofilaments dissociate but large amounts of proteins are nevertheless retained within the fiber. These proteins are contained within vacuoles and, after treatment with SDS, migrate in electric fields in approximately the same manner as proteins from undisturbed muscle. The drug has only minor effects on non-degenerating muscles (Lockshin, in preparation; Lockshin and Beaulaton, in preparation). The chloroquine apparently uncouples the preliminary alteration from final proteolysis.

The rates of disappearance of various enzyme activities differ, and some activities begin to drop as early as 4 days prior to emergence. We do not yet understand either the control of this process or its meaning, but we may be seeing premonitory changes which initiate the intracellular degradation. In this context, it is very interesting to note that there is a marked rise in lactic dehydrogenase, especially an LDH-5-like isozyme, in the muscle beginning 1-2 days prior to emergence. Thus a shift in intermediary metabolism may innaugurate the lytic phase (Lockshin, Bidlack, and Beaulaton, submitted and in preparation).

Proteases have been detected in the muscles by the use of tritiated (acetyl) hemoglobin and benzoyl-arginine-naphthylamide (BANA) as substrates. The pH optima are 3.6 against hemoglobin and approximately 6 against BANA; both activities rise as degeneration begins, and neither appears to escape lysosomal membranes. The enzymes attacking hemoglobin may be differentiated by the microbial inhibitor pepstatin. Nanogram amounts of this oligopeptide inhibit 20 % of the acid protease of unemerged animals and 50 % of the acid protease from emerged animals in which the muscles are degenerating. The sensitivity of both the assay and the drug are such that the **in vivo** effect of pepstatin could be assessed. Even when sufficient pepstatin was iontophoresed into individual fibers to inhibit the acid protease, loss of birefringence was not impeded (6). We have not yet assayed the effect of leupeptin on the cathepsin B-type enzymes. Nevertheless, the evidence from the pepstatin experiments again argues that dissolution of the myofilaments does not involve lysosomal cathepsin D.

Although our quantitation is not yet very good, autoradiograms of electrophoresed muscle proteins following the injection into the animals of S-35 methionine allows us to make some comments about the synthesis of proteins in degenerating muscles. Incorporation of methionine into soluble proteins continues unabated as degeneration begins, and continues, in approximately the same patterns as those seen prior to emergence, for at least 12 and perhaps as much as 17 hours after emergence. (Resting potential and membrane input resistance collapse at approximately 15 hours, after which time the rate of dissolution of the fiber increases (1, Beaulaton and Lockshin, submitted).

For these several reasons we feel that we do not as yet understand the mechanisms of proteolysis in this tissue. It appears to be inappropriate to attempt an explanation on the basis of digestion to oligopeptides of hemoglobin at pH 3.6, or perhaps even on the basis of digestion of BANA at pH 6. Our evidence at present suggests that we should not look for the release of amino acids but for the disappearance of intact proteins, which we may assess by measurement of enzymatic or immunologic activity. Our evidence furthermore suggests that our effectors--be they proteolytic enzymes or other--should exist in the cytoplasm, rather than within organelles, and that they should be active in a pH range near neutrality. We shall be undertaking this search in the near future. Once we can identify these enzymes, we shall be able to investigate the control mechanisms alluded to earlier.

REFERENCES

1. Lockshin, R.A., J. Insect Physiol. 19, 2359-2372 (1973).
2. Lockshin, R.A., Develop. Biol. 42, 28-39 (1975).
3. Lockshin, R.A. and J. Beaulaton, J. Ultrastr. Res. 46, 43-62 (1974).
4. Lockshin, R.A. and J. Beaulaton, J. Ultrastr. Res. 46, 63-78 (1974).
5. Lockshin, R.A. and J. Beaulaton, Life Sciences 15, 1549-1565 (1974).
6. Lockshin, R.A., Life Sciences, in press.

SEPARATION OF SOME PEPTIDE HYDROLASES FROM SKELETAL MUSCLE

M.F. Hardy, M.E. Parsons and R.J. Pennington

Regional Neurological Centre, General Hospital, Newcastle upon Tyne, U.K.

We have been studying peptide hydrolases in skeletal muscle in an attempt to elucidate their possible role in protein breakdown in normal and diseased muscle. The present paper outlines the separation of some of these enzymes from rat skeletal muscle.

In the first part of the work, combined anterior tibialis and gastrocnemius muscles were extracted with 10 mM K-PO_4 buffer, pH 7.2, containing 0.2 M NaCl + 1 mM dithiothreitol. The insoluble residue obtained after centrifugation was re-extracted with the same solution containing Triton X-100. This latter extraction was found to be necessary for the release of cathepsin C (DAP I)[1] as measured by the hydrolysis of gly-phe-NA.[1]

The two supernatants were dialysed, pooled, and submitted to chromatography on DEAE-Sephadex A-50, previously equilibrated with the extraction medium. Proteins were eluted with the medium followed by a linear NaCl gradient (0.2 M–– 0.5 M) in the same medium. The hydrolysis of several aminopeptidase substrates was measured in the column fractions (Fig. 1 (a), (b), (c)).

---

[1] Abbreviations: DAP, dipeptidyl aminopeptidase; NA, 2-naphthylamide; BANA, benzoyl-arginine 2-naphthylamide; BAPA, benzoyl-arginine 4-nitroanilide; BAA, benzoyl-arginine amide.

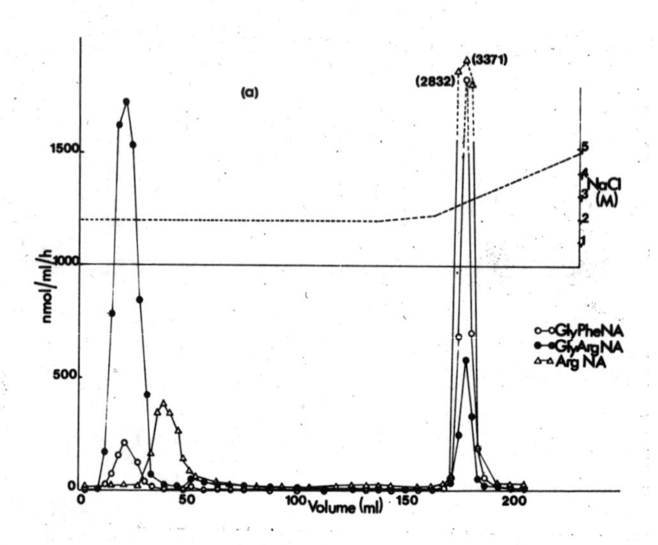

Fig. 1.(a). Separation of rat muscle amonopeptidases on DEAE - Sephadex.

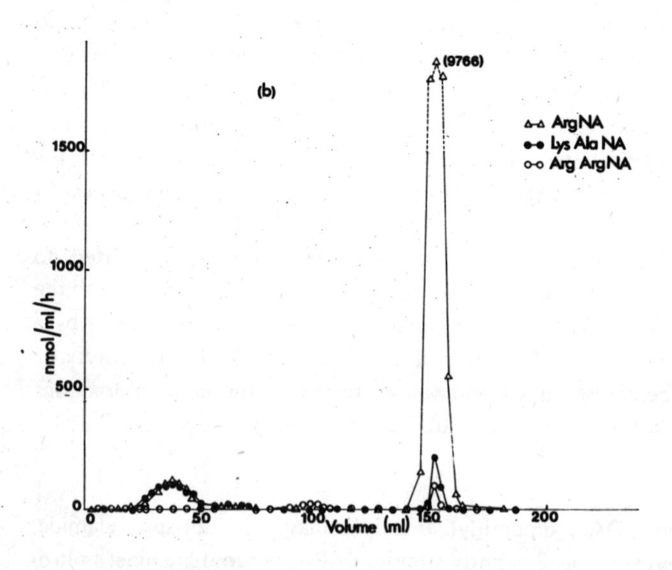

Fig. 1.(b). Separation of rat muscle aminopeptidases on DEAE -Sephadex.

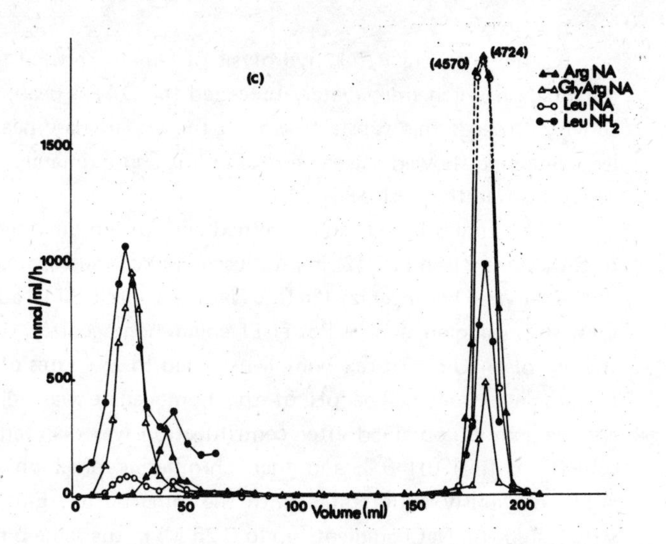

Fig. 1(c). Separation of rat muscle aminopeptidases on DEAE – Sephadex.

Fig. 1 (a) shows that enzyme activity hydrolysing arg-NA[1] was eluted in one sharp peak plus a second much smaller peak. The major peak was inhibited by 0.1 mM puromycin but the minor peak was not, thus indicating this major peak to be arylamidase. A separate peak of activity was obtained for gly-phe-NA hydrolysis but most activity on this substrate was displayed in the arylamidase peak. Hydrolysis of gly-arg-NA followed that of gly-phe-NA, but there was a much greater proportion of the activity in the early peak, which we presume to be DAP I. Hydrolysis of gly-phe amide has been demonstrated in this peak but not in the arylamidase peak. It therefore appears that gly-phe amide is a more specific substrate for DAP I than is the corresponding naphthylamide.

Hydrolysis of arg-arg-NA, the substrate for DAP III (1) showed two small peaks of activity, one coinciding with the main arylamidase peak (Fig. 1 (b)). Hydrolysis of the reported substrate for DAP II, lys-ala-NA (1), also showed two peaks of activity, both coinciding with the peaks of arg-NA splitting. We consider the first of these two coincident peaks to represent two distinct enzymes, since when the Triton-extracted supernatant was chromatographed alone, lys-ala-NA but not arg-NA was hydrolysed by this peak.

As shown in Fig. 1 (c), hydrolysis of leu-NA showed two minor peaks, the first of which immediately preceded the DAF I peak, but 90 % of the activity towards this substrate was in the arylamidase peak. Hydrolysis of leucine amide showed the same profile but approximately 50 % of the total activity was in this first peak.

Other studies have been concerned with the chromatographic properties of cathepsins B1 and B2 (2) in rat muscle. Mixed muscles from the hind limb of the rat were homogenized in 3 % NaCl — 1 mM EDTA, adjusted to pH 1.8 with HCl, using an all-glass Potter-Elvehjem homogenizer. (It was found that the use of an Ultra-Turrax homogenizer led to low rates of BAA hydrolysis in muscle extracts). The pH of the homogenate was adjusted to 4. The soluble extract obtained after centrifugation was dialysed against 10 mM Na-PO_4 buffer, pH 6.9, and then chromatographed on DEAE-Sephadex A-50, previously equilibrated with the same buffer. Elution was effected with a stepwise NaCl gradient (up to 0.25 M) in the same buffer.

Cathepsin B1, measured using BANA[1] as substrate, was separated into four components. The major component, which was not retained on the column at this pH, was rechromatographed on CM-cellulose at pH 5 and eluted with a linear NaCl gradient (0 — 0.4 M). No further fractionation was achieved, although there was some indication of a minor peak. The same four components were observed in denervated muscle, in which there is a considerable increase in cathepsin B1 activity (3). There was enhanced activity in all the fractions but the contribution of the minor peaks to the total activity was considerably increased.

Gel filtration of muscle extracts on Sephadex G-75 (Fig. 2) gave one major component which hydrolysed BANA; it had a molecular weight of 24-27000. This peak was greatly increased after denervation; the increase accounted for the increase in activity in the whole muscle extract. The minor peak (Fig. 2) was eluted at the void volume and was unaltered in denervated muscle. This peak was coincident with a peak of activity hydrolysing gly-arg-NA, and is presumed to be due to DAP I.

The hydrolysis of BAA[1] was measured also in the fractions, using a sensitive fluorimetric procedure (4). Two components hydrolysing BAA were obtained after gel filtration of muscle extracts (Fig. 2). The smaller peak had a molecular weight of 25000 and coincided with the major peak of BANA hydrolysis, which also showed slight activity towards BAPA[1]. The major component had a molecular weight of 47000 and showed no activity towards BANA and BAPA. It has a substrate specificity and molecular weight characteristic of cathepsin B2. Preliminary experiments indicate that this peak also hydrolyses benzoyl-glycyl-arginine, which would confirm the recently reported observation (5) of the carboxypeptidase nature of cathepsin B2.

Hydrolysis of BAA by homogenates of extensor digitorum longus and soleus muscles was measured after denervation, and it was found that the hydrolysis of this substrate also greatly increases and that the increase follows a time course similar to that observed with the corresponding naphthylamide (3). It is evident, from the gel filtration experiments, that the major part of the BAA hydrolysis in whole muscle extracts is due to the activity of cathepsin B2, and therefore the large increases observed after denervation in the level of BAA hydrolysis can only be explained on the basis of an increase in the activity of cathepsin B2 as well as cathepsin B1.

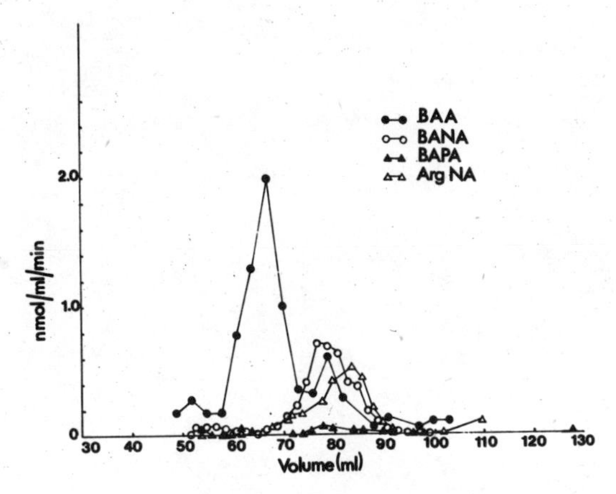

Fig. 2. Separation of rat muscle cathepsin B1 and B2 on Sephadex G—75.

ACKNOWLEDGEMENTS

This work was supported by the Medical Research Council and the Muscular Dystrophy Group of Great Britain.

REFERENCES

1. McDonald, J.K., Callahan, P.X., Ellis, S. and Smith, R.E., in Tissue Proteinases, (A.J. Barrett and J.T. Dingle, eds.), N. Holland, Amsterdam (1971), pp. 69-107.
2. Otto, K., in Tissue Proteinases, (A.J. Barrett and J.T. Dingle, eds), N. Holland, Amsterdam (1971), pp. 1–29.
3. Pluskal, M.G. and Pennington, R.J., Biochem. Soc. Trans. 1, 1307-1310 (1973).
4. Taylor, S.L., Ninjoor, V., Dowd, D.M. and Tappel, A.L., Anal. Biochem.60, 153-162 (1974).
5. Ninjoor, V., Taylor, S.L. and Tappel, A.L., Biochim. Biophys. Acta, 370, 308-321 (1974).

PROTEINASES THAT DIGEST THE CONNECTIVE TISSUE MATRIX

J. Frederick Woessner, Asher I. Sapolsky, Theresa Morales and Hideaki Nagase

Departments of Biochemistry and Medicine, University of Miami School of Medicine, Miami, Florida, USA 33152.

The connective tissue matrix offers a special problem in the field of protein breakdown by cell proteinases. The matrix proteins of interest all lie outside the cells that are to digest these proteins. The first step of proteolysis requires that the cells release proteases to the extracellular space. However, the liberation of proteolytic enzymes from the direct control of the cell poses serious dangers to the organism, since the enzymes might continue to digest indefinitely and they might digest proteins that were not intended for destruction. These considerations suggest that the proteases of interest with respect to matrix breakdown may be highly specific and that they may be governed by a series of control mechanisms.

The connective tissue matrix consists largely of proteins with varying amounts of attached carbohydrate. The main components are collagen, protein – polysaccharide complexes, glycoproteins and elastin. This paper will consider the digestion of collagen and the protein- polysaccharide complexes. Four enzymes will be discussed, two for each component.

---

*Research supported by Grant AM 16940 and HD 06773, U.S. Public Health Service.

CATHEPSIN D. Our studies of cathepsin D have gone in two directions: efforts to increase the purification and characterization of the specificity of uterine cathepsin D and exploration of the involvement of this enzyme in the breakdown of protein-polysaccharide complexes in cartilage. We previously reported purification of cathepsin D from the bovine uterus to 150 units/mg protein (1). This required, as a last step, cutting out slices from disc electrophoresis gels. It is now possible to achieve similar purification using DEAE-Sephadex chromatography. The enzyme is applied in 10 mM phosphate buffer, pH 8.8, and eluted with very gradually increasing concentrations of phosphate. Further purification is achieved by passing the active peaks from DEAE thru a hexanediamine–Sepharose column at pH 4. Enzyme is then eluted at pH 6. The final purification approaches 22--240 units/mg protein. The purified enzyme is completely inhibited by pepstatin and diazoacetyl-norleucine methylester. The following peptides are cleaved by this preparation (at the broken bonds):

Gly-Ala-Leu--Tyr LeuVal
Bz-Arg-Gly-Phe--Phe-Leu-4MBNA
Phe-Phe--Tyr-Thr
Leu--Tyr-NH_2
Ac-Phe--Tyr I_2
Phenyl Sulfite

The hexapeptide, based on a sequence in the B chain of insulin, was reported earlier (2) and remains the most rapidly cleaved synthetic substrate. Shortening the chain to a blocked pentapeptide reduces the rate considerably. Digestion of the dipeptides is not complete in 18 hours. BZ-Arg-Gly-Phe-Phe-Leu-4-methoxy-β-naphthylamide is split at the same rate by both pepsin and cathepsin D at pH 3.5. Kcat is 0.30 sec^{-1}, which compares favorably to the values reported by Ferguson **et al.** (3) for some hexapeptide substrates. Phenyl sulfite is also split at the same rate by the two enzymes. There is no doubt that pepsin and cathepsin D are closely related in their mechanism of action.

Cathepsin D has been shown to digest protein -polysaccharide light complex and proteoglycan subunit (PGS) from bovine nasal cartilage. The latter is the substrate of choice since it appears to represent the fundamental building block of proteoglycans. It consists of a long protein backbone with side chains of chondroitin and keratan sulfate attached through serine residues. Cartilage is rich in cathepsin D; this enzyme has been extensively purified from human patellar cartilage (4), rabbit ear cartilage, and chick embryo limb cartilage (5). Impure preparations of cartilage cathepsin D actively degrade PGS at physiological pH, but purified preparations have no action above pH 6 (5). In human osteoarthritis, cathepsin D is elevated 2 to 3-fold in the margin of cartilage lesions (6). Recent studies using the specific inhibitor pepstatin show that only 50 % of the apparent cathepsin D activity is inhibitable. The remaining activity at pH 5 appears to be due to an enzyme resembling the cathepsin F described by Barrett. It is not inhibited by thiol reagents or EDTA, but is blocked by whole egg white.

The activity of cathepsin D appears to be controlled by its requirement for an acid pH in the range 4-5. The cartilage matrix fluid pH is about 7.2. Therefore, the escape of this enzyme from its normal localization in the lysosomes would probably do little damage to the matrix. It probably functions chiefly in the digestion of materials brought into the interior of the cell. It has also been shown by Poole et al. (7) that cathepsin D escapes from cells and forms a thin halo around the cell in certain types of cartilage damage. Possibly the cell can secrete enough acid to favor cathepsin D action immediately adjacent to the cell.

NEUTRAL PROTEASES. Since cathepsin D appears to be unable to account for the breakdown of the cartilage matrix at any appreciable distance from the cell, attention was turned to the possible existence of neutral proteases in cartilage. Early efforts to extract from cartilage such enzymes that could degrade PGS uniformly resulted in failure. However, it was known that impure cathepsin D preparations from cartilage attacked PGS at neutral pH, so neutral activity had to be present. If crude homogenates were carried through a purification step such as gel filtration or ammonium sulfate fractionation, neutral protease activity could then be detected. Work is in progress on several types of cartilage, and neutral activity digesting PGS has been purified several hundred-fold from human patellae (4) and bovine nasal septum.

A critical problem in this work is that of finding a convenient assay for the digestion of PGS. Our approach has been to precipitate PGS with cetylpyridinium chloride. These two compounds interact chiefly by electrostatic interactions; therefore, raising the ionic strength of the medium will tend to redissolve the complex. Tri-chloroacetic acid is a suitable electrolyte, while at the same time it tends to precipitate proteins. At the proper concentrations (1 % CPC and 1.7 % TCA) PGS remains precipitated, but smaller digestion products of PGS are redissolved and can be determined as uronic acid. These principles have been developed into an assay method (8) which gives a linear increase in uronic acid color as the concentration of enzyme increases logarithmically.

The neutral PGS-digesting activity of human patellar cartilage has been separated from cathepsin D, but still possesses proteolytic action on other substrates such as casein and histone at pH 7. It is thought that these 3 neutral proteolytic activities are distinct since casein digestion is inhibited by 0.01 M cysteine, and histone digestion by aurothiomalate, whereas PGS digestion is unaffected by these compounds (4).

The effects of some inhibitors on the human and bovine enzymes are shown in Table I:

TABLE I

Effect of Various Compounds on PGS Digestion by Partially Purified Neutral Proteases of Human Patella and Bovine Nasal Septum

Compound	mM Conc.	Inhibition Human	Bovine
EDTA	5	50	85
Penicillamine	10	45	–
Dithiothreitol	5	55	–
Whole egg white	(10 mg)	60	–
Trasylol	5	0	0
Ovomucoid	(1 mg)	0	0
Soybean trypsin inhibitor	5	0	0
$ZnCl_2$	2	–	70
Iodoacetate	5	0	0

The bovine enzyme is much more strongly inhibited by EDTA. There is some suggestion that the human enzyme may be resolvable into two components, only one of which is sensitive to EDTA. The digestion products of PGS have been examined by molecular sieve chromatography on

Sepharose 4B. The size distribution of products is very close to that produced by trypsin. These fragments are larger than single chains of chondroitin sulfate, indicating that the protease cannot attack every segment of protein between adjacent polysaccharide chains.

Very little is known yet about the cartilage neutral proteases. Their subcellular distribution is not known, nor is it certain that they are ever released into the extracellular space other than at cell death. The difficulty of detecting activity in crude homogenates of cartilage suggests that the cartilage may contain some inhibitory substance to aid in the control of these enzymes. Sorgente **et al.** (9) have shown that the cartilage contains small molecular weight inhibitors of trypsin and chymotrypsin.

PZ PEPTIDASE. There are two possible pathways of collagen breakdown: extracellular and intracellular. Evidence exists for the functioning of both pathways in several tissues such as the involuting uterus. Extracellular collagen breakdown is believed to be initiated by collagenase. The name collagenase has given rise to some confusion because it was first applied to an enzyme from Clostridium histolyticum. Later, it was used for animal enzymes that digest collagen. However, the two types of collagenase have completely different specificities. The bacterial enzyme attacks peptides of the structure Pro-X-Gly-Pro. A useful substrate for the study of this enzyme is 4-phenylazobenzyloxy-carbonyl-Pro-Leu-Gly-Pro-D-Arg or PZ-Peptide (10). It was later found that animal tissue extracts could digest this synthetic peptide. However, the animal PZ-peptidase activity is separable from collagenase and has no action on native collagen. Various workers have speculated that this enzyme may come into play as soon as collagenase has cleaved collagen to digest the fragments of collagen.

In order to test some of these points, we have purified PZ-peptidase from chick embryos. It was noted that 13-14 day embryos had high level of this enzyme in their skin and that there was also active degradation of collagen at this time (11). In fact, the entire embryo proved to be rich in this enzyme. A purification scheme has been developed involving acid dialysis, gel filtration, CM and DEAE chromatography, hydrophobic chromatography and isoelectric focusing. The final preparation is over 600-fold purified and is homogeneous by electrophoresis at two pH's and with SDS added.

PZ-peptidase is apparently a metal enzyme. Its activity is enhanced about 40 % by Mg, Ca, Sr, but it cannot be inhibited more than about 40 % by EDTA and other chelating agents. In addition its activity is favored by dithiothreitol and completely blocked by p-chromercuribenzoate. Iodoacetate has no effect. The molecular weight is about 77000 by SDS gel electrophoresis.

The specificity of PZ-peptidase is not fully elucidated. It has no action on proteins such as casein or albumin. It does not digest collagen, collagen α-chains, or gelatin. A series of cyanogen bromide peptides of collagen was examined: α1-CB3 (147 residues), α1-CB4 (47 residues) and α1-CB2 (33 residues) were not digested, nor were $(Gly\text{-}Pro\text{-}Pro)_{10}$ and $(Gly\text{-}Pro\text{-}Pro)_5$. Therefore, it seems unlikely that this enzyme could act on fragments produced by collagenase action.

However, a smaller peptide of collagen is cleaved by PZ-peptidase; namelly, αl-(II)-CB6-C2:

Gly-Pro-Arg-Gly-Pro-Hyp-Gly-Pro-Ala-
Gly-Ala-Hyp↓-Gly-Pro-Gln-Gly-Phe

It can be seen from this example that the enzyme is not similar to bacterial collagenase, since it ignores a number of sequences of the type Pro-X-Gly-Pro. Furthermore, it will split smaller peptides such as Pro-Leu-Gly-Gly, and Leu-Gly-Gly at the Leu-Gly bond. However, it has no leucine aminopeptidase action on Leu-p-nitroanilide.

If this peptidase has a role in collagen breakdown, it must be limited to the final stages when peptides have been reduced to 15-20 residues or shorter. Since there is no detectable effect on proteins, control of this endopeptidase may not be critical. It could only come into action after collagenase and other proteases had partially digested larger proteins to suitable oligopeptides.

COLLAGENASE. We are studying this central enzyme of collagen degradation in the rat uterus during postpartum involution. In the process of involution 85 % of the uterine collagen is degraded within a period of 4 days. If collagenase is involved in collagen breakdown, this would be the tissue of choice in which to study the enzyme. A method has been developed for the assay of collagenase directly in uterine homogenates (12). This assay depends on the finding that collagenase and collagen fibers both go to the insoluble fraction when homogenates are centrifuged at 6000 xg, whereas, inhibitors of cllagenase go to the supernatant. Using this assay it was shown that collagenase levels rose to a peak at 24-48 hours postpartum, then declined as involution ceased. The collagenase activity appears to be under hormonal control. High doses of estradiol (100 μg/day) or progesterone (100 mg/day) inhibit collagen breakdown about 50 % and reduce the level of collagenase proportionally (13).

Serum α-2-macroglobulin and α-1-antitrypsin have been reported to inhibit collagenase. It was thought that some of these inhibitors might have been carried over into the uterine pellet. If trypsin were added to the pellet, it might bind to these inhibitors before they inactivated collagenase. This would enhance the measureable level of collagenase. The results of experiments to test this are shown in Table II:

TABLE II

Addition to Pellet Assay	Relative Collagen Digestion
None	70
Trypsin, 100 μg, 3 min, then SBTI, 360 μg	100
Soybean Trypsin Inhibitor, 360 μg	10

As anticipated, trypsin treatment for 3 min. at 37° enhanced the collagenase activity, in this case about 40 %. Totally unexpected was the almost complete inhibition of collagenase by soybean trypsin inhibitor alone, since this factor has no known action on collagenase. It must be concluded that the uterus contains a latent or hidden form of collagenase that is activated during the assay period by a trypsin-like enzyme in the uterus. This enzyme is not needed if trypsin is added for 3 minutes, since SBTI added at this point has no effect.

A series of trypsin inhibitors was tested with the following results. Inhibition of the collagenase-activating enzyme was produced by two types of soybean trypsin inhibitor, lima bean trypsin inhibitor, Trasylol, benzamidine, snail inhibitor, and α-1-antitrypsin. On the other hand, no inhibition was observed with tosyllysyl chloromethyl ketone, phenylmethyl sulfonylfluoride, ovomucoid or iodoacetate. It was of particular interest to note that α-1-antitrypsin (human) had no effect on collagenase once it was activated, the effect was only on the activation step.

Trypsin could not be replaced by chymotrypsin or elastase as an activator, nor could the tissue trypsin-like enzyme be inhibited by tosylphenylalanyl chloromethyl ketone.

The next step was to develop a method of extracting the collagenase from the insoluble pellets. This was accomplished by heating suspended pellets to 60° in Tris buffer containing 0.1 M $CaCl_2$. Indirect lines of evidence suggest that about 65-70 % of the activity can be recovered in this way. Fortunately, active and inactive forms of collagenase as well as activating enzyme all come out in this ectract.

The extracts have been fractionated on Sephadex G-150 columns. The inactive collagenase emerges first with a molecular weight of 70,000 daltons. If the extract is activated with trypsin, the peak is shifted to 45,000 daltons. If the extract is refrigerated for 1-2 days the activating enzyme converts the latent collagenase to an active form of MW approximately 60-65,000. Treatment with trypsin increases the activity of the active collagenase about 30-40 %, while reducing its molecular weight by 15-20,000. The following scheme diagrams these changes.

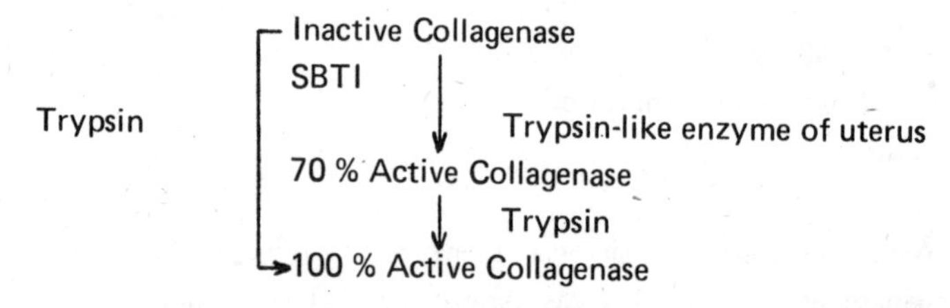

It is believed that the inactive form of collagenase is a proenzyme form. This is supported by recent findings of Birkedal-Hansen **et al.** (14) who find a related phenomenon in collagenase produced by cultured bovine gingiva. If these ideas are correct, then collagenase activity is governed by an extensive system of controls. It is first secreted as a proenzyme which requires a second protease for activation. The active collagenase is rapidly inhibited by serum and possibly by tissue inhibitors. Finally, steroid hormone levels govern the levels of collagenase.

REFERENCES

1. Sapolsky, A.I., and Woessner, J.F., Jr., J.Biol. Chem. 247, 2069 (1972).
2. Woessner, J.F., Jr. in Tissue Proteinases, (A.J. Barrett and J.T. Dingle, eds.), North Holland, Amsterdam (1971), p. 291.
3. Ferguson, J.B., Andrews, J.R., Voynick, I.M., and Fruton, J.S., J.Biol. Chem 248, 6701 (1973).

4. Sapolsky, A.I., Howell, D.S., and Woessner, J.F., Jr., J. Clin. Invest. 53, 1044 (1974).
5. Woessner, J.F., Jr., J. Biol. Chem. 248, 1634 (1973).
6. Sapolsky, A.I., Altman, R.D., Woessner, J.F., and Howell, D.S., J. Clin. Invest. 52, 624 (1973).
7. Poole, A.R., Hembry, R.M., and Dingle, J.T., J. Cell Science14, 139 (1974).
8. Sapolsky, A.I., Woessner, J.F., Jr., and Howell, D.S., Analyt. Bioch., in press (1975).
9. Sorgente, N., Eisenstein, R., and Kuettner, K., Biochim. Biophys. Acta, in press (1975).
10. Wunsch, E., and Heidrich, H-G., Z. Physiol. Chem. 333, 149 (1963).
11. Woessner, J.F., Jr., in Chemistry and Molecular Biology of the Intercellular Matrix, (E.A. Balazs, ed.), Academic Press, New York (1970), Vol. 3, p. 1663.
12. Ryan, J.N., and Woessner, J.F., Jr., Biochem. Biophys. Res. Commun. 44, 144 (1971).
13. Ryan, J.N., and Woessner, J.F., Jr., Nature 248, 526 (1974).
14. Birkedal-Hansen, H., Taylor, R.D., and Fullmer, H. M., J. Dent. Res. 54, 66 (1975).

CATHEPSINS FROM EXPERIMENTAL GRANULOMA AND THEIR ACTION ON COLLAGEN

S. Stražiščar, T. Zvonar, V. Turk

Department of Biochemistry, J. Stefan Institute, University of Ljubljana, Ljubljana, Yugoslavia.

In spite of several intensive studies, the degradation of fibrous collagen by tissue proteinases has not been elucidated in detail, (1,2). It has been shown that the majority of tissue collagenases are poorly active in degrading polymeric collagen. On the other hand some cathepsins were found to be quite active toward fibrous collagen (3,4,5).

Our interest in this problem was awakened when studying the activities of tissue proteinases in degrading various protein subst ates during evolutionary phases of experimental granuloma. We observed that proteolytic activities toward different substrates were dependent on the cell composition of granuloma respecting the cell types and the degree of their differentiation. The activities in degrading fibrous collagen and hemoglobin were in fair correlation in all evolutionary phases of granuloma, and the pH optimum of both reactions were at pH 3.5. (6)

These facts raised the question whether both reactions were catalyzed by the same enzyme or if there were two enzymes involved. It has been known that most hemoglobinolytic activity is due to cathepsin D, (3) but as to whether cathepsin D has collagenolytic activity or not, there has been no general agreement (7). We tried to answer this question by purifying the enzymes. We purified cathepsin D from experimental granuloma in an electrophoretically homogenous fraction. (8) It was practically without collagenolytic activity. (9)

Some authors thought that collagen in acid media and at the temperature of 37°C denatures to gelatine and is easily degraded by non-specific cathepsins. We tried to degrade gelatine by purified cathepsin D, but again cathepsin D exhibited only low activity for degradation of gelatine, whereas crude extracts from granuloma showed marked activity for degrading gelatine. This reaction was not inhibited by pepstatin, whick inhibits cathepsin D.

Nevertheless, there is an enzyme appearing in the same fractions through the purification procedure of cathepsin D and which is capable of degrading fibrous collagen and gelatine.

The purification procedure involved acidification to pH 3.6, ammonium sulphate precipitation, ion exchange chromatography on DEAE cellulose, gel filtration on sephadexes and electrophoresis on polyacrylamide gel. Proteolytic activities were followed in each purification step toward hemoglobin, fibrous collagen, gelatine and BANA.

The bulk of activities toward hemoglobin and fibrous collagen were in the same fractions on the ion exchangers as well as on several sephadexes (Fig. 1). Cathepsin D was finally purified using polyacrylamide gel electrophoresis, but the activity of collagenolytic cathepsin was lost after this procedure. Probably the enzyme was denatured.

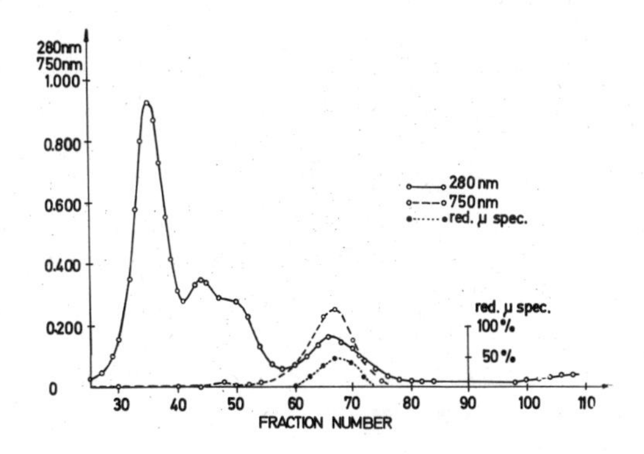

Fig. 1: Gel filtration of granuloma extract on Sephadex G-100. Dimensions of column – 70x6 cm. Flow rate of the eluent 35 ml/h O.D. 280nm – proteins in fractions. O.D. 750 nm proteolytic activity toward hemoglobin at pH 3,5. Reduction μ spec. – degradation of gelatine.

Using the active fractions from Sephadex G-100 we observed that collagen fibres were depolymerized and gamma, beta and alpha chains were the product of reaction (Fig. 2). These chains were further degraded to smaller polypeptides. Reaction products were followed using polyacrylamide gel electrophoresis. Shortened beta chains were clearly discernable on the gels as well as some shorter chains following alpha chains. The reaction was activated by cysteine. This fact suggested the possibility that the active agent might be of cathepsin B type.

According to the molecular weight of several cathepsins (cathepsin B2: 50 000, cathepsin D: 40 000 and cathepsin B1: 25 000) they would be eluted in subsequent fractions on Sephadex G-100, when suitable conditions were used. In our experiments the activities degrading hemoglobin and collagen appeared in the same fractions. If the activation of the reactions by cysteine was due to cathepsin B1 or B2, the curve, representing the activated reactions using subsequent fractions from Sephadex should be deformed or shifted to one or the other direction. But, as shown on the diagram, the curve representing the activated reactions toward hemoglobin retained its

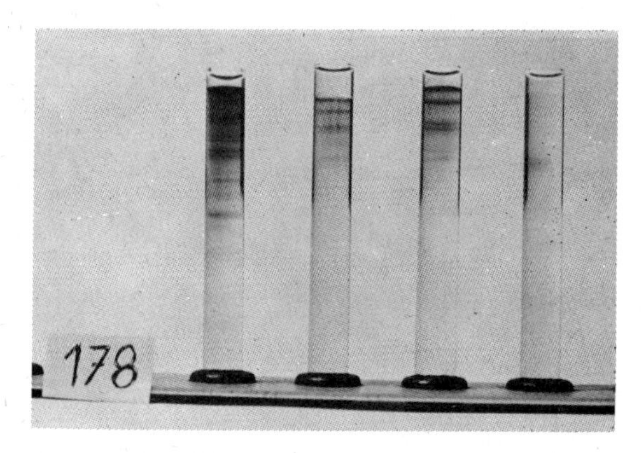

Fig. 2: Electrophoresis of the following reaction products:

1 — suspension of collagen fibres + enzyme sample,

2 — suspension of collagen fibres + thermally denatured enzyme sample,

3 — suspension of collagen fibres,

4 — enzyme sample incubated in buffer.

Experimental conditions in all cases were: pH 3.5 acetate buffer, temperature 37°C, incubation time 18 hours.

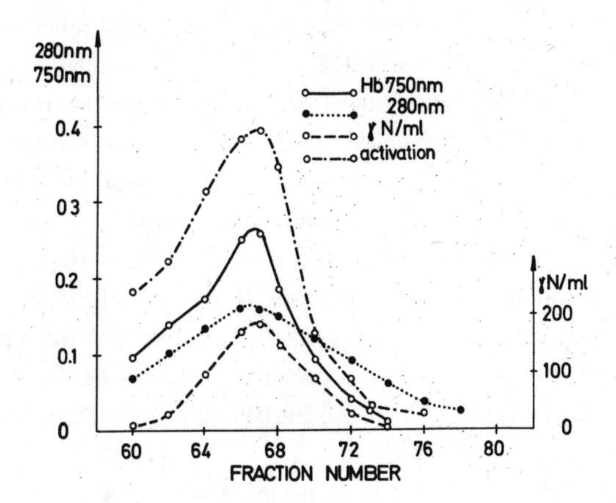

Fig. 3.A part of chromatogram from Fig. 1. Activities to degrade fibrous collagen (γN/ml) are added and activated (cystein) reactions toward hemoglobin are presented.

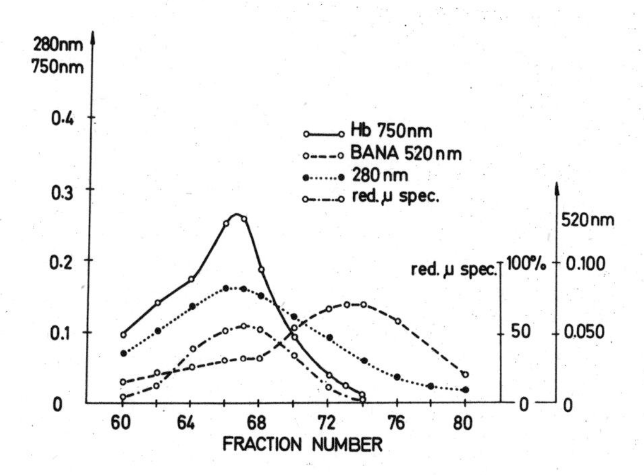

Fig. 4: Activities toward BANA in sequential fractions after gel filtration on Sephadex G-100. Same detail as in Fig. 3.

general shape and was not shifted to any direction (Fig. 3). The same phenomena were observed in the collagenolytic reaction. These facts strongly suggest that the active agent is not cathepsin B.

When these experiments were in progress it became clear that cathepsin B1 is able to degrade fibrous collagen and (4,5) we tested its activity toward BANA, the substrate degraded by cathepsin B1 and not by cathepsin B2 or cathepsin D, in the same fractions as the activities toward hemoglobin and fibrous collagen. Our results show (Fig. 4) that the activities toward BANA are in other fractions than the collagenolytic activity. Surprisingly there is not very much activity against gelatine and fibrous collagen in the fractions where cathepsin B1 is present. This may be due to the fact that we did not use either EDTA in the purification procedure, nor cysteine to activate the reactions.

On the basis of experimental results we can conclude that, under the described experimental conditions, cathepsin B1 plays only a marginal role in the degradation of fibrous collagen or gelatine. So far, we can say that the collagenolytic cathepsin is not identical with cathepsin D nor with cathepsin B1.

The elution diagram on Sephadex G-100 shows that the collageno-lytically active enzyme has a molecular weight similar to cathepsin D and higher than that of cathepsin B1. It is possible that beside collagenolytic activity it also has hemoglobinolytic activity activated by cystein.

In conclusion, we can say that there is no doubt that a special collagenolytic cathepsin exists, but a lot of work has to yet be done to elucidate adequately this problem.

ACKNOWLEDGEMENT

This work was supported by a research grant from the Boris Kidrič Foundation of the SRS and partly by the National Science Foundation, USA, grant no. 31389.

REFRENCES:

1. Woessner J.F. Jr.,Clinical. Orthoped. Nr. 96, 310 – 326, (1973)
2. Harris E.D. Jr. and Krane S.M., New England J. Med. 291, 557-563 (1974).
3. Bazin S. and Delaunay A.,Ann. Inst. Pasteur120, 50-71, (1971).
4. Burleigh M.C., Barrett, A.J., Lazarus, G.S., Biochem. J. 137, 387-398, (1974).
5. Etherington D.J.,Biochem. J. 153, 199-209, (1976).
6. Stražiščar Š. and Lebez D., Jugoslav. Physiol. Pharmacol. Acta 9 283-291, (1973).
7. Woessner J.F. Jr. In: Tissue proteinases (Barrett A.J. and Dingle J.T., eds.) North-Holland Publ. Co. Amsterdam-London (1971) pp. 291-311.
8. Stražiščar Š. and Turk V.,Zdrav. Vestn. 42, 579-581, (1973).
9. Stražiščar Š. and Turk V, Zdrav. Vestn.43, 685-689, (1974).

CATHEPSIN D OF RAT THORACIC DUCT LYMPHOCYTES: UNUSUAL INTRACELLULAR LOCALIZATION AND BIOCHEMICAL PROPERTIES

William E. Bowers*, John Panagides, Carl F. Beyer and Nagasumi Yago

The Rockefeller University, New York, New York 10021

Our studies have dealt with cathepsin D in rat thoracic duct lymphocytes, in which it displays an intracellular localization different from the other lysosomal acid hydrolases. Further studies on this enzyme have led to the finding that it has interesting biochemical properties different from the cathepsin D of many other rat tissues and other species.

METHODS

Preparation of homogenates of rat thoracic duct lymphocytes, fractionation by differential and isopycnic centrifugation, and calculations were carried out according to procedures described by Bowers (1), Beaufay et al. (2), and Leighton et al. (3). Enzymes were assayed according to Bowers et al. (4) and Yago and Bowers (5). Chromatography on G-100 Sephadex, partial purification of cathepsin D, inhibition by pepstatin and by antisera have been described fully (5).

---

*I wish to express my deep appreciation to Dr. Brian Poole for presenting this talk at the Symposiun.

RESULTS AND DISCUSSION

Fractionation: Homogenates of rat thoracic duct lymphocytes (TDL) were fractionated by differential centrifugation into three fractions: a nuclear fraction, a high-speed pellet fraction, and a high-speed supernatant. All of the lysosomal acid hydrolases contributed roughly 15 % of their total activity to the nuclear fraction, but cathepsin D differed from the other enzymes in its distribution between the high-speed pellet and supernatant fractions. Much more cathepsin D activity (approximately 65 %) was recovered in the high-speed supernatant than for the other lysosomal enzymes (approximately 15-20 %) which were associated mainly with the high-speed pellet fraction (1). These results indicated that there were at least two populations of acid hydrolase-containing particles, and so a more comprehensive study was undertaken.

Nine acid hydrolases were studied, and Fig. 1 records the distribution of these enzymes after isopycinc centrifugation of a postnuclear extract of rat TDL. They show, in confirmation of the results obtained after differential centrifugation, that most of the total activity for all the enzymes, except cathepsin D and to a lesser extent acid phosphatase, was sedimentable with a peak at a modal density of 1.18. The small portion of sedimentable cathepsin D activity was distributed in a heterogeneous fashion, and perhaps belonged to the same particles bearing the other acid hydrolases. However, most of the cathepsin D activity was unsedimentable (1).

The TDL preparation is very pure and consists almost entirely of lymphocytes. We were concerned, however, that a small number of macrophages, not easily identifiable among the vast majority of lymphocytes, contaminated our TDL preparation and contributed significant enzyme activities, especially cathepsin D. In order to eliminate this possibility, rat TDL were cultured for 2 hours to allow enough time for any macrophages to adhere to the plastic surface of the Petri dish, and then the non- adherent cells were removed and fractionated by isopycnic centrifugation. Although a few adherent cells were observed, the activity of the non-adherent cells for all enzymes on a per cell basis did not change, and the enzyme distributions obtained after isopycinc centrifugation of a postnuclear extract of non-adherent cells were identical to those shown in Fig. 1 (1). These results indicated that enzyme activities derived almost entirely from rat TDL.

Rat TDL themselves are comprised of approximately 95 % small lymphocytes and 5 % large lymphocytes. It was possible that small lymphocytes contributed most of the normal lysosomal acid hydrolases and that large lymphocytes contained most of the cathepsin D, or vice versa. In order to test this point, we devised a method of zonal centrifugation that fractionates cells mainly according to size. After zonal centrifugation, the cells showed a nearly symmetrical distribution that was slightly skewed toward the fractions situated most distal to the axis of rotation. Large lymphocytes occupied the skewed part of the distribution, as was shown by a variety of methods. The major question was whether or not the distributions for the lysosomal enzymes and cathepsin D would follow either that of the small lymphocytes or that of large lymphocytes. For all of the

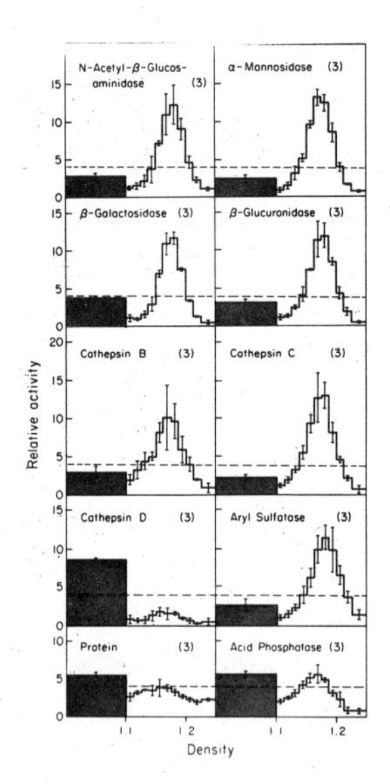

Fig. 1 Distribution of acid hydrolases and protein after isopycnic centrifugation in sucrose gradients of postnuclear extracts of rat thoracic duct lymphocytes. The shaded block with a density below 1.10 has an arbitrary density interval of 0.1 and represents the position of the sample layer. The dotted line indicates the histogram obtained if enzyme activity or protein were uniformly distributed. Each histogram shows the average of results with standard deviation, and the number of experiments is given between parentheses.

enzymes, including cathepsin D, there was no preferential association with large lymphocytes, and in fact, when small lymphocytes were purified by zonal centrifugation and then fractionated by isopicnic centrifugation, the distributions found were nearly identical to those seen in Fig. 1. Thus, cathepsin D and the other lysosomal enzymes both belonged mainly to small lymphocytes (6).

Further experiments were designed to investigate whether cathepsin D was associated either with thymus-derived T-lymphocytes or with bone marrow-derived B-lymphocytes, which are both present in a preparation of rat TDL, although T cells predominate. The answer is that the enzyme composition of T and B small lymphocytes – for both cathepsin D and for the other acid hydrolases – is about the same.

Thus, on the cellular level it appeared that cathepsin D and the other acid hydrolases belonged to the same cell. Recently we have found that mouse leukemia L1210 cells, which presumably represent a homogeneous, clonally – derived population of lymphocytes, also contained both normal sedimantable lysosomes and cathepsin D that was recovered mainly in an unsedimentable form after fractionation.

Although normal lysosomes and cathepsin D existed together in the same cell, the major question concerned the intracellular localization of cathepsin D. Recovery of this enzyme in a soluble, unsedimentable form after fractionation of rat TDL homogenates suggested two possibilities: 1) that cathepsin D was a soluble enzyme present in the cytosol; or 2) that it was contained inside particles that break during homogenization. In fact, the homogenization conditions that we used to break open rat TDL were quite severe; cells were exposed to hypotonic conditions for a few minutes and then vigorously homogenized with a tightfitting pestle. We therefore devised a more gentle method of disrupting the cell membrane and used antibody and complement for this purpose. Rat TDL were incubated briefly in rabbit anti-rat TDL antiserum, and then the cells were washed to remove excess unbound antibody. Complement was added, resulting in a lysis of more than 99 % of the TDL. When homogenates of antibody-complement lysed TDL were fractionated by differential centrifugation, the results shown in Table I were obtained. Nearly 90 % of the total activity for pyrophosphatase, an enzyme known to be present in the cytosol, was recovered in a high-speed supernatant fraction. If cathepsin D were also a cytosol enzyme, then it should have made a major contribution to the supernatant fraction. It did not, and we therefore ruled out a cytosol localization for cathepsin D.

Rather unexpectedly, however, we found that most of the total cathepsin D activity was associated with the nuclear fraction. This finding was in marked contrast to the localization of a typical lysosomal enzyme, N-acetyl-β-glucosaminidase, which was recovered mostly in a high-speed pellet.

Even more cathepsin D was found to be associated with the nuclear fraction after homogenization of antibody – complement lysed TDL using a loose-fitting pestle. As much as 75-80 % of the total activity of cathepsin D sedimented with this fraction, and, although this sedimentable activity was released by the addition of Triton X-100, we have thus far been unable to dissociate any cathepsin D-containing particles from the nuclei. At present, we do not know definitely where the enzyme is located intracellularly.

TABLE 1

Fractionation by differential centrifugation of a homogenate prepared from rat TDL lysed by antibody and complement.

Enzyme	Fraction	% of Total Activity
N-acetyl-β-glucosaminidase	Nuclear	17.1 ± 4.8
	High-speed pellet	63.6 ± 4.8
	Supernatant	19.3 ± 3.8
β-Glucuronidase	Nuclear	17.6 ± 4.6
	High-speed pellet	64.2 ± 4.0
	Supernatant	18.2 ± 4.2
Cathepsin D	Nuclear	52.1 ± 5.6
	High-speed pellet	15.7 ± 4.0
	Supernatant	32.7 ± 4.0
Pyrophosphatase	Nuclear	7.4 ± 3.8
	High-speed pellet	3.1 ± 1.4
	Supernatant	89.5 ± 4.0

Values are means ± standard deviation for 4 separate fractionations.

Because the intracellular localization of cathepsin D was so different from that of the other lysosomal enzymes, we initiated a study on the biochemical properties of this enzyme. A number of unusual properties were found, the first of which is illustrated in Fig. 2. An antiserum prepared in rabbits against rat liver soluble lysosomal enzymes inhibited to a considerable extent the enzyme activity of a crude extract of rat liver cathepsin D. Although the pH optimum for cathepsin D is 3.6, the assay was carried out at pH 5 in order to avoid a dissociation of antigen and antibody that occurs at lower pH values. Surprisingly, cathepsin D activity in an extract of rat TDL was only marginally affected by the same antiserum, even though both rat liver and rat TDL cathepsin D preparations showed identical pH curves with acid-denatured bovine hemoglobin as substrate (5).

When an extract of rat TDL was chromatographed on Sephadex G-100, the results shown in Fig. 3 were obtained. Whereas rat liver cathepsin D showed a single peak corresponding to an apparent molecular weight of 45 000, rat TDL extract contained two forms of cathepsin D. One form chromatographed in the same manner as the rat liver enzyme, but a second higher molecular weight form of approximately 95 000 was also found (5).

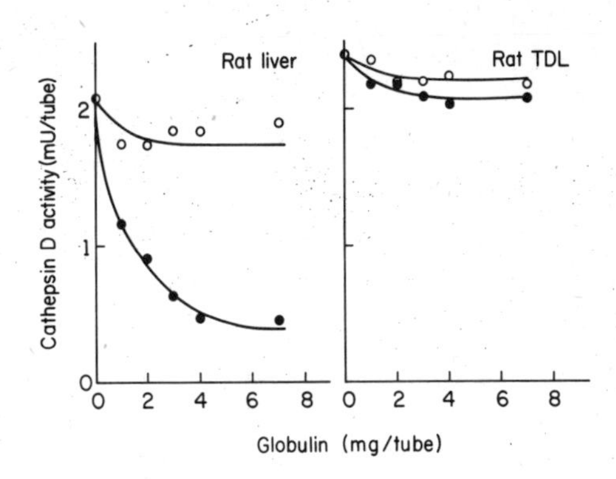

Fig. 2. Effect of globulins from sera of control and of immunized rabbits on cathepsin D activity in high-speed extracts of rat liver and of rat TDL. High-speed extracts of rat liver and of rat TDL were preincubated with control globulins from unimmunized rabbits (o) or globulins from antisera of rabbits immunized against soluble rat liver lysosomal enzymes (●) for 1 hour at 37°. Enzyme activity was measured at pH 5.0 according to the description of Yago and Bowers (5).

Owing to the similar chromatographic patterns on Sephadex G-100, the low molecular weight enzyme cannot be distinguished from rat liver cathepsin D. However, as is seen in Fig. 4, the sensitivity to inhibition by pepstatin for the lymphocyte enzymes differed markedly from that of rat liver cathepsin D. The lymphocyte enzymes were inhibited to the extent of 5 mU per nanogram of pepstatin, whereas only 1 mU of rat liver cathepsin D was inhibited by 1 nanogram of pepstatin. Because the lower molecular weight form of cathepsin D differs in its sensitivity to inhibition by pepstatin, it will be referred to as L-enzyme and the high molecular weight enzyme will correspondingly be denoted as H-enzyme.

The difference in sensitivity to inhibition by pepstatin was not due to the presence of pepstatin-binding contaminats in the rat liver cathepsin D preparation because the degree of inhibition did not change during the course of a 200-fold purification. Moreover, the degree of inhibition by pepstatin for rat liver cathepsin D was identical to that of rat kidney, rat adrenal, and rat fibroblast cathepsin D, as well as to that of several other

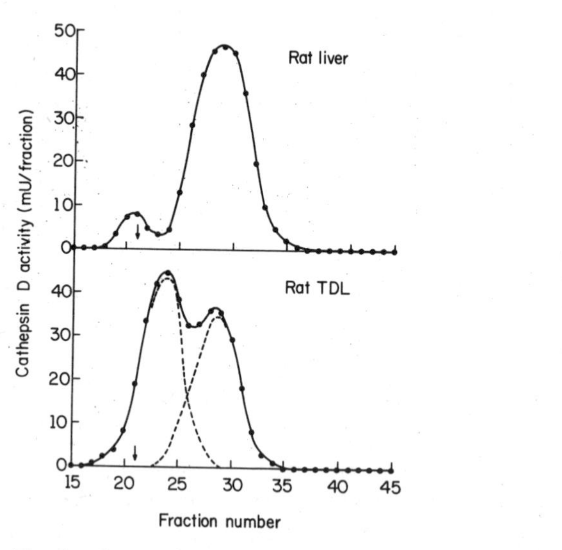

Fig. 3. Comparison of the chromatographic behavior of cathepsin D from rat liver and rat TDL. High speed extracts of rat liver and of rat TDL were subjected to chromatography on Sephadex G-100. The void volume is indicated by the arrow. The two curves depicted by the dashed lines are those resolved by a Dupont 310 curve resolver.

tissues from different species. Highly pepstatin – sensitive cathepsin D occurred only in rat TDL and in rat lymphoid tissues, and the sensitivity to inhibition did not change during partial purification. In rabbit, both liver and spleen cathepsin D were inhibited to the same extent, which is similar to the degree of inhibition of rat liver cathepsin D.

An important question concerned the relationship between H-enzyme and L-enzyme. H-enzyme, after chromatography on G-100 Sephadex as well as on G-200 Sephadex, reproducibly eluted at a position corresponding to an apparent molecular weight of 95 000, which was slightly more than twice

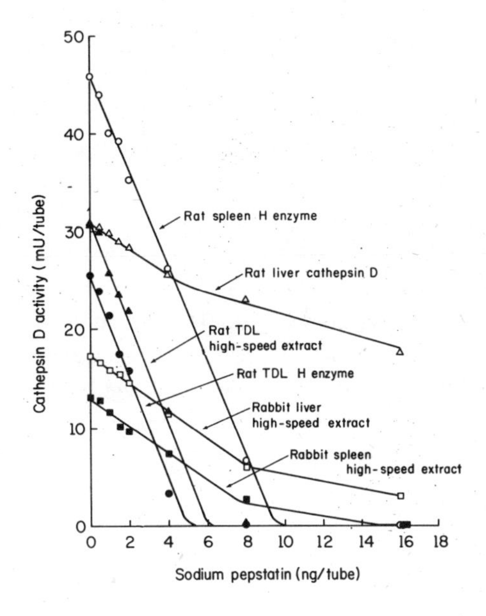

Fig. 4. Effect of pepstatin on cathepsin D activities of various tissues in rat and rabbit. Enzyme preparations were preincubated with varying amounts of sodium pepstatin. Symbols in the figure are: o, rat spleen H-enzyme (762 mU/mg protein); △ rat liver cathepsin D (4,300 mU/mg protein); ▲ rat TDL high-speed extract (103 mU/mg protein); ●, rat TDL H-enzyme (927 mU/mg protein); ◻, rabbit liver high-speed extract (18.4 mU/mg protein; and ■, rabbit spleen high-speed extract (11.7 mU/mg protein).

that of the L-enzyme. If a portion of the pooled preparation of H-enzyme was treated with β-mercaptoethanol, it converted to a low molecular weight form. There was no loss in enzyme activity accompanying this conversion, and recovery from the column was nearly 100 %. The low molecular weight enzyme formed after ß-mercaptoethanol treatment was found to be L-enzyme, due to its sensitivity to pepstatin.

H- and L-enzymes have been found in rat TDL and in rat lymphoid tissues, but not in other rat tissues. The enzymes have also been found in other rodents – mouse lymphoid tissues, hamster spleen – but not in horse spleen, rabbit thymus, spleen and liver, calf thymus spleen, or human tonsils.

Our arguments for considering H- and L-enzymes as a type of cathepsin D are the following: 1) they act best on acid-denatured hemoglobin as substrate, on which their pH curves are identical to that of rat liver cathepsin D; 2) they are inhibited by pepstatin, a specific inhibitor of cathepsin D and E, even though their sensitivity to inhibition is greater; 3) they are not inhibited by thiol reactive inhibitors, such as TPCK or TLCK, or by more specific cathepsin B inhibitors such as leupeptin; 4) H-enzyme in particular is not cathepsin E due to its lower activity on human serum albumin as substrate than on hemoglobin, and because after treatment of H-enzyme with β-mercaptoethanol only L-enzyme was found; and 5) other known cathepsins, such as A,B1, and C, have intracellular distributions that are totally lysosomal in rat TDL, in contrast to the H- and L-enzymes which, as mentioned before, are mostly recovered in a soluble, unsedimentable form. Therefore, although we do not yet have information about the specificity of these two enzymes, it seems most appropriate to consider them as cathepsin D-type proteases.

ACKNOWELDGEMENT

We gratefully acknowledge the excellent technical assistance of Ms. Elizabeth Dolci, Ms. Kathy Kohn, and Mr. Yi-Sheng John Lin and the superb typing of Ms. Anne McDermott and Mrs. Anna Polowetzky. This work was supported by Grant IM-67 from the American Cancer Society, Grants HD-05065, and CA-16875 from the U.S. Public Health Services and Grant GB-35258-X from the National Science Foundation.

REFERENCES

1. Bowers, W.E., J. Exp. Med. 136, 1394 (1972).
2. Beaufay, H., Bendall, D.S. Baudhuin, P., Wattiaux, R., and C. de Duve, Biochem. J. 73, 628 (1959).
3. Leighton, F., Poole, B., Beaufay, H., Baudhuin, P., Coffey, J.W., Fowler, S., and C. de Duve, J. Cell Biol. 37, 482 (1968).
4. Bowers, W.E., Finkenstaedt, J.T., and C. de Duve, J. Cell Biol. 32, 325 (1967).
5. Yago, N., and Bowers, W.E., J. Biol. Chem., in press (1975).
6. Bowers, W.E., J. Cell Biol. 59, 177 (1973).

PURIFICATION AND SOME PROPERTIES OF NATIVE AND IMMOBILIZED CATHEPSIN D

V. Turk, I. Urh, I. Kregar, J. Babnik, F. Gubenšek and R. Smith

Department of Biochemistry, J. Stefan Institute, University of Ljubljana, Ljubljana, Yugoslavia

Cathepsin D (E.C. 3.4.23.5) is a carboxyl proteinase existing intracellularly in many animal tissues and is considered to be one of the major enzymes in lysosomes, responsible for protein breakdown within the cell. It has been suggested that cathepsin D plays an important role in muscular dystrophy (1), and in the pathology of connective tissue diseases, particularly rheumatoid arthritis (2). This enzyme is responsible for the major part of the breakdown of proteoglycans in cartilage degradation (3,4), as well as for the autolytic degradation of cartilage (5,6). Cathepsin D could be also responsible for the breakdown of basic myelin protein forming encephalytogenic fragments (7). Finally, this enzyme may play a significant role in cardiac pathology (8) and tumour growth process (9).

Although there is much evidence that cathepsin D (and other lysosomal enzymes) is involved in physiological and pathological processes in tissues, there is still a lack of knowledge on this important enzyme.

Cathepsin D was first partially purified and characterized as an acid proteinase from bovine spleen (10). Considerable progress was made by Press et al. (11), who demonstrated the independent existence of cathepsin D free of cathepsins A, B and C, and found a similarity in the specificity between cathepsin D and pepsin, using the B chain of oxydized insulin as substrate. Since then cathepsin D has been investigated in many laboratories. The many isolations of this enzyme from different sources in a state of high purity are a common feature of such studies, but there are important differences between results, such as the number of multiple forms, isoelectric points, polypeptide composition and molecular weight.

The difficulties in preparing other proteinases and peptidases such as bovine factors X_1 and X_2 (12), bovine carboxypeptidase and its zymogen (13), and pepsin (14,15) are well known. These enzymes must be isolated with due regard for concommitant autolysis or proteolysis by other enzymes present in the tissues. Surprisingly little attention has been paid to this problem in the isolation of cathepsin D. Recent studies have shown that the isolated enzyme sometimes contains a variety of polypeptide chains which is best explained by partial degradation of the enzyme (16–19). Our experimental data suggest that the differences in polypeptide composition are the results of degradation of cathepsin D in vitro during isolation by classical procedures (20).

This paper briefly describes an isolation procedure for cathepsin D by affinity chromatography on hemoglobin agarose resin and some of the more important properties of pure cathepsin D. Immobilized cathepsin D was also prepared and some of its basic properties are presented.

EXPERIMENTAL

Purification procedure: Fresh bovine spleens were collected shortly after the animals were killed and frozen at -25^oC. All subsequent operations were carried out at 4^oC or below. The tissues were freed of large blood vessels and homogenized in an equal weight of water. The neutral homogenate was centrifuged and the supernatant was concentrated by ammonium sulphate or acetone precipitation. The precipitate was collected by centrifugation, dissolved in a small volume of water and dialyzed overnight against water. Undissolved material was discarded following centrifugation. The resulting protein solution containing 100–150mg protein/ml was adjusted to pH 3.5 with sodium acetate buffer containing 1 M sodium chloride, and was immediately applied to a column of hemoglobin agarose resin, which was equilibrated with the same buffer. Sepharose – hemoglobin resin was prepared following the method of Cuatrecasas (21). After application of the sample, the column was washed with 0.1 M acetate buffer containing 1 M NaCl at pH 5.0, and finally the active enzyme fractions were eluted with 0.1 M Tris buffer at pH 8.6 containing 1 M NaCl. The enzyme solution was concentrated by ultrafiltration and applied to a Sephadex G–100 column, equilibrated with the pH 8.6 buffer. The active enzyme fractions were pooled. As far as to this step, the purification of cathepsin D has been recently described in detail by Smith and Turk (20). As the final step of the purification, preparative disc electrophoresis or chromatography on DEAE cellulose column was employed for the separation of the three multiple forms of cathepsin D.

Preparative polyacrylamide gel electrophoresis was performed in a Canalco apparatus (Rockville, Md., USA). The stacker and separation gels (10 % acrylamide) were prepared according to the manufacturer's description, whereas the sample gel was exchanged by a dense sucrose solution. At the beginning 5 mA current was applied and when the protein bands appeared in the separation gel, the current was increased to 60 mA. The elution buffer, Tris–glycine, pH 8.5, was 10–times diluted cathode buffer.

Another method for the separation of the three multiple forms of cathepsin D was DEAE cellulose chromatography in 0.025 M Tris buffer using a linear 0.1 M NaCl gradient (see the paper by I. Kregar et al.: Isolation of cathepsin D by affinity chromatography on immobilized pepstatin, in this book).

The assay of proteolytic activity on hemoglobin substrate, isoelectric focusing, polyacrylamide gel electrophoresis and sodium dodecyl sulphate electrophoresis have all been described recently (20).

Analytical ultracentrifugation was carried out in a Phywe ultracentrifuge. Ultracentrifuge runs were carried out in a pH 7.3 potassium phosphate buffer (containing 0.1 M NaCl) of ionic strength 0.11.

Antibodies against pure cathepsin D were prepared by subcutaneous injection of rabbits with a mixture of equal volumes of enzyme solution (2.5 mg/ml of protein) and Freund's complete adjuvant. Three injections were made over a period of 15 days and rabbit blood was taken 10 days after the last injection. Immunoelectrophoresis, immunodiffusion analysis and inhibition of cathepsin D by rabbit antisera were used as described by Dingle et al. (5).

Amino terminal analysis was carried out by the dansylation method of Zanetta et al. (22).

Immobilized cathepsin D was prepared by chemical coupling of pure cathepsin D to CNBr–activated Sepharose 4B, following the general method of Axen and Ernback (23) with only minor modifications.

RESULTS AND DISCUSSION

Cathepsin D was purified from bovine spleen using precipitation methods, affinity chromatography on hemoglobin–agarose resin and gel filtration as previously described in detail (20). The final separation of the three multiple forms of cathepsin D by preparative gel electrophoresis is shown in Fig. 1. It is evident that a complete separation of multiple forms was achieved by this method. Fig. 2 shows the electrophoretic patterns of the separated multiple forms. All forms contain only one band which differs in its electrophoretic mobility.

Sodium dodecyl sulphate gel electrophoresis of the separated enzyme forms (Fig. 3) shows one polypeptide chain with a molecular weight of 42.000 for the first two forms. The third form contains an additional band of 28.000 molecular weight. In some experiments we also noted a band with molecular weight of 14.000. It seems likely that a rapid enough preparation, a complete purification in the cold and an extremely short time at acid pH would give a single protein band having a molecular weight of 42.000. Polypeptides of lower molecular weight are the result of in vitro degradation

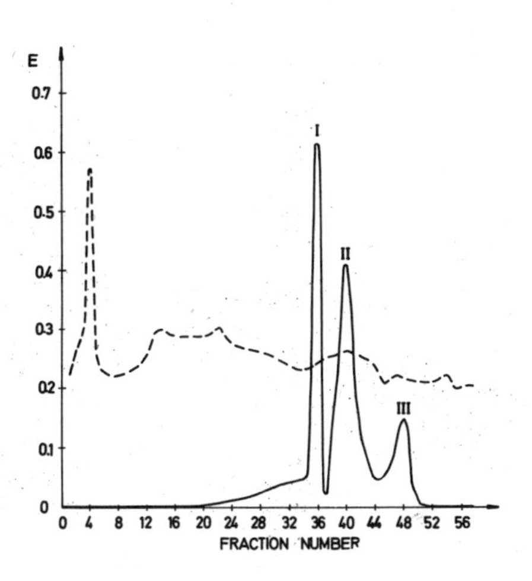

Fig. 1. Separation of multiple forms of cathepsin D by preparative polyacrylamide gel electrophoresis. ---- protein content measured as E_{280}, —— proteolytic activity on hemoglobin at pH 3.5.

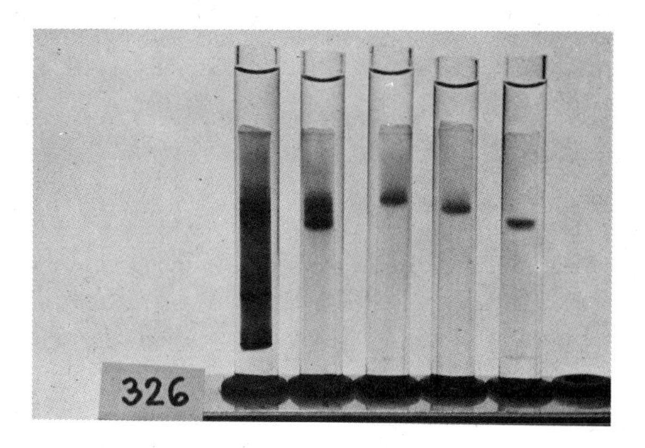

Fig. 2. Polyacrylamide gel electrophoresis at pH 8.5. Illustrated from left to right are: a) ammonium sulphate fraction, b) active fractions after Sephadex G–100 chromatography, c) form I; d) form II; e) form III.

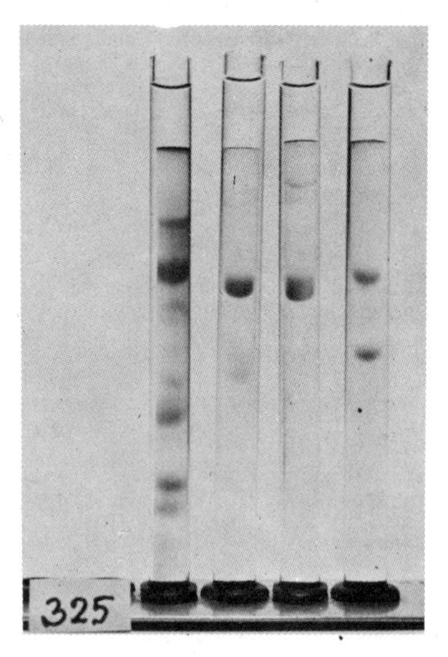

Fig. 3. Dodecylsuphate – polyacrylamide gel electrophoresis of separated enzyme forms. Illustrated from left to right are: a) protein standards (bovine serum albumin, ovalbumin, chymotrypsinogen A Lysozyme); b) form I; c) form II d) form III.

of the native enzyme during the isolation procedure, (20). Some of our previous stadies (17–20) show that classical purification procedures lead to the formation of several polypeptides of lower molecular weight and the most dominant is usually a polypeptide with molecular weight of 28.000. Our very early experiments clearly show that autolysis lasting several hours to 24 hours at 37°C at acid pH leads to the destruction of the cathepsin D molecule, forming low molecular weight polypeptides. Also 24 hrs autolysis in the cold (at 4°C) at pH 3.5 causes a similar effect (20). The molecular weight of all multiple forms was found to be 42.000. Similar results were obtained also by ultracentrifugal analysis. The sedimentation coefficient ($s^o_{20,w}$) is 3.9 s and the diffusion constant ($D^o_{20,w}$) is 8.8×10^{-7} cm^2/sec. When the sedimentation and diffusion constants were substituted in the Svedberg equation, a molecular weight of about 42.000 was obtained.

Antiserum against the pure first form of cathepsin D showed a single precipitin line with different extracts—supernatants of bovine spleeen, liver and thymus. Antiserum was used for the determination of its inhibitory capacity. Our results show complete inhibition of two forms and only slightly lower inhibition of the third form. These results can be compared with the data reported by Dingle et al. (5).

Isoelectric focusing also yielded three values of the isoelectric point for the multiple forms: 5.6, 5.9 and 6.4.

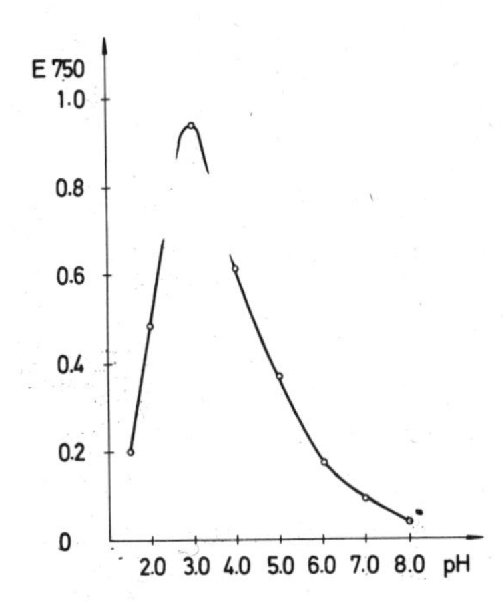

Fig. 4. pH optimum of immobilized cathepsin D.

Glycine was the only reacting NH_2—terminal residue that could be detected in the first and second form, which contains only one polypeptide chain. The third form contains glycine and another amino acid, probably serine. The preliminary amino acid composition of the multiple forms was almost identical.

The specificity of cathepsin D was unknown for a long time. The studies of Press et al. (11) on the action of cathepsin D on the B chain of oxydized insulin indicated a preferential attack at almost the same bonds that are cleaved by pepsin: Leu_{15}—Tyr_{16}, Tyr_{16}—Leu_{17}, Phe_{24}—Phe_{25} and Phe_{25}—Tyr_{26} bonds. These observations were largely confirmed when Keilova (24) prepared terminal heptapeptide Gly—Phe—Phe—Tyr—Thr—Pro—Lys, corresponding to residues 23—29 of the B chain of insulin and found that this peptide is cleaved by both cathepsin D and pepsin at the Phe—Phe and Phe—Tyr bonds. Also Woessner (25) reported that the peptide Glu-Ala—Leu—Tyr—Leu—Val is cleaved rapidly at the Leu—Tyr bond by cathepsin D in a very short time. In addition, Fruton and coworkers, (26,27) found that cathepsin D, gastric pepsin and Rhizopus pepsin acted on a series of peptide substrates by a similar mechanism, cleaving the senstive L—phenylalanyl—L—phenylalanyl bond. And finally, recently it was shown that brain cathepsin D selectively cleaves the two Phe—Phe bonds of the myelin basic protein (28).

The following synthetic substrates (the generous gift of Dr. Leon Barstow of the Department of Chemistry, University of Arizona, Tucson, USA) were used in our experiments:

A Gly—Phe—↓Phe—↓Tyr—Thr—Pro—Lys

B Lys—Pro—Thr—Tyr—↓Phe—↓Phe

C Lys—Pro—Thr—Tyr—↓Phe

It is surprising that the last two peptides were cleaved: Phe—Phe (B) and Tyr—Phe (C), because we normally think that cathepsin D is an endopeptidase, but similar observations were also reported by Woessner (25).

Another proof which indicates considerable homology between acid proteinases (cathepsin D, pepsin, acid proteinase of Aspergillus awamori, renin etc.) is the inhibition of these enzymes by diazo compounds. N—diazoacetyl glycine ethylester (the generous gift of Dr. T. Giraldi from University of Trieste, Italy), in the presence of cupric ions, inhibited cathepsin D and pepsin (pH 5.5, 0.01 M cupric acetate, 2 % hemoglobin substrate at pH 3.5, incubated 10 min at 37°C). The inhibitor was bound in the presence of cupric ions to cathepsin D and pepsin at the equimolar ratio. Another inhibitor of carboxyl proteinases that was tested was pentapeptide pepstatin (generous gift of Prof. H. Umezawa of the Institute of Microbial Chemistry, Kamiosaki 3–14–23, Shinagawa–Ku, Tokyo, Japan) and we found complete inhibition of cathepsin D at the molar ratio 1:1.

The last part of our investigations were experiments on the binding of pure cathepsin D to an insoluble matrix and some properties of immobilized cathepsin D. Biologically active proteins such as enzymes and antibodies artificially bound to insoluble carriers are of theoretical and practical interest. Protein carrier conjugates may serve as model system for proteins embedded in biological membranes or other native complex structures. They can also be used in affinity chromatography for the isolation of naturally occurring enzyme inhibitors from crude extracts or partially purified solutions. Very recently an immobilized rat liver cathepsin D (29) was prepared by coupling to enzacryl polyacetal (EPA–cathepsin) and to CNBr activated Sepharose. EPA–cathepsin was active toward synthetic hexapeptide substrate $(Gly{-}Phe{-}Leu)_2$ and did not split hemoglobin at all. Insoluble cathepsin D synthesized by covalent fixation of the enzyme to the activated sepharose was active toward hemoglobin with maximum activity at pH 7.0, and was completely inactive at acid pH. The proteolytic activity of the sepharose coupled cathepsin D was not inhibited by pepstatin, whereas the native enzyme was inhibited completely. They concluded that the finding that cathepsin D could be active at neutral pH shows some „light of understanding of their physiological role".

Because these very interesting results indicated the new role of cathepsin D in the metabolism of cell proteins, we decided to prepare immobilized cathepsin D by coupling the enzyme to activated sepharose by their method (29) which is also similar to the general method of coupling enzymes to sepharose (23).

Fig. 4 shows the pH optimum of the immobilized enzyme. The pH optimum of immobilized cathepsin D is still in the acid region at pH 3.2 to 3.5. The same results were obtained by commercial CNBr activated sepharose (Pharmacia, Uppsala, Sweden) and that freshly prepared in our laboratory. Only the prolonged low activity at neutral pH is observed. The native enzyme shows the same pH optimum and no activity in neutral pH region. Our experiments also show that pepstatin completely blocks the activity of immobilized cathepsin D.

The question arises as to why these differences exist. The most probable answer is that the cathepsin D prepared by Kazakhova and Orekhovich (29) was impure (purification procedure from 1963 and 1967, cited in ref. 29). In our experiments we used completely pure cathepsin D and therefore we can conclude that cathepsin D fixed on agarose resin retains its basic properties, such as the already mentioned pH optimum and inhibition with pepstatin.

The data presented here provide additional evidence that it is very important to develop a successful purification method which will give pure enzyme, and affinity chromatography offers this possibility. There should now be a very large opportunity to study cathepsin D structure and its function, in order to define its true physiological role in the metabolism of cell.

ACKNOWLEDGEMENTS

We thank Mrs. M. Pregelj, Mrs. A. Burkeljc, Mrs. N. Pelicon and Mr. K. Lindič for their excellent technical assistance. This work was supported by a grant from the Research Community of Slovenia and in part by a grant from the NSF, USA, Grant No. GF–31389.

REFERENCES

1. Weinstock, I.M. and Iodice, A.A. In: Lysosomes in Biology and Pathology (J.T. Dingle and H.B. Fell, eds.) North Holland, Amsterdam (1969) Vol. 1, p. 450.
2. Page–Thomas, D.P. In: Lysosomes in Biology and Pathology (J.T. Dingle and H.B. Fell, eds.), North Holland, Amsterdam (1969) Vol. 2, p. 87.
3. Dingle, J.T. In: Tissue Proteinases (A.J. Barrett and J.T. Dingle, eds.) North Holland, Amsterdam (1971), p. 313.
4. Sapolsky, A.I., Altman, R.D., Woessner, J.F. and Howell, D.S. J.Clin. Invest. 52, 624 (1973).
5. Dingle, J.T. Barrett, A.J. and Weston, P.D. Biochem. J. 123, 1 (1971).
6. Morrison, R.I.G., Barrett, A.J., Dingle, J.T. and Prior, D. Biochim. Biophys. Acta 302, 411 (1973).
7. Benuck, M., Marks, N. and Hashim, G.A. Europ. J. Biochem. 52, 615 (1975).

8. Wildenthal, K. In: Lysosomes in Biology and Pathology. (J.T. Dingle and R.T. Dean, eds) North Holland, Amsterdam, (1975) Vol. 4, p. 167.
9. Poole, A.R. In: Lysosomes in Biology and Pathology. (J.T. Dingle, ed.) North Holland, Amsterdam (1973) Vol. 3, p. 303.
10. Anson, M.L. J. Gen. Physiol. 22, 79 (1939).
11. Press, E.M., Porter, R.R. and Cebra, J. Biochem. J. 74, 501 (1960).
12. Fujikawa, K., Legaz, M.E. and Davie, E.W. Biochemistry 11, 4882 (1972).
13. Uren, J.R. and Neurath, H. Biochemistry 11, 4483 (1972).
14. Rajagopalan, T.G., Moore, S. and Stein, W.H. J. Biol. Chem. 241, 4940 (1966).
15. Fruton, J.S. In: The Enzymes (P.D. Boyer, ed.) Acad. Press New York (1971) Vol. 3, p. 119.
16. Sapolsky, A.I. and Woessner, J.F. J.Biol. Chem. 247, 2069 (1972).
17. Turk, V., Kregar, I., Gubenšek, F. Smith, R. and Lapanje, S. In: Intracellular Protein Catabolism (H. Hanson and P. Bohley, eds.) Johann Ambrosius Barth, Leipzig (1974/6) p. 260.
18. Kregar, I., Turk, V., Gubenšek, F., Smith, R. Croat. Chim. Acta 46, 129 (1974).
19. Turk, V., Kregar, I., Gubenšek, F. and Babnik, J. Biol. Vest. 23, 107 (1975).
20. Smith, R. and Turk, V. Europ. J. Biochem. 48, 245 (1974).
21. Cuatrecasas, P. Proc. Nat. Acad. Sci. 63, 450 (1969).
22. Zanetta, J.P., Vincendon, G., Mandel, P. and Gombos, G. J. Chromatog. 51, 441 (1970).
23. Axen, R. and Ernback, S. Europ. J. Biochem. 18, 351 (1971).
24. Keilova, H. In: Tissue Proteinases (A.J. Barrett and J.T. Dingle, eds.) North Holland, Amsterdam (1971), p. 45.
25. Woessner, J.F. In: Tissue Proteinases (A.J. Barrett and J.T. Dingle, eds.) North Holland, Amsterdam (1971), p. 291.
26. Ferguson, J.B. Andrews, J.R., Voynick, I.M. and Fruton, J.S. J. Biol. Chem. 248, 6701 (1973).
27. Sampath–Kumar, P.S. and Fruton, J.S. Proc. Nat. Acad. Sci. 71, 1070 (1974).
28. Benuck, M., Marks, N. and Hashim, G.A. Europ. J.Biochem. 52, 615 (1975).
29. Kazakova, O.V., Orekhovich, V.N. Internat. J. Peptide Prot. Res. 7, 23 (1975).

ISOLATION OF CATHEPSIN D BY AFFINTY CHROMATOGRAPHY ON IMMOBILIZED PEPSTATIN

I. Kregar, I. Urh, R. Smith, H. Umezawa[+] and V. Turk

Department of Biochemistry, J. Stefan Institute, University of Ljubljana, Ljubljana, Yugoslavia and [+]Institute of Microbial Chemistry, 3–14–23, Kamiosaki, Shinagawa-ku, Tokyo, Japan.

Affinity chromatography with an immobilized substrate proved to be a successful tool in the preparation of pure and undegraded cathepsin D (1). Immobilized reversible inhibitors are even more suitable for the isolation because the enzyme remains inactive during the affinity chromatography procedure. Pepstatin is known as a strong inhibitor of carboxyl proteases (2) and we wanted to make use of its inhibitory property for the isolation of cathepsin D. In this report, a method for isolation of cathepsin D is described using affinity chromatography on pepstatin-agarose.

Immobilized pepstatin was prepared as follows: 60 mg of pepstatin (Institute of Microbial Chemistry, Japan) were dissolved in a small volume of methanol and then mixed with 30 ml of diluted NaOH solution. A suspension of washed AH-Sepharose 4B (Pharmacia, Sweden), 3 g of dry weight, was added to the pepstatin solution, followed by the addition of 0.5 g of 1-ethyl-3N-dimethylcarbodiimide (Calbiochem, USA). pH was maintained at 6.2. After 2 hrs, another portion of carbodiimide was added, pH controlled for 1 hr and then the mixture was left overnight in the coldroom. On the next day the resin was washed extensively with water and buffers used later in affinity chromatography. The amount of bound pepstatin was approximately 2 uM/g of dry resin, as determined by amino acid analysis.

Calf thymus was used as a source of enzyme. It was obtained in the slaughterhouse and brought to the laboratory on ice. A 33 % homogenate was prepared in distilled water containing 1 mM sodium azide. It was centrifuged and in the neutral extract proteins were precipitated with ammonium sulphate to 70 % saturation. After removal of salt by dialysis, the protein solution was mixed with concentrated buffer of appropriate pH. The precipitate was discarded and the supernatant was applied to the pepstatin agarose column. Binding of the enzyme on the resin was tested at pH 3.5 and 5.0. It has been found that more enzyme is bound at pH 3.5, but 35 % of activity was lost when the sample was acidified to pH 3.5. At pH 5.0 the loss of activity was only 4 %, but there was 20 % of unbound enzyme. Altogether 10 % higher yields were obtained at pH 5.0. Fig. 1 shows the elution diagram from the pepstatin-agarose column when the enzyme was bound at pH 5.0. It can be seen that 20 % of activity appeared in the first peak. The bulk of activity was eluted with a buffer of pH 8.6.

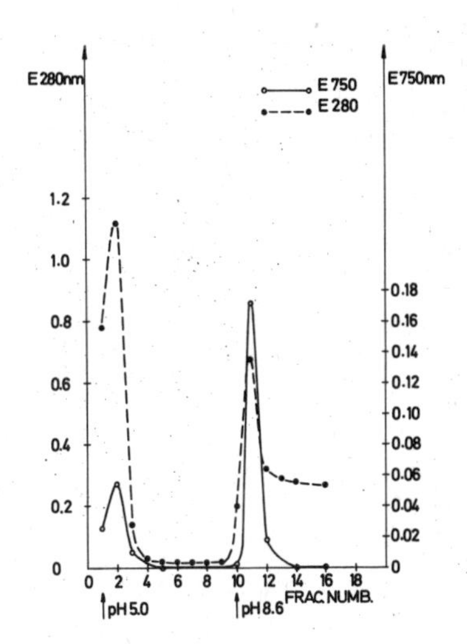

Fig. 1. Pepstatin-Sepharose chromatography of O – 70 % ammonium sulphate precipitated fraction from thymus. —— proteolytic activity on hemoglobin at pH 3.5;––– protein measured as E_{280}. Dimensions of the column were 6,5 x 1 cm, flow rate was 15 ml/h.

Active fractions in the second peak were pooled, concentrated by Amicon ultrafilter and subjected to gel chromatography on Sephadex G-100. Inactive material was eluted in the first peak, separated from active proteins as evident from Fig. 2. The purity of fractions was controlled by polyacrylamide gel electrophoresis. After Sephadex chromatography, cathepsin D preparations were obtained consisting of three active forms.

Separation of the active forms was achieved on DEAE cellulose using 0.025 M Tris buffer and a linear gradient toward 0.1 M NaCl. The elution profile of the separation is shown in Fig. 3. One can see that the first form was completely separated, whereas the second and the third were separated to a lesser extent. Gel electrophoresis of the obtained fractions confirms the successful separation of enzyme forms (Fig. 4).

Electrophoresis of active forms in the presence of sodium dodecyl sulphate shows that the first form consists of only one polypeptide chain having a molecular weight of 42 000. The second and the third form also contain a polypeptide of molecular weight of 27 000.

Affinity columns with immobilized pepstatin have been successfully employed in renin purification by Murakami et al. (3,4) and by Corvol et al.

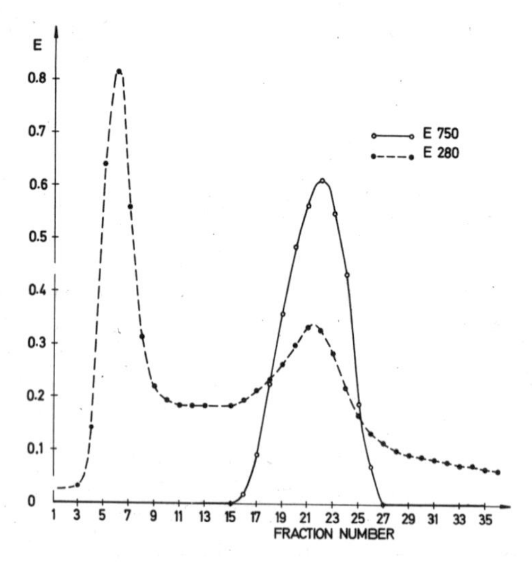

Fig. 2. Sephadex G-100 chromatography of the enzyme from the affinity chromatography column.—— proteolytic activity on hemoglobin at pH 3.5; ––– protein measured as E_{280}. Dimensions of the column were 100 x 2 cm, flow rate was 15 ml/h.

(5). Similar columns with immobilized peptides containing D-amino acids were used by Stepanov for the purification of pepsin and awamorin (6).

Our results show that there are no differences in the isolated cathepsin D, as far as the number of active forms and their polypeptide composition, by using either hemoglobin-agarose or pepstatin-agarose in the purification procedure. The advantage of the latter is in the higher binding of the enzyme to the column and therefore in better yields.

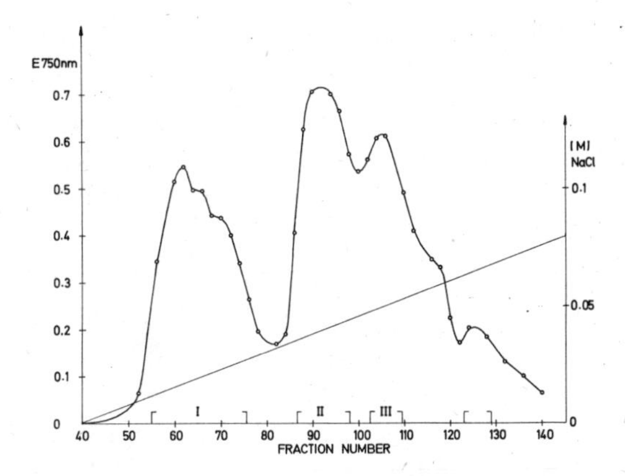

Fig. 3. Separation of active fractions from Sephadex G-100 on DEAE cellulose column (30 x 2 cm). 0.2 M Tris in 0.01 M NaCl was used for the elution and later a linear gradient toward 0.08 M NaCl was applied. Flow rate was 33 ml/h.

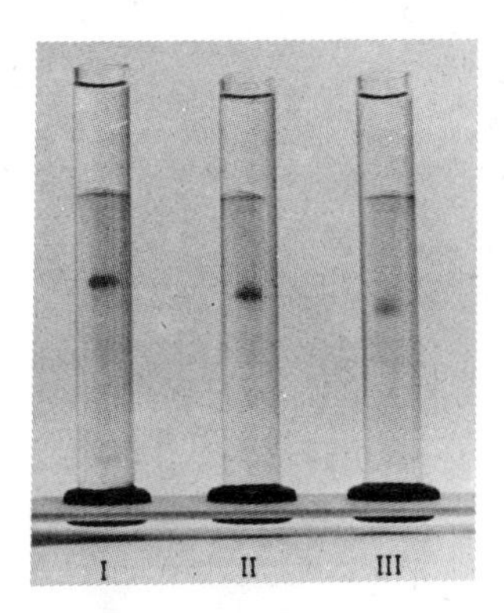

Fig. 4. Polyacrylamide-gel electrophoresis at pH 9.5 of separated multiple forms of cathepsin D (I, II and III from DEAE cellulose).

ACKNOWLEDGEMENT

We thank Mrs. M. Pregelj, Mrs. A. Burkeljc and Mr. K. Lindič for their excellent technical assistance. This work was supported by a grant from the Research Council of Slovenia and in part by a grant from the NSF, USA, Grant no. GF-31389.

REFERENCES

1. Smith, R., Turk, V., Europ. J. Biochem. 48, 245-254 (1974).
2. Umezawa, H., Aoyagi, T., Morishima, H., Matsuzaki, M., Hamada, H., Takeuchi, T., J. Antibiotics, 23, 259-262 (1972).
3. Murakamı, K., Inagami, T., Michelakis, A.M., Cohen, S., Biochem. Biophys. Res. Comm. 54, 482-487 (1973).
4. Murakami, K., Inagami, T., Biochem. Biophys. Res. Comm.62 757-763 (1975).
5. Corvol, P., Devaux, C., Menard, J. FEBS Letters 34, 189-192, (1973).
6. Stepanov, V.M., Lavrenova, G.I., Slavinskaya, M.M., Biokhimiya 39, 384-387 (1974).

Note added in proof: While this paper was in press, a paper on cathepsin D purification by affinity chromatography on immobilized pepstatin was also published by Kazakova and Orekhovich in Biokhimiya 40, 969 (1975).

SYNTHETIC OCTAPEPTIDE INHIBITOR OF CATHEPSIN D

F. Gubenšek, L. Barstow[+], I. Kregar and V. Turk

Department of Biochemistry, J. Stefan Institute, University of Ljubljana and [+]Department of Chemistry, University of Arizona, Tucson, Az., USA

Some five years ago Keilova (1) reported on low molecular weight substrates and inhibitors of cathepsin D. In the light of the recently introduced technique of affinity chromatography, her data became interesting again. A low molecular weight peptide inhibitor containing a D-amino acid could serve as a ligand in the isolation of cathepsin D and other acid proteinases.

At the moment pepstatin (2) seems to be the most useful ligand for this purpose; however, it is not commercially available and its synthesis is not yet known. A suitable synthetic inhibitor would be a valuable tool in the purification procedure.

Parikh and Quatrecasas (3), led by similar reasoning, succeeded in synthesising an enanthiomeric analogue of renin substrate, which proved to be a strong inhibitor and suitable ligand. Stepanov et al. (4) also reported the successful use of enanthiomeric substrate analogues of acid proteinases as affinity ligands. Their inhibitors were di– and tri-peptides, the latter being the better ligand.

An inhibitor to be used as a ligand for affinity chromatography should bind the enzyme appreciably better than protein impurities present in the mixture to be purified. However, too strong binding seems to be inappropriate, as it requires too drastic conditions for the recovery of the bound enzyme. It should be also resistant against proteinase activity, although this is not imperative, as it may bind the enzyme relatively well quite far away from its pH optimum.

We determined the inhibitory properties of a low molecular weight peptide containing the D-isomer of phenylalanine, for which we could expect inhibitory properties on the basis of the results of Keilova (1). Among the five hexapeptides she tested, three containing single D-phenyl alanine showed appreciable inhibition of hydrolysis of her hexapeptide substrate of cathepsin D.

A similar octapeptide was synthetized by Merrifield solid phase synthesis: Gly-Glu-Gly-Phe-Leu-Gly-D-Phe-Leu. From the data of Keilova we could expect that this octapeptide would be cleaved only to a minor degree by cathepsin D and that it should be a competitive inhibitor. In order to increase the solubility of the peptide, a polar residue was added on the N-terminal side of the peptide. The elongation would also maintain the active part of the peptide far enough from the solid support and thus facilitate the binding.

According to our expectations the peptide was only slightly cleaved by cathepsin D and appreciably by pepsin, as determined by thin-layer chromatography after 2 and 24 hours of hydrolysis at 37°C.

Cathepsin D used in these experiments was a highly purified sample from bovine thymus containing only the three active forms found by gel electrophoresis (5). Pepsin was a 3-times crystallized commercial preparation

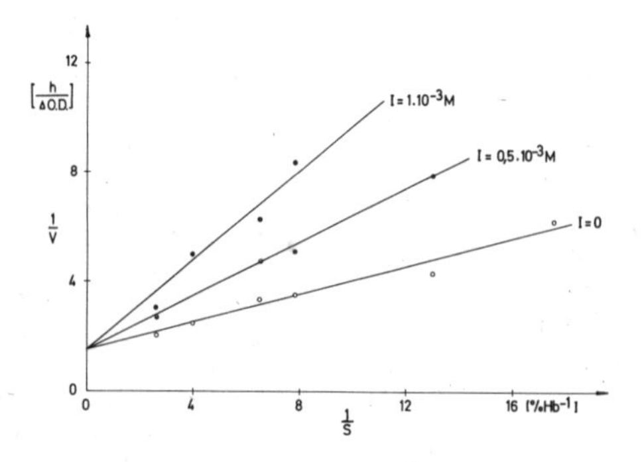

Fig. 1. The Lineweaver—Burk plot. Substrate concentration is expresed as % of hemoglobin, and velocity of reaction as OD_{750} per hour. Concentration of inhibitor as indicated.

of the Nutritional Biochemical Co., USA. Proteolytic activity was measured against hemoglobin at pH 3.5 for cathepsin D and at pH 2.0 for pepsin, using the modified method of Anson (6).

The octapeptide inhibited hydrolysis of 2 % hemoglobin by cathepsin D by about 40 % at a concentration of 1.2×10^{-3} M. In order to obtain a more detailed picture of the inhibition, a Lineaveaver-Burk plot was made using two different concentrations of inhibitor (Fig. 1.). The data obtained indicate a true competitive inhibition. The dissociation constant of the inhibitor was found to be $K_i = 5.2 \times 10^{-4}$ M. From the same plot we also obtained the Michaelis constant for the highly purified cathepsin D from thymus, which was found to be $K_M = 2.7 \times 10^{-5}$ M hemoglobin, and is in good agreement with some earlier data for this enzyme obtained in our laboratory. The comparison of the constant on the molar scale shows that cathepsin D binds hemoglobin much more strongly than the octapeptide inhibitor, assuming that the two constants reflect the affinity for the enzyme.

No significant inhibition of pepsin was obtained using the same concentrations of inhibitor, i.e. 0.5 and 1.0 mM. This difference in inhibitory properties could reflect the difference in the secondary enzyme – inhibitor interactions distant from the active centre, which seem not to play an important role in the weakly specific pepsin.

We can conclude from these experiments that the octapeptide inhibitor has too low an affinity for cathepsin D to permit an effective binding of the enzyme to immobilized inhibitor, unless its binding properties were favorably changed on covalent attachment to the solid phase. This is a possibility which remains to be proved by additional experiments. The good results with the synthetic inhibitors of pepsin, awamorin and renin, however, encourage a further search for a more suitable synthetic peptide inhibitor of cathepsin D.

ACKNOWLEDGEMENT

We thank Mrs. K. Leonardis for her excellent technical assistance. This work was supported by a grant from the Research Council of Slovenia and in part by a grant from the NSF, USA, grant no. 31389.

REFERENCES

1. Keilova, H., Blaha, K. and Keil, B., Eur. J. Biochem. 4, 442-447 (1968).
2. Ikezawa, H., Aoyagi, T., Takeuchi, T., Umezawa, H., J. Antibiotics 24, 488-490 (1971).
3. Parikh, I., Cuatrecasas, P., Biochem. Biophys. Res. Comm. 54, 1356-1361 (1973).
4. Stepanov, V.M., Lavrenova, G.I. and Slavinskaya, M.M., Biokhimiya 39, 384-387 (1974).
5. Smith, R., Turk, V., Eur. J. Biochem. 48, 245-254 (1974).
6. Anson, M.L., J.Gen.Physiol.22, 79-89 (1939).

LUNG PROTEASES AND PROTEASE INHIBITORS

Jeffrey Ihnen and George Kalnitsky

Department of Biochemistry, University of Iowa Iowa City, Iowa 52242 U.S.A.

SUMMARY

Homogenates of guinea pig lung show significant protease activity at pH values from 2.2 to 9.9. Perfusion results in increased activity in both acid and alkaline ranges. Protease activity can be extracted by repeated homogenization and sonication of the pellet. Some of the neutral and alkaline proteases require a dialyzable co-factor and are inhibited by 1.2 mM Ag^+ and Cu^{++}. 2-mercaptoethanol increases activity of the acid cathepsins and of an enzyme at pH 9.9, while DFP inhibits at both acid and alkaline pH. Lung is very active as shown by the „profile" of its total protease activity over a wide pH range. Serum can inhibit protease activity of lung extracts through the whole pH range tested.

INTRODUCTION

Lung is a very active metabolic organ in support of its main function of gaseous exchange, yet very little is known about the proteases in this tissue (1). Many years ago, it was reported that lung contains a protease which is optimally active at pH 3-4 (2,3) and that lung extracts exhibit aminopeptidase, tri- and di-peptidase activities (4,5). A more thorough investigation of lung protease activity, carried out in 1955, showed that lung extracts contained 2 pH optima, one at pH 4 and the other at pH 8 (5,6). The pH 4 activity or Proteinase I, was slightly purified (6) and is probably cathepsin D. More recently, cathepsins BI and B2 (7) and the dipeptidyl-aminopeptidases I, II, III and IV (8) have been shown to be present in lung. In addition, lung is very active in the metabolism of vasopeptides such as angiotensin and bradykinin (9), and lung proteases and inhibitors are intimately related to the clotting mechanism (1). Thus, it is known that this organ contains proteolytic enzymes, but, with a few exceptions such as cathepsin D and converting enzyme, nothing more is known of the properties of any of these enzymes in this tissue nor of the actual proteolytic capability of this organ.

This report presents evidence that guinea pig lung has significant proteolytic activity across a wide range of pH and that this activity can be inhibited by blood serum. Included also is a brief examination of some properties of these proteases.

METHODS

Autolysis of Lungs and Protease Assay

Lungs were removed from guinea pigs (English Colony short-haired strain or, in a few casses, the Hartlesy Strain) after guillotining or anaesthetizing the animal with ether. The lungs were homogenized (5 % w:v) in 0.33 M sucrose-1 mM EDTA in a Potter-Elvehjem homogenizer driven by a Tri-R (model Stir-R) motor, and equal volumes of this homogenate (or various fractions thereof) were then mixed with Universal buffer (10) which had been adjusted to the appropriate pH by the addition of 0.2 M NaOH. At times 0 and 4.5 hours at 37° C, 1 ml aliquots were removed and 3 ml of 10 % trichloracetic acid were added to precipitate the proteins and stop the reaction. After centrifugation and filtration through Whatman No. 4 paper, the increase in absorbance of the clear supernatant over that of the blank was measured at 280 nm. Readings were proportional to the amounts of aliquots taken. Absorbance is the increase in absorbance of the aliquot multiplied by the volume of the solution and represents proteolytic activity of the whole lung.

Assays Using Substrates

The substrate assay, employing denatured hemoglobin at pH 2-5 (11) and Hammersten quality casein from pH 6-10 (12), was also used to measure proteolytic activities of lung homogenates and fractions, as well as of cell extracts. For the assay, 0.5 ml enzyme preparation was added to 1.6 ml of 1 % substrate and allowed to incubate for 30 minutes at 37° after which the reaction was stopped by the addition of 3.5 ml of 10 % trichloracetic acid. The solution was then centrifuged, filtered and the absorbance read at 280 nm. For the controls, TCA was added to the substrate prior to the addition of the enzyme preparation.

Perfusion of Lung Tissue

After the guinea ping had been anaesthetized, the thoracic cavity was opened and the perfusing solution of 0.9 % saline was introduced into the right ventricle of the heart. Flow was initiated and the 2 atria were opened to allow drainage. After the lungs had turned a pale white – requiring at least 100 ml of saline – the lungs were removed and handled as described above.

Differential Centrifugation

The homogenate was spun at 700 g for 10 minutes to give a nuclear fraction. The supernatant was then centrifuged at 25,000 g for 30 minutes to give mitochondrial-lysosomal and supernatant fractions. Pellets were suspenden by homogenization in sucrose-EDTA.

Enzyme Extracts

The proteolytic enzymes were extracted by repeated homogenization and sonication of the pellet, as described under Results. The sonication was accomplished with a Sonified Cell Disrupter (Model W140) obtained from Heat Systems-Ultrasonics, Inc. Centrifugation was carried out using a Sorvall RC 2-B centrifuge and an SS-34 rotor.

All protein concentrations were determined by the Biuret method (13) using bovine serum albumin as the standard.

Each compound which was tested for its effects on the proteases was added to the enzyme extract and the pH was adjusted to the desired values. After a 15 minute incubation (37°) 0,5 ml of the mixture was then used in the substrate assay as described above.

Blood was obtained from the animals by heart puncture, allowed to coagulate at 4° for several hours and then centrifuged. The serum was decanted and stored at 4° until used.

RESULTS AND DISCUSSION

Autolysis Studies

Guinea pig lung homogenates, allowed to autolyze for 4,5 hours, showed significant proteolytic activity throughout the pH range tested, from pH 3 through 10 (Figure 1, lower curve). Since blood contains protease inhibitors, the lungs were perfused to remove trapped blood, then homogenized and allowed to autolyze, as before. The resultant protease activity (Figure 1, upper curve) is significantly increased at pH 4–5, and 7–10.

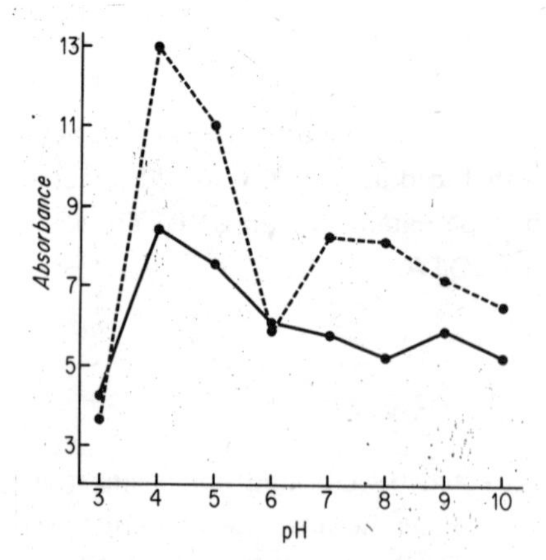

Figure 1. Protease activity of guinea pig lung homogenates as determined by autolysis. Each point is the average of four experiments carried out in duplicate. ●---● perfused lung; ●——● non-perfused lung.

Perfused and non-perfused lung homogenates were fractionated into nuclear, mitochondrial-lysosomal and supernatnat fractions. Each fraction was allowed to autolyze at the indicated pH values and then assayed for protease activity. Perfusion did not increase the activities of the nuclear or mitochondrial-lysosomal fractions, but it did increase the activities in the supernatant fraction (Figure 2). Thus, perfusion appears to remove blood protease inhibitor(s) from the supernatant fraction.

Solubilization of Lung Proteases

Extraction of proteases from lung homogenates was accomplished by a series of three homogenizations and sonications of the pellet (Figure 3). The extracts were assayed with denatured hemoglobin at pH 4,0 and with casein at neutral and alkaline pH's as desribed under methods. This procedure solubilized about 85 % of the total proteolytic activity at pH 4, 6.5 and 8.5. The remainder was in the pellet and was readily assayed. The extraction resulted in little or no loss of activity at pH 8.5 and 6.9. However, the activity recovered at pH 4,0 was 210 % of that assayed in the original homogenate. Both the proteolytic and acid phosphatase activities at pH 4.0 are released at approximately the same rate by this solubilization procedure, indicating the probable lysosomal origin of this protease activity.

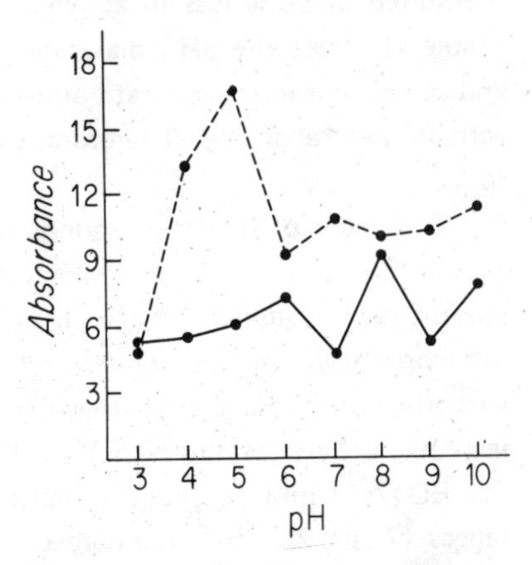

Figure 2. Protease activity of supernatant fractions obtained from guinea pig lung homogenates. ●- - -● perfused lung; ●——● non-perfused lung. Other conditions as in Figure 1.

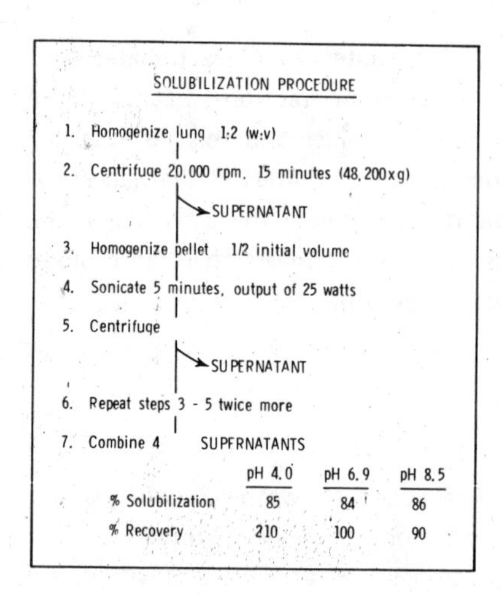

Figure 3. Solubilization of guinea pig lung proteases.

Some Properties of the Proteases in the Lung Extract

Dialysis had little or no effect on the activity of the acid proteases, but it resulted in 50 % loss in activity at pH 6.0 and 65-70 % loss at pH 7-10 (Table 1). When the pH 7 dialysate was concentrated (from 1 liter to 10 ml) and 1 ml of the concentrate added to 4 ml of extract, most of the lost activity was recovered. Therefore, some dialyzable component is necessary for activity.

A number of thiol compounds were tested for their effects on protease activity in the lung extract. 2-mercaptoethanol was found to be a better -SH reagent than cysteine, glutathione or dithioerythritol. Mercaptoethanol (at 1 mM concentration) clearly increased the activity at pH 9.9 (Table 2) but had virtually no effect at any other pH. 5 mM mercaptoethanol resulted in an increased activity at acid pH values; 10 mM concentration was optimal.

EDTA (1 mM) had some inhibitory effect in the neutral and alkaline pH ranges (Table 2). From the dialysis experiments, this would be expected if metal ions were necessary for activity. The slight increase in activity with EDTA in the acid region may be due to the presence of small amounts of inhibitory metal ions.

TABLE 1

Effect of Dialysis on Lung Protease Activity.

EFFECT OF DIALYSIS ON LUNG PROTEASE ACTIVITY

pH	% Recovery of Activity	pH	% Recovery of Activity
2.2	89	6.1	49
3.1	97	7.0	35 (89)*
4.0	95	8.0	25
5.0	92	8.8	23
		9.9	35

*on addition of concentrated dialysate to the dialyzed enzyme preparation

The lung extract was dialyzed against 50 mM phosphate buffer at pH 7.0 for 22 hours, after which aliquots were adjusted to the indicated pH values and assayed for activity against denatured hemoglobin or casein.

Of the metal ions tested, Ag^+ and Cu^{++} showed the greatest inhibition in the alkaline range (Table 2). Zn^{++} was much less inhibitory, as were the other metal ions tested. Added NaCl had no effect at any of the pH values tested.

Table 3 summarizes the effects of several classical inhibitors on the lung protease extracts. DFP inhibits in both the acid and alkaline regions, indicating the presence of serine proteinases. TLCK shows some inhibitory effects in the acid region only. This may indicate the presence of an -SH enzyme (14) or of a trypsin-like enzyme (15), ZPBK which inhibits bacterial and mold proteases (16,17) also shows some inhibitory properties in the acid region. Iodoacetate can readily inhibit -SH enzymes and, with this extract, it appears to inhibit at pH values where addition of 2-mercaptoethanol increases protease activity.

TABLE 2

% Inhibition or (Activation) of Lung Protease Activity by –SH, EDTA or Heavy Metals.

% Inhibition or (Activation) of Lung Protease Activity by -SH, EDTA, or Heavy Metals

			Metal Ions (1.2 mM)				
ph	-SH	EDTA (I mM)	Ag^{+}	Cu^{2+}	Zn^{2+}	Pb^{2+}	Cd^{2+}
2.2	(100)*	(100)	—	—	—	—	—
3.1	(40)*	(28)	—	—	—	—	—
4.0	(48)*	(21)	—	—	—	—	—
5.0	(100)*	(20)	—	70	20	25	—
6.1	—	20	77	85	45	25	37
7.0	—	32	90	82	33	52	28
8.0	—	30	92	68	—	18	—
8.8	—	35	98	45	51	—	51
9.9	(75)+	37					

* /10 mM   + 1 mM

The substrate assay was used as described in the section under methods. Universal buffer (10) was used in the experiments with 2-mercaptoethanol and EDTA. With the metals, the following buffers (50 mM) were used: glycine, pH 2,3; acetate, pH 4,5; 2 (N-morpholino) ethane sulfonic acid, pH 6,7; and tris (hydroxymethyl) methyl aminopropane sulfonic acid, pH 8,9.

Development of Normal „Profiles"

We have developed a „profile" of the total protease activity present in guinea pig lung by perfusing the lung, extracting the proteases, and adding 2-mercaptoethanol to the extract (Figure 4). Significant activity is present at all the pH values tested, with pronounced activity in the acid, neutral and alkaline ranges. Thus, lung is a very active metabolic organ, with respect to both the total proteolytic enzyme activity and the pH range over which these enzymes are active.

Table 3. % Inhibition of Proteases in Guinea Pig Lung Extract by Various Compounds.

% Inhibition of Proteases in Guinea Pig Lung Extract by Various Compounds

pH	DFP	TLCK	ZPBK	IAc
2.2	40	40	—	20
3.1	26	27	20	29
4.0	35	34	27	41
5.0	27	16	21	37
6.1	—	—	—	—
7.0	—	—	—	—
8.0	25	—	—	21
8.8	24	—	—	—
9.9	19	—	—	25

DFP, TLCK (N^2tosyllysine chloromethylketone) and ZPBK (carbobenzoxylalanine bromomethylketone), 1×10^{-4} M; IAC (iodoacetate) 4×10^{-3}M. ZPBK was dissolved in 70 % methanol; the final methanol concentration in the substrate assay was 1.6 %.

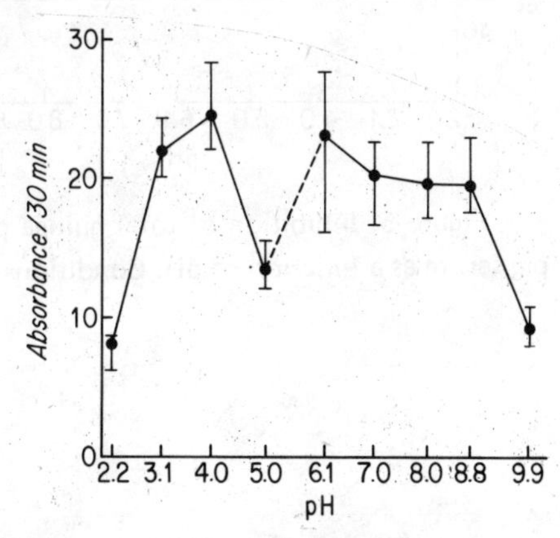

Figure 4. Total protease activity of guinea pig lung extracts as a function of pH. Each point is the average of six experiments carried out in duplicate. The vertical bars indicate the range of the values obtained, 2-mercaptoethanol was added to 1 mM concentration at pH 9.9 and 10 mM at pH 2.2 through 5.0. Other conditions were as described in substrate assay in methods section.

Since perfusion experiments indicated that blood serum inhibited lung proteases at several pH values (Figures 1 and 2) we also developed a second „profile" – for serum inhibition of lung protease activity. This was carried out by incubating equal volumes of normal guinea pig serum and normal total lung extract before addition of the substrates (denatured hemoglobin or casein). This „profile" expressing the inhibition capability of serum from normal animals, is shown in Figure 5. Significant inhibition of protease activity in the lung extract is obtained throughout the entire pH range tested. Thus, blood protease inhibitors can inhibit proteases in lung, under the conditions of these experiments. Whether this actually does happen, **in vivo,** remains to be demonstrated.

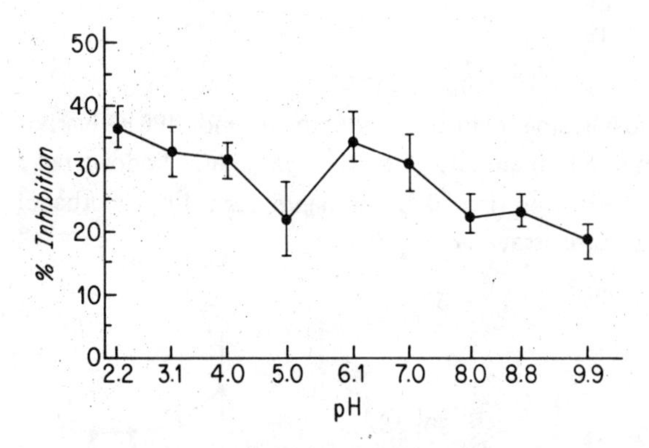

Figure 5. Inhibition of total guinea pig lung protease activity by guinea pig serum as a function of pH. Conditions as in Figure 4.

ACKNOWLEDGMENT

This work was supported by grants from the National Institutes of Health (HL 16920-01) and the Environmental Protection Agency contract number EPA 68-02-0746 with the University of Iowa.

REFERENCES

1. Heinemann, H.O. and Fishman, A.P., Physiol, Rev, 49,1–47 (1969).
2. Nye, R. N., J. Exp. Med. 35, 153-160 (1922).
3. Weiss, C., Arch. Path. 33, 182-187 (1942).
4. Fruton, J.S., J. Biol. Chem 166, 721-738(1946).
5. Dannenberg, A.M., Jr. and Smith, E.L., J. Biol. Chem 215, 45-54 (1955).
6. Dannenberg, A.M., Jr. and Smith, E. L., J. Biol. Chem. 215, 55-66 (1955).
7. Otto, K., in Tissue Proteinases (A.J. Barrett and J.T. Dingle, eds.), North Holland, Amsterdam (1971), PP. 1-28.
8. McDonald, J.K., Calllahan, P.X. Ellis, S. and Smith, R.E. in Tissue Proteinases, (A.J. Barrett and J.T. Dingle, eds.), North Holland, Amsterdam (1971), PP. 69-107.
9. Sanders, G.E. and Huggins, C.G. Ann. Rev. Pharmacol. 12, 227-264 (1972).
10. Long, C., Biochemists Handbook, E. and F. Spon, London (1961), pp 41-42.
11. Anson M.L., J. Gen. Physiol. 22, 79–89 (1938).
12. Kunitz, M.J. J. Gen. Physiol. 30, 291-310 (1947).
13. Gornall, A.C., Bardawill, C.J. and David, M.M. J. Biol. Chem. 177 751-766 (1949).
14. Whitaker, J.R. and Perez Villasenor, J., Arch. Biochem. Biophys. 124, 70–78 (1968).
15. Shaw, E., Mares-Guia, M. and Cohen, W. Biochem. 4, 2219-2224 (1965).
16. Shaw, E. and Ruscica, J., J. Biol. Chem. 243, 6312-6313 (1968).
17. Dworschack, R.T., Wyborny, L. and Kalnitsky, G., Arch. Biochem. Biophys. 159, 463-467 (1973).

THE INFLUENCE OF CATHEPSIN D INHIBITORS (PEPSTATIN AND ANTICATHEPTIC ANTIBODIES) ON SERUM CATHEPSIN AND LYSOZYME ACTIVITY OF ^{60}Co IRRADIATED ANIMALS

Jan Štefanovič, Miroslav Ferenčik and Olga Absolonova

Research Laboratory of Immunology, Department of Medical Microbiology and Immunology, Faculty of Medicine, Comenius University, Bratislava, ČSSR

SUMMARY

In lethally irradiated (^{60}Co) chickens and rabbits serum lysozyme levels were found to decrease gradually down to undetectable values. Serum catheptic activity increased sharply in chickens after ionizing irradiation and then remained at normal or slightly elevated values. Changes in activities of both studied lysosomal enzymes observed after ionizing irradiation, could be compensated to a certain extent by application of cathepsin D inhibitors – pepstatin and particularly anticatheptic immunoglobulin G.

INTRODUCTION

Phagocytes of irradiated men and animals are, besides other defects, particularly characterized by reduced bactericidal activity (1-4). We have found that after x-irradiation of rabbits the antibacterial activity of polymorphonuclear leucocytes decreases, whereas catheptic activity increases (5,6), and that increased concentrations of cathepsin D in leucocyte extracts can inactivate total antibacterial activity as well as the activity of individual antibacterial substances such as myeloperoxidase and catalase (7-9). Increased cathepsin D and decreased antibacterial activity was also found in leucocytes of patients irradiated by terapeutic doses of x-rays (2,10) and in the bone marrow of irradiated rats (11).

The aim of the present study is to investigate if, besides changes in cathepsin D activity, irradiation of experimental animals results also in changes of lysozyme levels, and if these potential changes could be influenced by pepstatin or specific anticatheptic antibody.

MATERIALS AND METHODS

Whole body irradiation with a single dose of 1,000 R in rabbits (weighing 2.5 – 3.0 kg) and 1,500 R in chickens (1.2 – 1.5 kg) was applied by means of Chisobalt apparatus (^{60}Co – source of gamma radiation).

Pepstatin, iso-valeryl-L-valyl-L-valyl-4-amino-3- hydroxy-6-methyl-heptanoyl-L-alanyl-4-amino-3-hydroxy-6-methyl-heptanoic acid, was kindly supplied by Prof. H. Umezawa and Banyu Pharm. Co. (Tokyo).

Antiserum against chicken cathepsin D was prepared by repeated immunization of rabbits with 1,250 times purified cathepsin D isolated from chicken spleens (12) and emulsified in complete Freund's adjuvans (Difco). The inhibitory capacity of anticatheptic sera was determined according to Dingle **et al.** (13). 8 ul of the most effective antiserum inhibited one unit of cathepsin D (its definition is to be found in ref. 12). The IgG fraction (anticatheptic IgG) was prepared from this antiserum by batch with DEAE Sephadex A-50 (14). The same procedure was applied in preparing IgG fraction from normal rabbit serum. In **in vitro** experiments, 1 mg of rabbit anticatheptic IgG completely inhibited 11.2 units of chicken cathepsin D. Normal rabbit IgG had no inhibitory effect on cathepsin D activity.

Lysozyme was determined by the method of Gavin (15) and serum cathepsin activity was assayed by autodegradation of fresh serum at pH 3.5 according to Homolka in our own modification (16).

RESULTS AND DISCUSSION

Three groups of rabbits (10 animals per group) were used in the first experiment. The mean lysozyme values in sera of these animals are shown in Figure 1.

In the second experiment, chickens were divided up into 5 groups (8 animals per group). Figure 2 shows average serum lysozyme and Figure 3 serum cathepsin levels in individual groups of chickens at different time intervals following irradiation.

The presented results demonstrate in conformity with other authors (17-20), that after lethal irradiation of both rabbits and chickens, serum lysozyme changes are of the same character — decrease of levels down to undetectable values. This decrease was delayed in our experiments after application of pepstatin and particularly of anticatheptic antibodies. The sharp increase in serum cathepsin activity during the first hours following irradiation of chickens was also compensated to a certain extent after administration of pepstatin and anticatheptic heterologous IgG. The above results indicate that ionizing radiation effects changes in release of lysosomal enzymes and that the changes concerning cathepsin and lysozyme activities can be influenced by cathepsin D inhibitors.

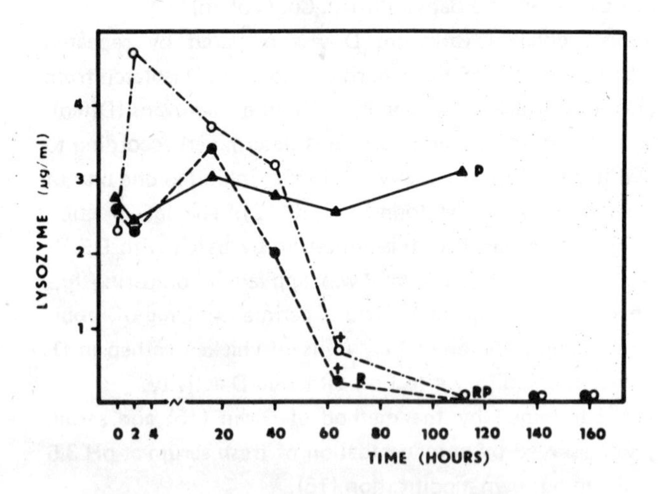

Figure 1. Lysozyme levels in sera of irradiated and unirradiated rabbits.

Two hours before irradiation and then 2, 16, 40, 64, 112, and 160 hours after irradiation animals in group R were administered 4 ml of saline and animals in group RP 4 ml (100µg) of pepstatin. Group P received pepstatin in the same doses and intervals as group RP, but was not irradiated.

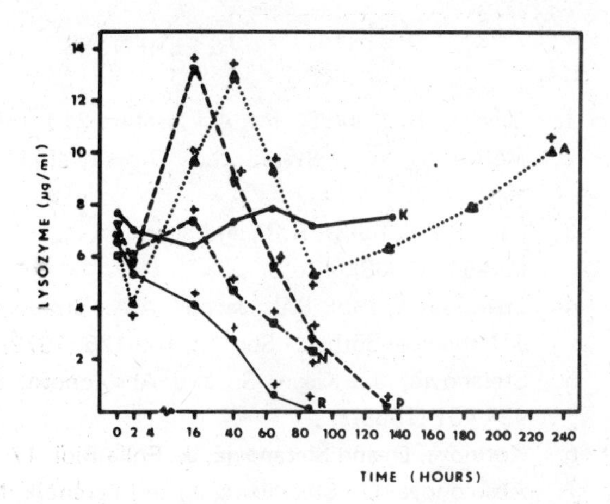

Figure 2. Lysozyme levels in sera of irradiated and unirradiated chickens.

Group R – animals only irradiated; group P – animals were given 25 μg (1 ml) of pepstatin intravenously 2 hours before and 2 and 16 hours after irradiation; group A – chickens received 2 ml (24 mg) of rabbit anticatheptic IgG 24 hours before irradiation; group N – the same as group A, but chickens were administered normal rabbit IgG; group K – control animals which were not given pepstatin or IgG and were not irradiated.

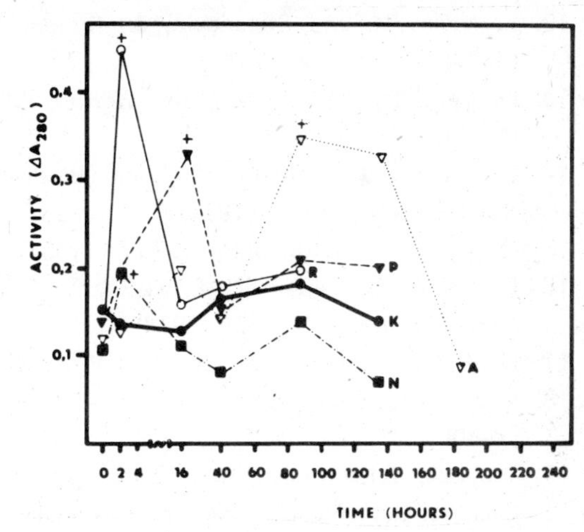

Figure 3. Cathepsin levels in sera of irradiated and unirradiated chickens.

The inividual groups – as in Figure 2.

REFERENCES

1. Selvaraj, R. J. and Sbarra, A.J., Nature 210, 158-161 (1966).
2. McRipley, R.J. Selvaraj, R.J., Glovsky, M.M., and Sbarra, A.J., Rad. Res. 31, 706–720 (1967).
3. Paul, B.B., Strauss, R.R., and Sbarra, A.J., J. Reticuloendothelial Soc. 5, 538-549 (1968).
4. Sbarra, A.J., Paul, B.B., Jacobs, A.A., Strauss, R.R., and Mitchel, G.W., J. Reticuloendothelial Soc. 12, 109-126 (1972).
5. Štefanovič, J., Kišon, Š., and Absolonova, O., Folia Microbiol. 13, 424-431 (1968).
6. Kotulova, D. and Štefanovič, J., Folia Biol. 17, 26-32 (1971).
7. Absolonova, O., Štefanovič, J., and Ferenčik, M., Bratisl. Lek. Listy 59, 676-681 (1973).
8. Ferenčik, M. and Štefanovič, J., Folia Microbiol. 18, 402-409 (1973).
9. Štefanovič, J. „Non-Specific" Factors Influencing Host Resistance (ed. W. Braun and J. Ungar), S. Karger, Basel (1973), pp. 62-67.
10. Kopitar, M., Škrk, J., Šavnik, L., Lebez, D., and Poniž, T., Strahlentherapie 129, 596–603 (1966).
11. Cotič, V. and Lebez, D., Yugosl. Physiol. Pharmacol. Acta 1, 141-142 (1965).
12. Ferenčik, M. and Štefanovič, J., Biologia (Bratislava) 29, 453-463 (1974).
13. Dingle, J.T., Barrett, A.J., and Weston, P.D., Biochem. J. 123, 1-13 (1971).
14. Baumstark, J.S., Laffin, R.J., and Bardwill, V.A., Arch. Biochem. Biophys. 108, 514-522 (1964).
15. Gavin, J.J., Appl. Microbiol. 5, 25-34 (1957).
16. Ferenčik, M., Štefanovič, J., Košutzky, J., and Absolonova, O., Vet. Med. (Praque) 19, 327-336 (1974).
17. Hook, W.A. and Muschel, L.H., Fed. Proc. 18, 573-580 (1959).
18. Hook, W.A., Carey, W.F., and Muschel, L.H., J. Immunol. 84, 569-575 (1960).
19. Trojickij, V.L., Kaulen, D.R., Tumanjan, M.A., Fridenstejn, A.J., and Cachava, O.V. Radiacionnaja Immunologija, Medicina, Moscow (1965), p. 85.
20. Mestecky, J., Jilek, M., and Marečkova, M., Folia Microbiol, 11, 179-183 (1966).

THE RELATIONSHIP BETWEEN CATHEPSIN D ACTIVITY AND ANTIBACTERIAL ACTIVITY IN SUBCELLULAR FRACTIONS OF RABBIT AND CHICKEN POLYMORPHONUCLEAR LEUCOCYTES

Miroslav Ferenčik and Jan Štefanovič

Research Laboratory of Immunology, Department of Medical Microbiology and Immunology, Faculty of Medicine, Comenius University, Bratislava, ČSSR

SUMMARY

Myeloperoxidase, catalase, lysozyme, cathepsin D and E, beta-D-glucuronidase, and acid phosphatase were found in subcellular fractions of rabbit polymorphonuclear leucocytes, while chicken leucocytes lacked myeloperoxidase, catalase, and cathepsin E. Pepstatin added to rabbit leucocytes before their disintegration caused the inhibition of cathepsin D and E and an increase of specific activity of myeloperoxidase, as well as of total antibacterial activity.

INTRODUCTION

Phagolysosome is the site of intracellular killing and destruction of phagocytozed microorganisms. In polymorphonuclear leucocytes (PMN) a number of hydrolytic enzymes and antibacterial substances, localized mainly in lysosomes, participate in this process (1-7). It seems that there is a dynamic equilibrium between hydrolytic enzymes and macromolecular antibacterial substances in the lysosomes and also in the phagocytic vacuole.

Some defects and damage of lysosomal membranes might bring on derangement of this balance, which becomes manifest mainly through a more rapid degradation of proteins with enzymatic or antibacterial activity by cathepsin D and other intracellular proteases (8-10).

To test this hypothesis, we determined the relative distribution of seven lysosomal enzymes in subcellular fractions of PMN and compared their activities with the antibacterial activity of each fraction. As leucocytes from two species were used, species-specific differences could be studied. Investigation of a possible effect of cathepsin D on the myeloperoxidase and total antibacterial activities was done through inhibition of cathepsin D activity by adding pepstatin before disintegration of leucocytes.

MATERIALS AND METHODS

Polymorphonuclear leucocytes were isolated from peritoneal exudates of rabbits and chickens (11, 12). After washing with 0.34 M sucrose, the leucocytes were homogenized for 180 sec. in glass homogenizer fitted with a motor driven teflon pestle (5.000 rev. per min.). Only suspensions of leucocytes containing at least 90 % PMN were taken for homogenization. The homogenates were fractionated to 5 fractions by differential centrifugation according to Welsh and Spitznagel (13).

Pepstatin was kindly supplied by Prof. H. Umezawa and Banyu Pharm. Co. (Tokyo).

The antibacterial activity was determined with a modification of Hirsch's technique (14).

Cathepsin D (E.C. 3.4.4.23., reference 11), cathepsin E (11), acid phosphatase (E.C. 3.1.3.2., ref. 15), lysozyme (E. C. 3.2.1.17., ref. 16), beta -D-glucuronidase (E.C. 3.2.1.31., ref. 17), catalase (E.C. 1.11.1.6., ref. 18), and myeloperoxidase (E.C. 1.11.1.7., ref. 19, 11) estimations were done on all fractions.

When studying the effect of cathepsin D on total antibacterial and lysosomal enzymes activities, rabbit peritoneal leucocytes were divided up into two halves. Pepstatin was added to the first half so as to reach a concentration of 1 μg/ml, and the same volume of saline solution was added to the other half. Both parts were submitted to alternate freezing and thawing 6 times, and then fractionated by centrifugation, using the procedure mentioned above.

RESULTS AND DISCUSSION

Distribution of studied lysosomal enzymes in subcellular fractions of rabbit and chicken polymorphonuclear leucocytes is shown in Figure 1. There are differences in the relative representation of enzymes in individual fractions and also differences in the enzymatic equipment of rabbit and chicken leucocytes. In rabbit PMN, the presence of all enzymes under study could be demonstated, whereas in chicken leucocytes, the presence of cathepsin E, myeloperoxidase, and catalase could not be detected by the applied methods.

Cathepsin D activity is compared with antibacterial and myeloperoxidase activity in individual fractions in Figure 2. The antibacterial activity is present in the first four fractions from rabbit leucocytes, while in chicken leucocytes it is found in the first fraction only. The results given in this Figure suggest that in rabbit PMN the specific activity of myelo-

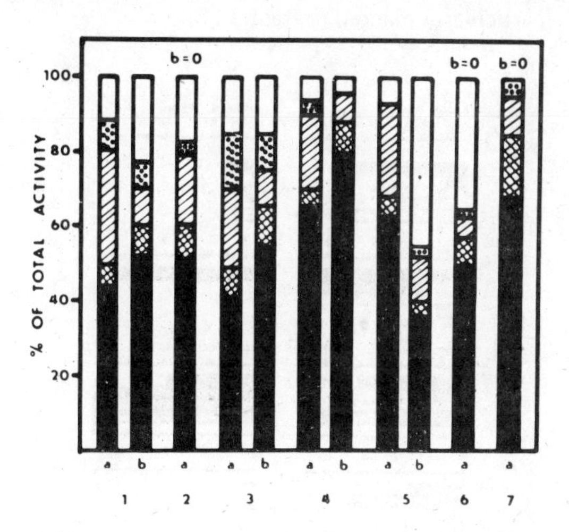

Figure 1. Relative activities of lysosomal enzymes in subcellular fractions of rabbit (columns a) and chicken (columns b) leucocytes.

Fractions: I – 150 x g (solid column), II – 800 x g (cross-hatched col.), III – 10,000 x g (cross-lined col.). IV – 50,000 x g (stippled col.), V – above 50,000 x g (open col.). Enzymes: 1 – cathepsin D, 2 – cathepsin E, 3 – acid (p-nitrophenyl) phosphatase, 4 – lysozyme, 5 – beta-D-glucuronidase, 6 – catalase, 7 – myeloperoxidase.

peroxidase is proportional to the bactericidal activity of fractions (when **Escherichia coli** is used as the test microorganism). The highest bactericidal activity was found in fraction 2, concomitantly with the highest activity of myeloperoxidase and the lowest cathepsin D activity.

The degree of antibacterial activity in individual fractions prepared from rabbit PMN is depicted in Figure 3.

A different distribution of specific enzymatic activities and of antibacterial activity is obtained when repeated freezing and thawing is used for disintegration of leucocytes instead of the glass homogenator. The most striking difference is in the increase of relative activities in the supernatant fraction (above 50.000 x g). Qualitative differences between the enzymatic equipment of rabbit and chicken PMN, however, remain under these conditions. The results obtained with pepstatin and without this cathepsin D and E inhibitor are presented in Figure 4. It can be seen that inhibition of cathepsin D, and also cathepsin E by pepstatin results in elevation of specific myeloperoxidase activity, as well as of antibacterial activity in all fractions (particularly marked in fraction 5).

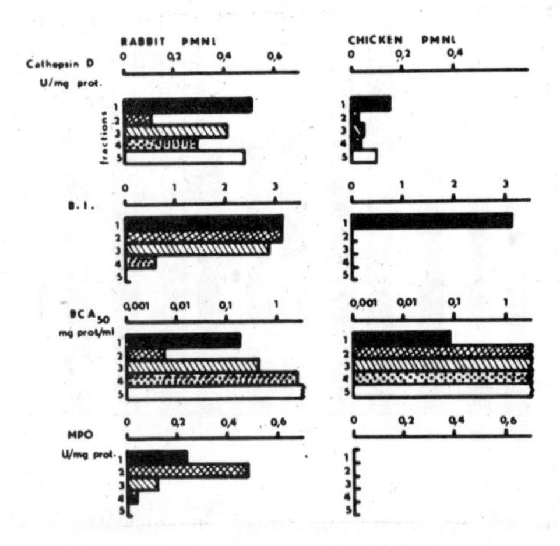

Figure 2. The relationship among cathepsin D, antibacterial and myeloperoxidase activities in subcellular fractions of rabbit and chicken polymorphonuclear leucocytes.

B.I. – bactericidal index (20).

BCA_{50} – bactericidal activity resulting in 50 % inhibition of growth of the test microorganism under standard conditions.

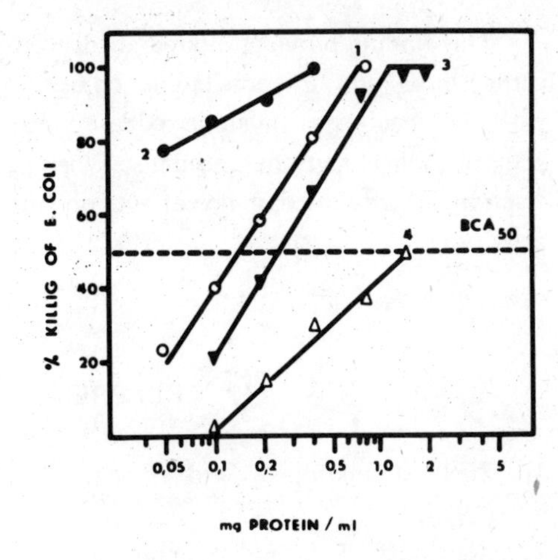

Figure 3. Bactericidal activities of the first four fractions from rabbit polymorphonuclear leucocytes.

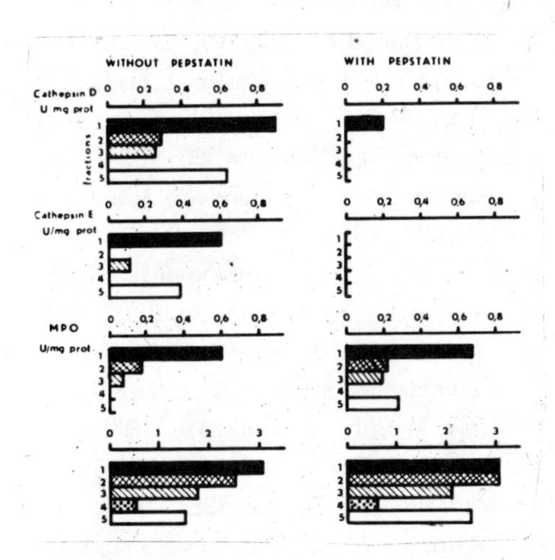

Figure 4. The effect of pepstatin on cathepsin D, cathepsin E, myeloperoxidase and total antibacterial activities in subcellular fractions of rabbit polymorphonuclear leucocytes.

This finding provides indirect evidence that after disintegration of PMN, cathepsin D and E (possibly in co-operation with other proteases) can partially inactivate myeloperoxidase and so reduce total antibacterial activity of leucocyte homogenates. When cathepsin D is not present in the fraction, pepstatin has no effect on myeloperoxidase or antibacterial activity.

REFERENCES

1. Zucker-Franklin, D. and Hirsch, J.G., J. Exp. Med. 120, 569-578 (1964).
2. Klebanoff, F.J. and Hamon, C.B., J. Reticuloendothelial Soc. 12, 170-196 (1972).
3. Sbarra, A.J., Paul, B.B., Jacobs, A.A., Strauss, R. R., and Mitchell, G.W., Jr., J. Reticuloendothelial Soc. 12, 109-126 (1972).
4. Baggiolini, M., Enzyme 13, 132-160 (1972).
5. Spitznagel, J.K. in Phagocytic Mechanisms in Health and Disease (R.C. Williams and H.H. Fudenberg, eds.), Georg Thieme Publ., Stuttgart (1972), pp. 83-106.
6. Spitznagel, J.K., Dalldorf, F.G., Leffell, M.S., Folds, J.D., Welsh, I.R.H., Cooney, M.H., and Martin, L.E., Lab. Invest. 30, 774-785 (1974).
7. Allen, R.C., Yevich, S.J., Orth, R.W., and Steele, R. H., Biochem. Biophys. Res. Commun. 60, 909-917 (1974).
8. Absolonova, O., Štefanovič, J., and Ferenčik, M., Bratisl. Lek. Listy 59, 676-681 (1973).
9. Ferenčik, M. and Štefanovič, J., Folia Microbiol. 18, 402-409 (1973).
10. Štefanovič, J. in „Non-Specific" Factors Influencing Host Resistance (W. Braun and J. Ungar, eds.), S. Karger, Basel (1973), pp. 62–67.
11. Ferenčik, M. Štefanovič, J., Absolonova, O., and Kotulova, D., J. Hyg. Epid. Microb. Immunol. 19, (1975) – in press.
12. Ferenčik, M., Štefanovič, J., Košutzky, J., and Absolonova, O., Vet. Med. (Prague) 19, 327-336 (1974).
13. Welsh, I.R.H. and Spitznagel, J.K., Infect. Immun. 4. 97-102 (1971).
14. Štefanovič, J. and Absolonova, O. Folia Microbiol., 11, 358-363 (1966).
15. Andersch, M.A. and Szczypinski, A.J., Amer. J. Clin. Pathol. 17, 571-574 (1947).

16. Gavin, J.J., Appl. Microbiol. 5, 25-34 (1957).
17. Fishman, W.H. in Methods of Enzymatic Analysis (H. U. Bergmeyer ed.) Acad. Press, New York (1963) pp. 869-871.
18. Sinha, A.K., Analyt. Biochem. 47, 389-394 (1972).
19. Hosoya, T., Kondo, Y., and Ut, N., J. Biochem. 52, 180-189 (1962).
20. Wardlaw, A.C. and Pillemer, L., J. Exp. Med. 103, 553-561 (1956).

ON THE SPECIFICITY OF CATHEPSIN B2

K. Otto, Husan Afroz, R. Müller, and P. Fuhge

Department of Physiological Chemistry, Division of Enzymology, University of Bonn, Bonn (G.F.R.)

Cathepsin B2 that evolved eight years ago together with cathepsin B1 (1) from the then cathepsin B – described in 1957 by Greenbaum and Fruton (2) – has, in the meantime, been purified to homogeneity (3). In gel electrophoresis, it displays two bands, close together and of similar appearance, both of them hydrolyzing benzoylarginineamide (Bz–Arg–NH_2). This substrate was the classic substrate for the former cathepsin B and still is a substrate for both cathepsin B1 and B2.

Like cathepsin B1, cathepsin B2 is also a thiol enzyme which requires a mercapto–compound and EDTA for full activity.

After the purification of cathepsin B2 had been accomplished – essentially more than 2 years ago (4) – we began to investigate its specificity with some oligo– and polypeptides, first with the oxidized B-chain of insulin. Recently, Tappel and coworkers (5) published a paper on cathepsin B2 and its specificity, in which they pronounced cathepsin B2 to be a carboxypeptidase.

Already a couple of years ago, when we were not yet certain about the purity of this enzyme – the interference of the many isoenzymes of cathepsin D worried us – we had tested its action on other enzymes as to a possible inactivating effect, with little success. This, together with only limited hemoglobin degradation, was not very indicative for an endopeptidase action of cathepsin B2. On the other hand, with the oxidized B-chain of insulin, there clearly were small peptides as degradation products – pointing again to endopeptidatic cleavages. Later, we realized that the oxidized B-chain is an unfavourable substrate, partly because of proline in position 28, partly due to the relatively ‚unspecific' amino acid sequence.

However, with glucagon as a substrate, cathepsin B2 displays pure carboxypeptidase activity. This is illustrated in Fig. 1, which shows the time course of the hydrolysis of this peptide hormone by cathepsin B2. Theoretically, of the 29 amino acids in glucagon, 27. can be released as single amino acids, the last dipeptide remaining unhydrolyzed. We found up to 25 moles of amino acid. A characteristic feature of glucagon hydrolysis is a lag period after the release of approximately 5 amino acids. Under the conditions used here, the hydrolysis of glucagon is completed after circa two hours. At certain intervals, samples were withdrawn and subjected to amino acid analysis. Thus, the appearance of the amino acids can be followed. (Fig. 2).

One can see the release of the singly—occurring amino acids Met^{27}, Trp^{25}, Val^{23}, Ala^{20}, Lys^{12}, Gly^{4}. Their time course follows exactly the sequence within the polypeptide. Similarly, the appearance of those amino acids which occur twice and with a certain distance in between can be demonstrated as two ascending steps; whereas those whose position is close together (e.g. Arg^{17} – Arg^{18}) behave as one single unit. Except for the time lag shown in Fig. 1, all amino acids seem to be split off equally well, including the basic ones. Proline does not occur in glucagon.

The pH optimum for glucagon hydrolysis is approximately 4.4 (acetate buffer); there is no trace of endopeptidase activity by cathepsin B2. The K_m value calculated from initial velocities (5 min) is about 0.4 mM.

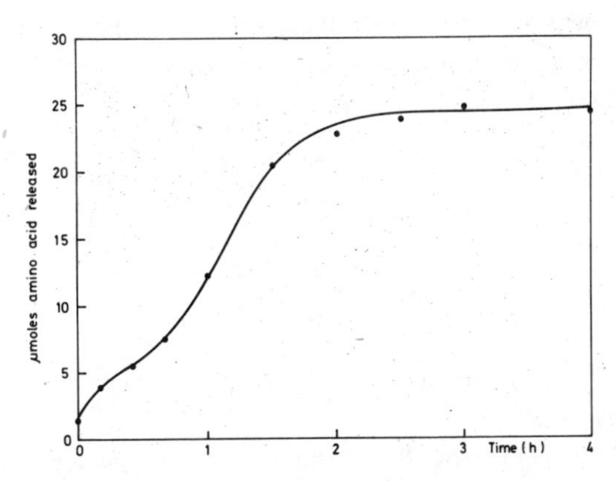

Fig. 1. Hydrolysis of 1μmole of glucagon by cathepsin B2 (75 μg) at pH 4,2

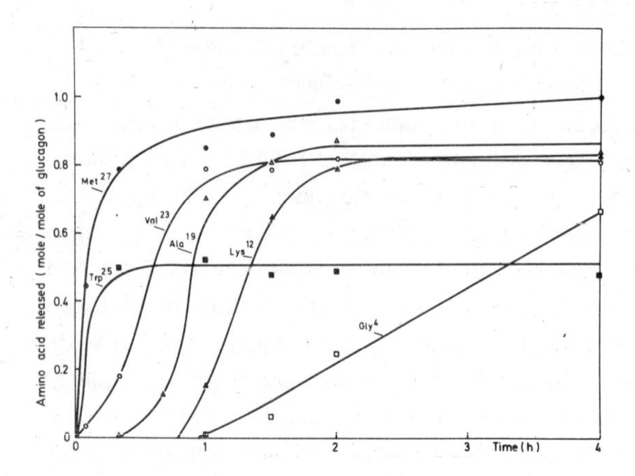

Fig. 2. Time course of the release of amino acids from glucagon by cathepsin B2

When turning to a simpler peptide that, at the same time, contains proline, e.g. bradykinin, then we can easily recognize with the ninhydrin technique that very probably only one amino acid residue is released. This could be verified with amino acid analysis which shows only the C-terminal arginine. Hydrolysis comes to a halt when approaching proline or the amino acid next to it. There is no endopeptidatic cleavage.

This mode of action is confirmed by the effect of cathepsin B2 on angiotensin I and angiotensin II. With angiotensin I, only Leu^{10} and His^{9} are split off as expected beforehand, Phe^{8} not at all. Proline, here too, stops further hydrolysis. Consequently, angiotensin II is not hydrolyzed at all, and there is no endopeptidase activity either, different from findings with cathepsin A by Logunov and Orekhovitch (6).

				5		AE			10					15		
Phe-	Val-	Asn-	Gln-	His-	Leu-	Cys-	Gly-	Ser-	His-	Leu-	Val-	Glu-	Ala-	Leu-	Tyr-	Leu-
0	0	46	46	46	46	85	100	100	92	92	92	100	77	77	0	0

	AE	20					25					30
-Val-	Cys-	Gly-	Glu-	Arg-	Gly-	Phe-	Phe-	Tyr-	Thr-	Pro-	Lys-	Ala
12	15	15	14	8	15	?	?	?	0	0	50	100

Fig. 3. Hydrolysis of aminoethylated B-chain of insulin by cathepsin B2 at pH 4,5 (12 h).
Relative amounts of some of the released amino acids are assigned arbitrarily.

A similar result is obtained with oxytocin whose carboxyl terminus consists of glycineamide. Apparently, neither the glycineamide nor the amide group alone can be split off. Tappel (5) has reported that synthetic peptides with glycine in the carboxyl-terminal position are not suitable substrates for cathepsin B2.

Turning back again to the oxidized B-chain of insulin with a proline residue in its antepenultimate position 28, we found that only the C-terminal alanine is split off and, in addition, a certain percentage of the penultimate lysine (pos. 29) too. But further analysis of the hydrolyzed amino acids shows a number of additional residues released. We must assume that in this case an additional endopeptidatic cleavage — or maybe even more than one — has occurred, followed by subsequent carboxypeptidatic hydrolysis. Quite similar results were obtained with the aminoethylated B–chain of insulin (fig. 3). Thus, besides Ala^{30} and Lys^{29}, we find amino acids of positions 3–13 and, in smaller quantities, those of positions 14–15 and, perhaps, some from positions 18–26.

We wanted to confirm this unexpected result with another suitable substrate, and we chose melittin, a peptide from bee venom and made up from 26 amino acid residues. It contains a proline residue in position 14 and at the carboxyl terminus the rarely encountered glutamic acid diamide. If cathepsin B2 were merely a carboxypeptidase and if it were to require a carboxyl group for its activity, then there should be no proteolytic degradation at all. As we already know from the hydrolysis of Bz–Arg–NH_2, this is not the case. Consequently, melittin is hydrolyzed to a

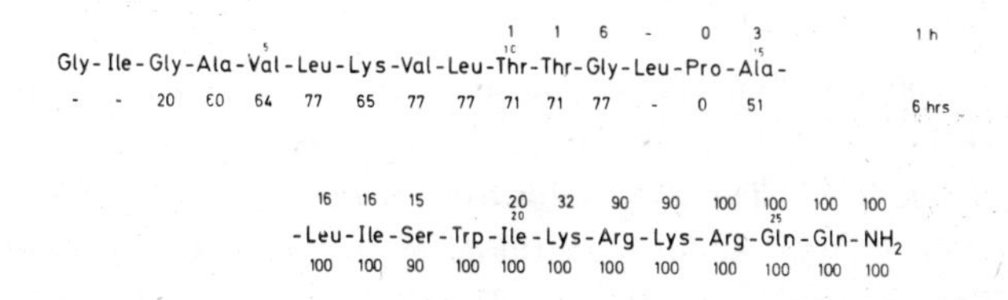

Fig. 4. Hydrolysis of melittin by cathepsin B2 at pH 5
Relative amounts of some of the released amino acids are assigned arbitrarily (Arg = 2x100).

considerable extent. There remains the question how the terminal glutamic acid diamide is split off and, equally important, whether or not an additional endopeptidatic cleavage occurs, thus being a starting point for subsequent exopeptidatic activity.

The experiments revealed that melittin is easily hydrolyzed, though perhaps not as quickly as is glucagon. Under the conditions used, the hydrolysis is not fully completed after six hours (Fig. 4).

With only 60 minutes hydrolysis time, mainly the amino acids from positions 16–26 are released, including 2 moles of glutamine (pos. 25 and 26). After six hours hydrolysis, amino acids from positions 3–12 become clearly visible, threonine and valine functioning as markers. When assigning a degree of 100 % cleavage to the carboxylterminal amino acids, the amino acids ‚beyond' proline are split off to a degree of about 70,%.

This points to an endopeptidase action of cathepsin B2, at least in a case like this when proline occurs in the molecule. At the same time, the question is answered whether or not cathepsin B2—as a carboxypeptidase – requires a free carboxyl group in its substrate. That appears not to be the case as proven here again. Nevertheless, cathepsin B2 may require amides of particular amino acids as the ‚carboxyl-terminus' since oxytocin with glycineamide was not attacked.

Finally, it could be shown that the amide nitrogen is indeed liberated as ammonia because 2 moles of glutamine appear together with 1 mole of ammonia. The latter could be determined like to amino acids in the analyzer, as well as by means of the glutamate dehydrogenase reaction with α–ketoglutarate and $NADH_2$. It is also possible to follow up the hydrolysis of the terminal amide in a cuvette with a coupled test system containing melittin, cathepsin B2, and the glutamate dehydrogenase test ingredients. With this arrangement, it can be shown that even at pH7 reasonably fast reaction rates are attainable.

Melittin contains an accumulation of lysine and arginine, together four residues. That does not appear to be a hindrance for the exopeptidatic action of cathepsin B2. The same holds true for the protamines clupeine and salmine, which contain about two thirds of their amino acid residues as arginine. They are both quickly hydrolyzed – as already mentioned by Tappel (7). As both of them contain some proline residues, hydrolysis is however, not complete, and an additional endopeptidatic cleavage might be expected here too. Due to the uniform amino acid sequences of these protamines, together with their possible inhomogeneity, such a cleavage is not easy to prove; the time course of the hydrolysis of these substances by cathepsin B2 – a final value is reached after a comparatively short time – is not in favour of such a possibility.

When we now turn to an acid polypeptide, namely to the oxidized A-chain of insulin, there is hardly any recognizable degradation by cathepsin B2, except perhaps for very prolonged times; less than one mole of amino acid is released. It must be the cysteic acid residues or the generally acid character which is responsible for this behaviour, as otherwise the amino acid sequence shows no uncommon features.

That this is probably so can be confirmed by using the aminoethylated A-chain of insulin. The now aminoethylated cysteine residues resemble lysine, thus turning this peptide into a more basic compound. This leads to a rather fast hydrolysis by cathepsin B2. As no proline residue is present in the A–chain, the degradation is merely a carboxypeptidatic one. Fig. 5 shows the time course of such a hydrolysis, together with an estimation of the release of the amino acid residues.

Since in the oxidized B-chain, cysteic acid is not an unsurmountable hindrance for cathepsin B2, the facile hydrolysis of the aminoethylated A–chain could be interpreted as rather the accumulation of cysteic acid residues than as an isolated residue of this kind.

Finally, we tested adrenocorticotrophin (ACTH), a polypeptide with 39 amino acid residues, among them four proline residues. One of these four is located close to the C-terminus (pos. 36), comparable in its location with the

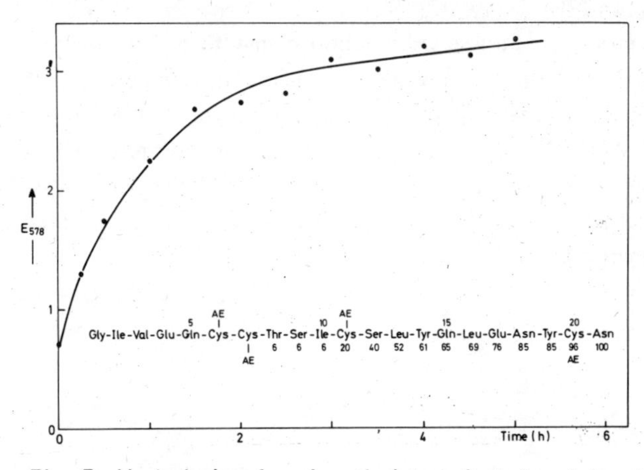

Fig. 5. Hydrolysis of aminoethylated A-chain of insulin (0,5 μmole) by cathepsin B2 (21 μg) at pH 4,5

B–chain of insulin. Here, too, we find that besides the final phenylalanine and the penultimate glutamic acid some other amino acids from the interior of the molecule are released. That, again, points to an endopeptidase effect. But due to the size and amino acid sequence of this hormone, it is not clear yet where the endopeptidatic cleavage (or cleavages?) are located.

Last but not least, a brief word on the pH–optima of all these hydrolyses. Generally, the reaction optimum was between pH 4 and 5, except perhaps for salmine which showed a slightly higher optimum. Those basic peptides – melittin also – clearly tend to the higher pH–optima. The pH–optimum of Bz–Arg–NH_2 (approximately 6) is in acordance with this rule. As mentioned before, melittin in combination with the glutamate dehydrogenase reaction shows substantial degradation rates with cathepsin B2 even at pH 7.

Summarizing our results we may state:

1. Cathepsin B2 is primarily a carboxypeptidase with a pH–optimum of between 4 and 5 depending on the substrate involved. Basic peptides show a tendency to higher values. The acidic oxidized A-chain of insulin is only slightly attacked.

2. Probably all the existing peptide bonds can be hydrolyzed, except for bonds following or preceding proline.

3. ,Cathepsin B2 shows amidase action, as exemplified already many years ago with Bz–Arg–NH_2 and now with the peptide melittin. This is not completely unique, as it has been observed with at least one other carboxypeptidase, too (8).

4. When the carboxypeptidase degradation comes to a halt due to a proline residue, as for example in the B-chain of insulin, in melittin, or in ACTH, an additional quasiendopeptidatic action ‚beyond' the proline residue can occur. Whether or not a particular amino acid constellation or a minimal chain length is required, we cannot say as yet. This endopeptidase effect could not be demonstrated with the relatively short-chained peptides bradykinin and angiotensin.

ACKNOWLEDGEMENTS

The authors are greatly indebted to Professor Dr. Horst Hanson (Halle, G.D.R.) for valuable advice. We gratefully acknowledge the technical assistance of Miss Gabriele Schwartzer. This investigation was supported by the Deutsche Forschungsgemeinschaft and by the Funds of the Chemische Industrie.

REFERENCES

1. Otto, K., Z. Physiol. Chem 348, 1449–1460 (1967)
2. Greenbaum, L.M. and Fruton, J.S., J. Biol. Chem. 226, 173–180 (1957)
3. Otto K. and Riesenkönig, H., Biochim. Biophys. Acta 379, 462–475 (1975)
4. Otto, K. in: Intracellular Protein Catabolism, (Hanson, H., Bohley, P, eds.), Verlag J. Ambrosius Barth, Leipzig, pp 252 – 259 (1974).
5. Ninjoor, V., Taylor, S.L., and Tappel, A.L., Biochim. Biophys. Acta 370, 308–321 (1974)
6. Logunov, A.I. and Orekhovitch, V.N., Biochem. Biophys. Res. Commun. 46, 1161–1168 (1972)
7. De Lumen, B.O. and Tappel, A.L., Biochim. Biophys. Acta 293, 217–225 (1973)
8. Hayashi , R., Moore, St., and Stein, W.H., J. Biol. Chem. 248, 2296–2302 (1973).

A THIOL DEPENDENT ACID PROTEASE

T. Turnšek, I. Kregar, D. Lebez, V. Turk

Department of Biochemistry, J. Stefan Institute, University of Ljubljana, Ljubljana, Yugoslavia

It is a well known fact that lysosomes have an adequate complement of proteolytic enzymes for intracellular protein degradation. However, little is known about the steps that occur during such proteolysis, the total number and nature of the enzymes involved or their possible concerted action. A pathway of protein hydrolysis by lysosomal enzymes was proposed (1,2) in which acid proteases attack proteins to produce oligopeptides that are then digested to completion by exopeptidases and dipeptidases. Our experiments were aimed to investigate the number and nature of acid endopeptidases present in calf lymph nodes. This tissue was used as a source of the cathepsins, because it has been suggested that, along with cathepsins B and C, at least two different acid proteases could be responsible for the hydrolysis of hemoglobin and albumin as reported by Stein and Fruton (3). Fräkki et. al. (4) measuring the digestion of endogenous tissue proteins, hemoglobin and casein by crude rat thymus and lymph nodes' extracts, proposed that the whole proteolytic activity consists of several separate enzymes with different pH optima.

EXPERIMENTAL

A 50 % aqueous homogenate was prepared from fresh minced tissue with a Waring blender. It was centrifuged at 5000 x g for 90 min in a Sorvall

RC 2 centrifuge and the supernatant was acidified to pH 3.5 centrifuged again at 5000 x g for 20 min; afterwards the proteins in the supernatant were precipitated with 80 % ammonium sulphate. Finally the proteases were precipitated with 30 – 60 % acetone, dissolved in 0.01 M acetate buffer of pH 5.5 and applied to a column of CM cellulose.

The protein content in eluted fractions was followed by measuring the absorbance at 280 nm.

The acid proteolytic activity was determined using hemoglobin as a substrate as described by Anson (5). 0.4 ml of the enzyme preparation were incubated with 2 % hemoglobin solution in acetate buffer at pH 3.5 or 3.0 and at 37°C. The reaction was stopped with 4 ml of 5 % trichloroacetic acid. The hydrolytic products in the filtrates were determined by measuring the colour differences at 750 nm between samples and blanks after the addition of 4 ml of 0.5 M NaOH and 1.2 ml of the Folin-Ciocalteau reagent. Blanks were prepared so that trichloroacetic acid was added to the enzyme solution prior to the addition of hemoglobin. Whenever the activity in the presence of dithiothreitol (DTE) was measured, the absorbance difference between samples and blanks was determined in the filtrates directly at 280 nm.

The amidase activity was measured according to Otto (6). 0.1 ml of the enzyme preparation was incubated for 60 min at 37°C with 1 ml of the substrate solution, which was composed of 20 mM Bz-Arg-NH_2, 1mM ethylenediaminotetraacetic acid (EDTA) and 1mM DTE in 0.1M acetate buffer, pH 5.7.

RESULTS

Our first experiments were made to examine the effect of thiol activators and inhibitors as well as pepstatin on the 30-60 % acetone fraction. The results are shown in Table I; one can see that Hg^{++} inhibited acid hydrolysis of hemoglobin by 30 %, while Ag^{+} had no effect. Pepstatin inhibited the activity up to 90 % and the addition of both, pepstatin and Hg^{++} inhibited it completely. Preincubation with EDTA and DTE increased the hemoglobin hydrolysis at acid pH. We see that in the mixture of cathepsins acid proteolytic activity is partly influenced by thiol reagents. After the inhibition of cathepsin D by pepstatin, the residual activity, which is due to other acid proteases, increased linearly with the time of incubation (Figure 1). Our experiments are very similar to those of Huisman et. al. (7). They

TABLE I.

THE INFLUENCE OF SOME EFFECTORS ON HEMOGLOBIN HYDROLYSIS AT pH 3.0 MEASURED IN 30 – 60 % ACETONE EXTRACT

To 0.25 ml of enzyme solution 0.15 ml of the effector solution was added to give the concentration in incubation mixture as indicated in the table. After 15 min preincubation at 37°C 2 ml of 2 % hemoglobin solution was added and incubated for 30 min at 37°C.

Added solution	A750 nm	A280nm	Residual activity	% inhibition
Acetate buffer pH 3.0	1.528+	0.300	(100)	(0)
$HgCl_2$ (5 mM)	1.144	–	73	27
$AgNO_3$ (2 mM)	1.640	–	103	(0)
Pepstatin (4 μM)	0.168	–	10	90
Pepstatin + $HgCl_2$	0.030	–	2	98
Pepstatin + $AgNO_3$	0.176	–	10	90
DTE (4 mM) + EDTA (2 mM)	–	0.370	123	(0)

+ The enzyme solution (A.U./mg protein = 1.86) was diluted and the values calculated for the dilution factor.

studied the role of individual cathepsins from rat liver lysosomes in protein digestion at pH 5.0, inhibiting them selectively by specific inhibitors; the degradation of hemoglobin and albumin occurred even when the cathepsins B1, B2, C and D were blocked by leupeptin, omission of chloride ions or by the presence of pepstatin. The authors ascribed the residual activity to some other thiol dependent protease. Contrary to our results, this activity was completely inhibited by Ag^+, but we applied a lower concentration of Ag^+ and measured hemoglobin digestion at pH 3.0 and not at pH 5.0.

The separation of acid proteinases was achieved by ion exchange chromatography followed by gel filtration. Three proteolytic active peaks were eluted from a CM– cellulose column with 0.01 M acetate buffer, pH 5.5 and a linear gradient of CH_3COOH to 0.2M conc. (Figure 2). The first peak exhibited amidase activity as well as hemoglobin splitting activity. Further purification on Sephadex G-100 has shown (8) that while in the second and third proteolytic peak only multiple forms of cathepsin D were present, different cathepsins were obtained from the first proteolytic peak (Figure 3). Two amidase activities were eluted in the range of molecular weights 23 000 and 51 000 as determined by Whitaker method (9). This activity was measured on Bz-Arg-NH_2, the common substrate of cathepsins B1 and B2, which indicates that we have separated the cathepsins B1 and B2. They were not studied further.

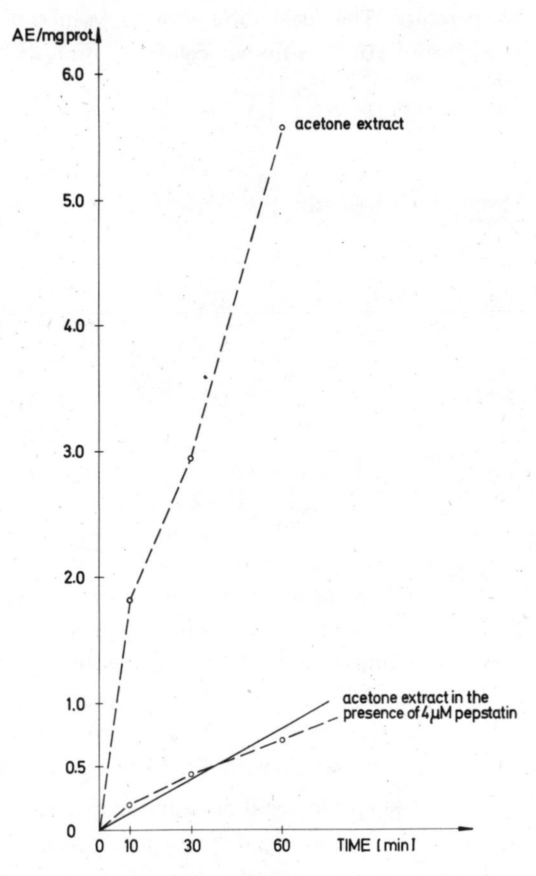

Fig. 1. Time dependence of proteolytic activity of acetone extract in the presence of 4 μM pepstatin and without pepstatin.

Two acid proteinases were also separated by gel filtration, having molecular weights of approximately 39 000 and 14 000. Further characterization of the proteinases proved (8) that the former was cathepsin D, while the latter — designated here as cathepsin S — had different characteristics. However, some of them were similar to those of cathepsin D; thus the protein substrate most susceptible to acid hydrolysis was hemoglobin, while the degradation of bovine serum albumin and human serum albumin was 50 % of hemoglobin hydrolysis; gamma globulin was not degraded at all. The pH optimum for hemoglobin hydrolysis was at pH 3.0. The enzyme preparation of cathepsin S was not stable in the absence of DTE, its proteolytic activity decreasing by 50 % when left for 24 hrs at room temperature. The main difference between cathepsin D and S lies in the influence of effectors on hemoglobin hydrolysis (Table II). Pepstatin, which

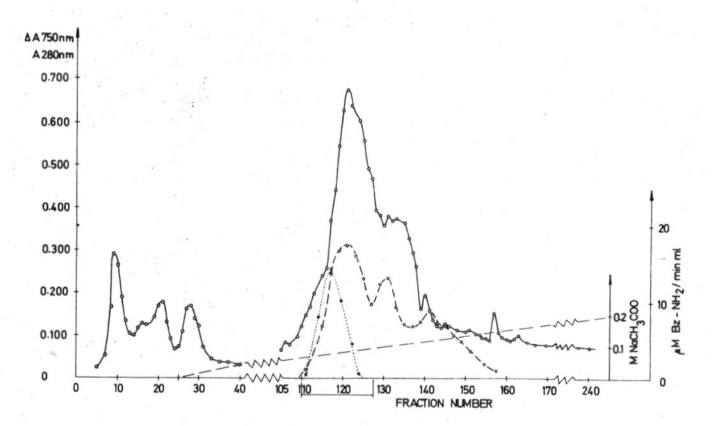

Fig. 2. Chromatography of acetone extract on CM cellulose. 28 ml of acetone extract containing 15.8 mg protein/ml were applied on a column (5 cm x 50 cm) and eluted with 0.01 M acetate buffer, pH 5.5. After the second protein peak the elution was continued using a linear gradient to 0.2 M concentration of CH_3COONa (pH 5.5). Flow rate: 65 ml/h. Fractions volume: 20 ml. —— A_{280} nm, --- ΔA_{750} nm ,acid proteolytic activity on hemoglobin at pH 3.5, — .... amidase activity on Bz–Arg–NH_2 at pH 5.3. Fractions were pooled as indicated. Reproduced from Turnšek et al. (22) by kind permission of Elsevier Publ. Co.

is strong inhibitor of cathepsin D, did not influence the cathepsin S activity. It was also not affected by $MgCl_2$ but partly inhibited by $CoCl_2$ and completely by $HgCl_2$ and iodoacetic acid. The addition of DTE increased the activity by 300 %. These results indicate that cathepsin S is a sulphydryl endopeptidase which is not sensitive to pepstatin.

The enzyme preparation obtained after gel chromatography was not homogenous as checked by disc electrophoresis. Four protein components were obtained. They were extracted from the gels with H_2O, concentrated and their proteolytic activities on hemoglobin at pH 3.0 was measured in the absence and in the presence of pepstatin. It was found that one component was not inhibited by pepstatin, probably representing the cathepsin S. Rechromatography of the isolated component under the same experimental conditions – 7.5 % polyacrylamide gel in Tris glycine buffer, pH 8.4 – showed that cathepsin S was obtained in the pure form.

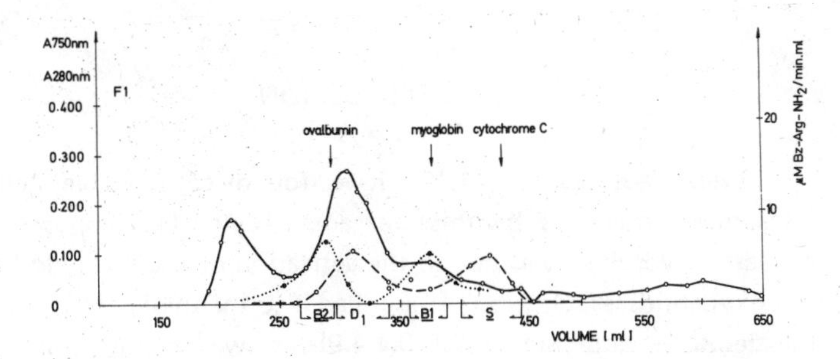

Fig. 3. Chromatography on Sephadex G-100. 4 ml of concentrated enzyme preparation containing 10.6 mg protein/ml were applied on a column (2.5 cm x 85 cm) of Sephadex G-100. Elution buffer was composed of 0.15 M sodium acetate, 0.1 M NaCl, 1 mM EDTA and 0.5 mM dithiothreitol, and the pH was adjusted to 4.6. Flow rate: 35 ml/h, fractions volume: 7-8 ml. —— A_{280}nm, ––– ΔA_{750} nm, acid proteolytic activity, ... amidase activity. Fractions were pooled as indicated. Reproduced from Turňšek et al. (22) by kind permission of Elsevier Publ. Co.

TABLE II

INFLUENCE OF SOME EFFECTORS ON HEMOGLOBIN HYDROLYSIS OF CATHEPSIN S

Added solution	Relative activity
Acetate buffer, pH 3.0	(100 %)
Pepstatin	100
$CoCl_2$	69
$MgCl_2$	100
$HgCl_2$	0
Iodoacetic acid	0
Dithiothreitol	310

0.2 ml of enzyme solution (0.66 mg protein/ml) were incubated with 0.2 ml of 5 mM effector solution in acetate buffer, pH 3.0 for 12 min at 37°C. Then 2 ml 2 % hemoglobin solution in the same buffer were added and incubated for 3 hrs at 37°C.

Pepstatin was added in 3 μM concentration.

DISCUSSION

Several authors (10,11,12) have found considerable activation of lysosomal proteinases by thiol reagents. Taylor (13) supposed that this increased hydrolysis was due to the activation of the sulfhydryl dependent carboxypeptidases. Others (14) ascribed it to the thiol dependent dipeptidyl transferase – cathepsin C (E.C.3.4.4.9)– as well as cathepsins B1 and B2 (E.C.3.4.22.1) which should act in concert with endopeptidases in lysosomal extracts or tissue homogenates (15,16,17). Rather crude preparations of cathepsin D, obtained early by Anson (18) and Press et. al (19), were also found to be partly activated by cysteine, although the pure form of cathepsin D, obtained later, was unaffected by standard activators and inhibitors of the „thiol, serine and metal" proteinases (14). Cathepsin D, purified from human erythrocytes was even 90 % inhibited by 10 mM DTE and 40 % by 1 mM DTE, suggesting an essential disulphide group, as reported by Reichelt et. al. (20).

We think that the sulfhydryl dependent endopeptidase cathepsin S, found in calf lymph nodes, could be responsible for the observed thiol dependent acid hydrolysis. It differs from known thiol dependent cathepsins B1 and B2 by its lower molecular weight—as determined by gel filtration—, its inability to split Bz—Arg—NH_2, as well as by the much lower pH optimum for the hydrolysis of hemoglobin. Cathepsin S differs from acid endopeptidases cathepsins D (E.C.3.4.23.5) and E by its inability to be inhibited by pepstatin. Recently Yago et. al. (21) also isolated an acid cathepsin from thoracic duct lymphocytes which had a molecular weight of about 80 000 and was very sensitive to pepstatin, so it differs characteristically from cathepsin S isolated from calf lymph nodes.

ACKNOWLEDGEMENTS

We thank Mrs. M. Božič, Mrs. J. Komar and Mrs. V. Štrukelj for their skilful technical assistance. We are especially grateful to Dr. H. Umezawa from Tokyo for his generous gift of pepstatin. This work was supported by a grant from the Research Council of Slovenia and in part by a grant from the NSF, USA, Grant No. GF-31389.

REFERENCES

1. Coffey, J.W., De Duve, C., J. Biol. Chem. 243, 3255-3263 (1968)
2. Tappel, A.L. In J.T. Dingle, H.B. Fell (eds.) Lysosomes in Biology and Pathology, vol. 2, pp 207—244, North—Holland Amsterdam—London (1969).
3. Stein, O., Fruton, J.S., J. Biol. Med.215, 163—170 (1960).
4. Fräki, J., Ruskaanen, O., Hopsu-Havu, V.K., Kuralainen, K., Hoppe Seyler's Z.Physiol. Chem. 354,933—943 (1973).
5. Anson, M.L., J.Gen. Physiol. 22,79—87 (1939).
6. Otto, K., Hoppe Seyler's Z.Physiol. Chem. 348,1449—1460 (1967).
7. Huisman, W., Lanting, L., Doddema, H.J., Bouma, J.M.W., Gruber, M., Biochim. Biophys. Acta 370,297-301 (1974).
8. Turnšek, T., Kregar, I., Lebez,D., Croat. Chem. Acta 47, 59—69 (1975).
9. Whitaker, J.R., Analyt. Chem. 35,1950—1953 (1963).

10. Huisman, W., Bouma, J.M.W., Gruber, M., Biochim. Biophys. Acta 297, 98–109 (1973).
11. Huisman, W., Bouma, J.M.W., Gruber, M., Biochim. Biophys. Acta 297, 93–97 (1973).
12. Goettlich–Riemann, W., Young, J.O., Tappel, A.L., Biochim. Biophys. Acta 243,137–146 (1971).
13. Taylor, S.L., Tappel, A.L., Biochim. Biophys. Acta 341,99–111 (1974).
14. Barrett, A.J. In J.T. Dingle (ed.) Lysosomes, pp 46–135, North–Holland Amsterdam–London (1972).
15. Huang, F.L., Tappel, A.L., Biochim. Biophys. Acta 236, 739–748 (1971).
16. Kato, T., Kojima, K., Murachi, T., Biochim, Biophys. Acta 289, 187–193 (1972).
17. De Lumen, B.O., Tappel, A.L., Biochim, Biophys. Acta 293,217–225 (1973).
18. Anson, M.L., J. Gen. Physiol. 23,695–704 (1939).
19. Press, E.M., Porter, R.R., Cebra, J. Biochem.J., 74,501–514 (1960).
20. Reichelt, D., Jacobsohn, E., Haschen, R.J., Biochim. Biophys. Acta 341, 15–26 (1974).
21. Yago, N., Bowers, W.E., Fed. Proc. 33,257 (1974).
22. Turnšek, T., Kregar, I., Lebez, D., Biochim. Biophys. Acta, 403, 514 (1975).

CATHEPSIN L AND CATHEPSIN B3 FROM RAT LIVER LYSOSOMES

H. Kirschke, J. Langner, B. Wiederanders, S. Ansorge, P. Bohley, and H. Hanson

Physiologisch-chemisches Institut der Martin–Luther–Universität, DDR-402 Halle (Saale), Holystraße 1, GDR

At the first symposium on „Protein catabolism" we mentioned our intentions to detect all soluble proteinases in rat liver lysosomes, which are active at pH 6-7 under in vitro conditions.

We start our enzyme preparation from a lysosomal extract containing only 0.12 % of mitochondrial and 0.15 % of microsomal proteins. Therefore we are sure that the isolated proteinases are from lysosomes.

The main part (80 %) of the proteolytic activity at pH 6-7 in the lysosomal extract is distributed among four proteinases with m. w. of 15-30000. Three of these cathepsins have been purified nearly 9000 fold over the specific activity in the homogenate. They are characterized by their physical and kinetic properties and their capability of hydrolysing low molecular and protein substrates. From 100 rat livers we got 1-3 mg cathepsin L, and 8-11 mg of both cathepsin B3 and B1. I will discuss the cathepsins L and B3, and I will compare them with the known cathepsin B1 from rat liver lysosomes. All these cathepsins are thiol-proteinases.

The specific activities show the diversity of the three cathepsins (Table 1). Cathepsin L is the most important proteinase degrading azocasein. This holds true also for other protein substrates such as glucagon, cytosol proteins, haemoglobin, and others.

Table 1

Specific activities and Km of cathepsins at pH 6.0
Incubation conditions: 20 mM phosphate buffer; 5 mM glutathion; 5 mM EDTA; 37° C; azocasein 2.5 %, Bz–ArgNH$C_{10}H_7$ 1 mM, LysNH$C_{10}H_7$ 1mM, Bz–ArgOEt 0.1 M.
Specific activity: μMoles/min/mg enzyme; *nMoles/min/mg enzyme.

	cath. B1	cath. B3	cath. L
Specific activity			
azocasein *	3.8	5.3	51.7
Bz–ArgNH$C_{10}H_7$	0.9	5.6	< 0.1
LysNH$C_{10}H_7$	0	4.8	0
Bz–ArgOEt	3.9	5.3	0
Km (Mol/l)			
azocasein	3.10^{-5}	1.10^{-4}	7.10^{-6}
Bz–ArgNH$C_{10}H_7$	1.10^{-3}	4.10^{-4}	–
LysNH$C_{10}H_7$	–	5.10^{-5}	–
Bz–ArgOEt	5.10^{-3}	4.10^{-2}	–

Additionally to its endopeptidase activity, cathepsin B3 possesses also aminopeptidase activity. Up to now we did not succeed in separating the proteinase from the aminopeptidase. There are some hints that one enzyme possesses these two catalytic activities. For instance we observed a reciprocal competitive inhibition of degradation of endo- and aminopeptidase substrates. Some other clues I will give later.

A preferential degradation of in vivo short lived cytosol proteins was clearly demonstrated for cathepsin L. Differences in the degradation of short and long lived cytosol proteins were also measured for cathepsin B1 and B3, but not so markedly. The data on Km show the great affinity of cathepsin L for protein substrates.

Leupeptin is a powerful inhibitor for cathepsin L and B1 (Fig. 1). The complete inhibition of cathepsin L and B1 in these experiments is caused by a tenfold excess of leupeptin, with respect to the molar ratio of enzyme and inhibitor. Cathepsin B3 does not show such a sensitivity to leupeptin. The best inhibitor of cathepsin B3 is the chloromethyl ketone of leucin, which has a very low inhibitory function for cathepsin L and B1. Exactly the same results were obtained from experiments using synthetic substrates instead of azocasein. The inhibitory constants of leupeptin and cathepsin B1 and B3 differ from each other (Table 2). The higher inhibitory constant of cathepsin B3 holds good not only for the endopeptidase substrate but also for the aminopeptidase substrate.

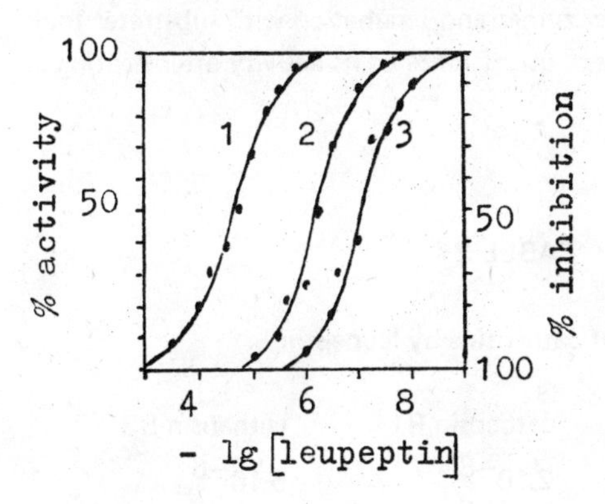

Fig. 1 Inhibition of cathepsins by leupeptin substrate: azocasein 2,5 %

pH 6,0

1 cathepsin B3 : 5.10^{-6}M
2 cathepsin B1 : 5.10^{-6}M
3 cathepsin L1 : 3.10^{-7}M

Disc electrophoresis of cathepsin L shows a lot of protein bands. A similar result was given by isoelectric focusing. But all eluates of these protein bands possess proteolytic activity. We think this is a question of enzyme molecules containing different amounts of neuraminic acid or another charged compound. After inhibition with leupeptin there are detectable only one main and one weak protein band by disc electrophoresis. The band of cathepsin B1 differs from that of cathepsin L. It is located more towards the anode, whereas cathepsin B3 lies more cathodically because of its isoelectric point of 7.5. Cathepsin B3 gives two protein bands — one of them is a double band. All of them possess both endopeptidase and aminopeptidase activity.

Also by measuring the temperature stability we are able to discriminate between the three cathepsins B1, B3, and L. The cathepsins were heated for

30 minutes at the given temperature (Fig. 2) without their substrates. Thereafter the dilution of the enzymes and incubation with substrates took place at 37°C. Cathepsin B3 kept intact 80 % of its activity after heating at 60°C.

TABLE 2

Inhibition of cathepsins by leupeptin

Ki (Moles/l)	cathepsin B1	cathepsin B3
Bz–Arg$NHC_{10}H_7$	2.10^{-8}	5.10^{-6}
Bz–ArgOEt	(5.10^{-8})	(8.10^{-6})
Lys$NHC_{10}H_7$	–	5.10^{-6}

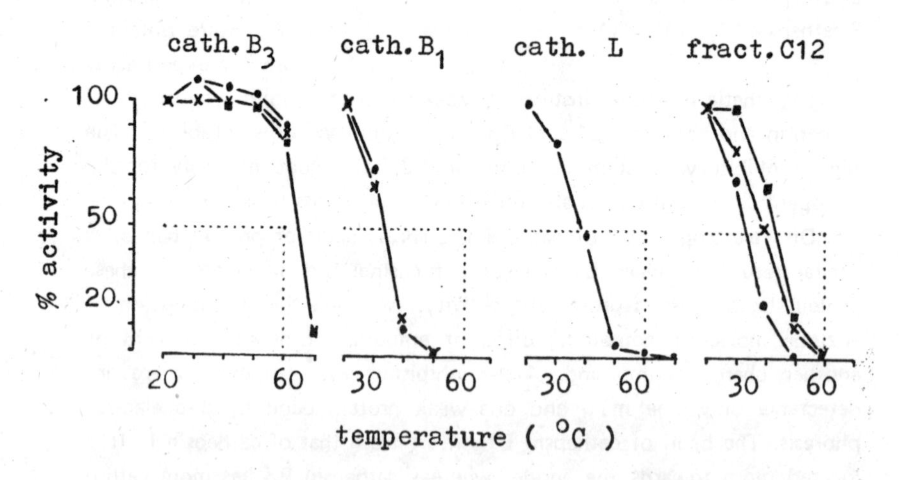

Fig. 2 Temperature stability of cathepsins. Cathepsins (~10^{-4}M) were incubated (30 min) alone in phosphate buffer pH 6.0 at the given temperature, followed by dilution and incubation with substrates, glutathione and EDTA at 37°C and pH 6.0. Substrates: –.– azocasein; –x– BANA; –·– LeuNa; subsidiary line . . . . . .

I would like to add, that this holds true not only for the hydrolysis of the proteinase substrates azocasein and BANA but also for the amino-peptidase substrate LeuNA. Cathepsin B1 possesses only 10 % of its activity after heating at 40°C. At this temperature cathepsin L preserved nearly 50 % of its activity. On the right hand of Fig. 2 you can see the temperature stability of a proteinase fraction which contains a lot of other enzymes and proteins. Nevertheless the stability of the proteolytic activity is low.

ACTIVITY OF ACID AND NEUTRAL PROTEINASES IN BONE MARROW OF BUSULFAN TREATED RATS

V. Cotič and D. Lebez

Department of Biochemistry, J. Stefan Institute, University of Ljubljana, Ljubljana, Yugoslavia.

Bone marrow is an organ which, under the influence of various factors, quickly changes its functional activity. This organ is known to contain a relatively high content of acid and neutral proteinases (1,2).

In our experiments we have investigated the influence of busulfan (myleran) on the function of bone marrow and changes of proteinase activity in this connection. Busulfan is a substance with an alkylating effect. Its action in bone marrow is seen as depressive effect on proliferation and is widely used for treatment of chronic myeloic leukemia. It is known that the depressive effect is evident already at low doses, first on granulocytopoiesis and later also on erythrocytopoiesis and thrombocytopoiesis.

Male Wistar rats were used in our experiments. Busulfan oil suspension was injected intramuscularly 4 consecutive days in a daily dose of 1 mg/kg.

Femoral bone marrow was taken and homogenized in Hanks solution. In the supernatant proteolytic activity was measured by modified Anson method (3) at pH 3.5 for the acid and at pH 6.5 for the neutral proteinases (4). The substrate was 2 % hemoglobin solution.

The number of erythrocytes and of reticulocytes was determined in peripheral blood after sacrification of each animal. These results are presented in Fig. 1.

The upper curve represents the number of erythrocytes. As you can see this number remained unchanged for 30 days after the first injection. The lower curve shows the number of reticulocytes. It is evident that the number of reticulocytes increases after 7 days. This means that erythropoiesis was stimulated during first few days of the experiment. Approaching the 30 th day the number slightly drops below the normal value. This means that the erythropoiesis was hindered between 15 th and 30 th day.

The time dependence of the number of leucocytes in peripheral blood following the injection of busulfan is shown in Fig. 2. It is evident that the dose applied did not provoke the depression of granulocytopoiesis. On the contrary, bone marrow is slightly stimulated at the beginning of experiment. The same effect was noticed and reported in some clinical cases (5). Therefore the usual therapy requires long term therapy with low doses.

The upper curve shows the number of cells in bone marrow. A moderate increase in the number of cells is evident in the second week.

The time dependence of the acid proteolytic activity is shown in Fig. 3. The lower curve represents the activity of acid proteinases in the tissue which remains constant during the first week. However, at the end of the

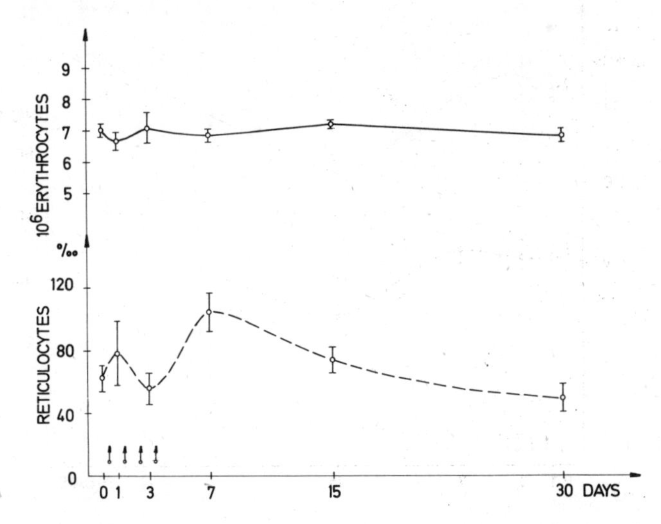

Fig. 1. Time dependent numbers of erythrocytes and reticulocytes in μl of peripheral blood of busulphan treated rats. ↑ Indicates time of application of busulphan.

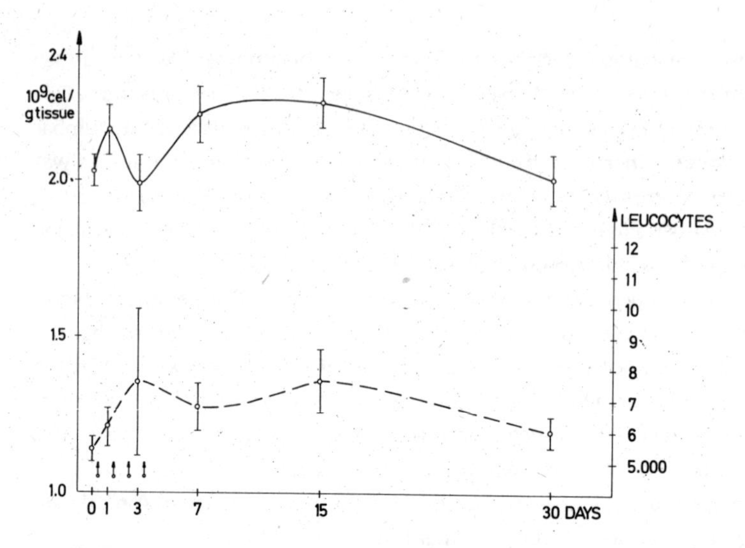

Fig. 2. Time dependent numbers of bone marrow cells in 1 g of tissue (——) and numbers of leucocytes in μl of peripheral blood (––––) of busulphan treated rats.

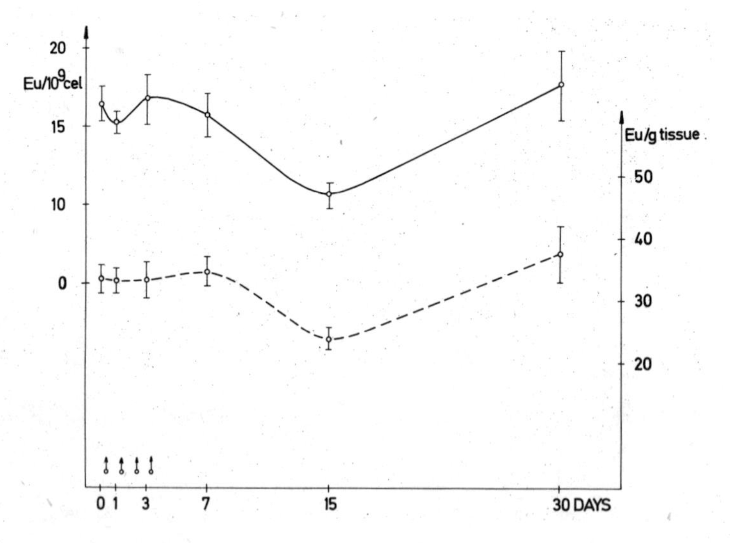

Fig. 3. Time dependent activity of acid proteinases in bone marrow of busulphan treated rats. (——) activity in 10^9 cells, (–––) activity in g of tissue.

second week a marked decrease was observed. Similar changes were observed in proteolytic activity per cell as shown in the upper curve.

It is interesting to know that a marked decrease in the activity of acid proteinases in our experiments coincides with the marked decrease in the number of bone marrow cells and in the number of leucocytes in the similar experiments of Niskanen who applied higher dose of busulfan (6). We suppose that our decrease of activity reflects the functional damage of the cells. This damage is too small, due to the lower dose, to provoke the change of the number of the cells.

In Fig. 4. we show the time dependence of the activity of neutral proteinases. The maximum of the activity was observed on the 3th day. After that the activity decreases and remains slightly under the normal value on the 15 th day.

The pronounced increase in activity of neutral proteinases coincides with the time of the excitation of bone marrow.

Fig. 5. shows even more clearly the difference in the behaviour of acid and neutral proteinases. The lower curve shows the proportion of the activity of acid and neutral proteinases. It is clearly shifted toward neutral proteinases in the first few days. Then it is near to the normal proportion. By the 15th day the neutral proteinases prevail again due to the marked decrease of acid proteinases in that time.

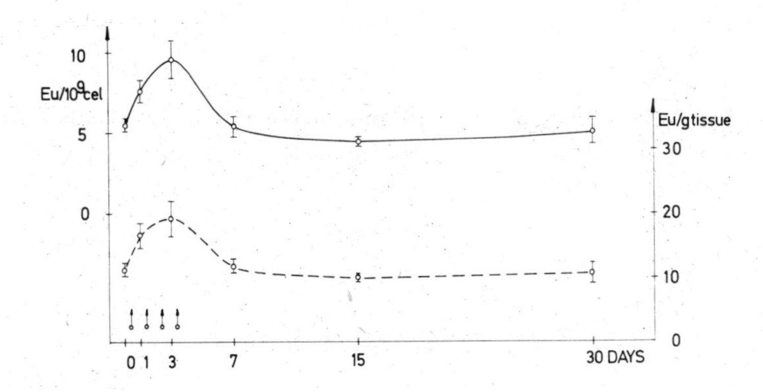

Fig. 4. Time dependent activity of neutral proteinases in bone marrow of busulphan treated rats. (——) activity in 10^9 cells, (– – –) activity in g of tissue.

The upper curve shows the concentration of soluble nitrogen compounds in bone marrow. A pronounced decrease in the nitrogen content is seen at the end of second week. Simultaneous increase in the number of cells suggests an increase of more mature cells which are smaller. This can be expected if we assume that busulfan causes damage to stem cells. Therefore the proliferation of younger cells is hindered for some time and in proliferation and maturation only slightly older cells take part.

Our results show that cells in various states of maturation contain distinct patterns of proteinases. They also suggest that in metabolic and proliferative activated cells activity of neutral proteinases increases whereas the activity of acid proteinases is increased in cells with higher catabolic processes (4).

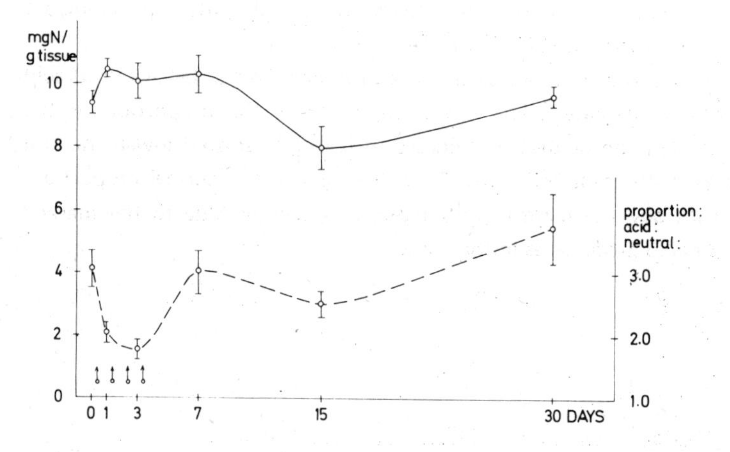

Fig. 5. Time dependent amount of soluble nitrogen compounds (——) and proportion of acid and neutral proteolytic activity in bone marrow of busulphan treated rats.

ACKNOWLEDGEMENTS

We thank Mrs. K. Leonardis and Mrs. M. Božič for their excellent technical assistence. This work was supported by a grant from the Research Council of Slovenia

REFERENCES

1. Lapresle, C. in Tissue proteinases (A.J. Barrett and J.T. Dingle edt.) pp 135-155. North Holland Publ. Co. Amsterdam – London (1971).
2. Cotič, V., Lebez, D., Iugosl. Physiol. Pharmacol. Acta 1, 141-142 (1965).
3. Anson, M.L.,J. Gen. Physiol. 22, 79-89 (1939).
4. Cotič, V. Ph. D. thesis, pp 1-168., Univerza v Ljubljani, Ljubljana (1974).
5. Ehrhart, H., Hörmann, W., Scheffel, G., and Armbröster, E.,Klin. Wschr. 48, 204-209 (1970)
6. Niskanen, E., Acta Pathol. Microbiol. Scand. 70 Suppl. 190, 1-93 (1967).

THE NEUTRAL PROTEINASES OF HUMAN SPLEEN

Alan J. Barrett

Tissue Physiology Department, Strangeways Research Laboratory, Cambridge CBI 4 RN, England.

Hedin (1) reported that bovine spleen preparations showed proteolytic activity in both acidic and alkaline conditions. Many years later, the acidic-acting peptide hydrolases of this tissue were the subject of several detailed studies, butt little atention was paid to the neutral – alkaline activities. Investigations such as those of LoSpalluto et al. (2) indicated the presence of neutral proteinases in human spleen, also, and the purpose of the present paper is to describe the results of work by Dr. Phyllis Starkey, with the author, on these enzymes.

Purification of the enzymes

Hedin (1) had found that moderate salt concentrations were required for the solubilisation of the alkaline proteinases of bovine spleen, and we found the same to apply to the human tissue. Our assays of proteolytic activity were made with azocasein as substrat (3) at pH 7.5, and tests with a variety of salts and detergents led to the selection of M-KCl containing 0.1 % Brij 35 for extraction of activity from the tissue. Insoluble material was separated from the extracts by centrifugation, and discarded.

The second stage of the purification procedure was autolysis overnight at pH 4.5 and 35°C. Full enzymic activity was retained, and the partial destruction of contaminating proteins facilitated their removal in subsequent steps. Again the preparation was centrifuged, and the pellet discarded.

The next purification step exploited the tendency of the neutral proteinases to precipitate at low ionic strength. The supernatant from autolysis was dialysed against 50 mM-sodium acetate buffer, pH 5.1. This resulted in the formation of a fine precipitate containing almost all of the neutral proteinase activity, whereas most of the protein remained in solution. The precipitate was collected and extracted with 2M-$MgCl_2$

The first ion-exchange step was with DEAE-cellulose. The enzymic activity remained in solution when the $MgCl_2$- extract was dialysed into 75 mM-Tris HCl, pH 7.8, containing 0.1 % Brij 35 (which was included in all of the column buffers), but approximately 90 % of the activity could be adsorbed on to DEAE-cellulose under these conditions. Elution with a gradient to 0.6 M-NaCl in the same buffer resulted in the elution of two separate peaks of neutral proteinase activity. The first peak was due to an elastase which could be assayed very conveniently with a synthetic substrate, benzyloxycarbonyl-L-alanine 2-naphthol ester (Z-Ala-2-ONap) that was synthesised for us by Dr. Graham Knight in our laboratory. The second peak of activity was due to an enzyme somewhat similar to chymotrypsin, which we have named „cathepsin G". Cathepsin G is routinely assayed by use of benzoyl-DL-phenylalanine 2-naphthol ester (Bz-DL-Phe-2-ONap). In both of the assays with synthetic substrates, the 2-naphthol released was detected by coupling with the diazonium salt, Fast Garnet GBC, to give a red colour ($\Delta\epsilon_{520}$ = 33,000).

The two neutral proteinases were further purified separately. The elastase pool was adjusted to pH 7.0 and diluted with 3 vols of water, for adsorption on CM-cellulose. The column was equilibrated with 50 mM-potassium phosphate buffer, pH 7.4, and subsequently eluted with a gradient of NaCl to 0.4M, in the same buffer. Finally, the elastase was run on a column of Sephadex G-75 in 50 mM-potassium phospate, pH 7.0, containing M-NaCl.

Cathepsin G, after elution from DEAE-cellulose, was dialysed against 0.20 M-potassium phosphate, pH 6.8, and run on a column of hydroxyapatite. Elution was with a gradient to 0.75 M-buffer of the same composition. The active material was run finally on Sephadex G-75 under the conditions described for elastase.

The total purification of elastase was 1300-fold and that of cathepsin G, 750-fold; yields were a few percent (i.e. several milligrams of each enzyme per kilogram of spleen).

The „stickiness" of spleen elastase and cathepsin G, their tendency to adhere to the surface of solid materials, was not entirely supressed even by the combination of M-NaCl with 0.1 % Brij 35, and therefore led to anomalous behaviour of the enzymes during purification. For example, there is no doubt that both are highly cationic proteins, so that their adsorption to DEAE-cellulose was unexpected. Also, they were more strongly retarded on Sephadex than was consistent with the molecular weight values determined by SDS-gel electrophoresis.

The elastase also was purified, on a small scale, by affinity chromatography on ovoinhibitor (from chicken egg white) covalently linked to Sepharose. The enzyme was eluted with a sodium formate buffer at pH 3.0

Properties of the proteinases

Electrophoresis. Both enzymes were run in gel electrophoresis with a discontinuous buffer system giving a running pH of 3.5 (4). The gels were stained for protein with Coomassie Brilliant Blue and also were stained for enzymic activity with naphthol ester substrates. The elastase appeared as a group of four fast-running bands, each with enzymic activity. An additional diffuse zone of less mobile material without enzymic activity appeared when enzyme samples had been allowed to autolyse.

Cathepsin G also showed four bands of protein and enzymic activity, and again autolysis led to the formation of slower running material without activity. The mobility of the active forms of the enzyme was greater than that of the elastase forms, and similar to the mobility of avian lysozyme.

In SDS-gel electrophoresis without reduction, elastase appeared as two zones corresponding to molecular weights of 27500 and 30000. Greater heterogeneity after reduction was attributable to autolysis. Without reduction, cathepsin G gaves two bands in the range 27000-30000, and reduction caused the appearance of a third, still within this range.

It was concluded from the results of electrophoresis that both the elastase and cathepsin G of human spleen occur as multiple forms differing slightly in charge and perhaps in molecular weight. The final preparations of each enzyme seemed to be essentially free from contamination by other proteins.

It did not prove possible to achieve good resolution in isoelectric focusing of the proteinases, but always the enzymes were detected at the alkaline ends of the gradients, indicating isoelectric points of 10 or above for both.

pH Optima. The elastase showed maximal activity against azo-casein and Z-Ala-2-ONap at approx. pH 8.7, whereas cathepsin G was most active against azo-casein and Bz-DL-Phe-2-2ONap at pH 7.5.

Synthetic substrates. The spleen proteinases were tested for activity against a variety of low molecular weight substrates. Neither was active against the nitroanilide or 2-naphthylamide of Bz-DL-Arg, substrates of trypsin. The elastase was rather more active than pancreatic elastase, on a weight basis, against Boc-Ala-ONap, Boc—Ala-2-ONap and Z-Ala-2-ONap; it showed no activity against derivatives of phenylalanine and tyrosine, but was active against chloroacetyl naphthol AS-D, a non-specific substrate of several proteinases and esterases.

Cathepsin G was active against Bz-Tyr-OEt, Bz-DL-Phe-2-ONap and Ac-Phe-2-ONap, although much less so than chymotrypsin. For Bz-Tyr-OEt, K_m was 2.6 mM (0.5 mM for chymotrypsin) and k_{cat} 3.1 s^{-1} (19.6 for chymotrypsin). Cathepsin G also was active against phenylpropionyl naphthol AS and chloroacetyl naphthol AS-D, but not against derivatives of alanine.

Protein substrates. In addition to azo-casein, both of the proteinases were active against cartilage proteoglycan, in solution. The action of the elastase on elastin was shown both with elastin-agarose plates (5) and by the radiochemical method of Takahashi et al. (6). On a weight basis, the elastinolytic activity was much less than that of pancreatic elastase, but nevertheless it was strong. The action of the pancreatic enzyme on elastin was much more powerfully inhibited by both EDTA (10 mM) and NaCl (1M) than that of the spleen enzyme.

Somewhat surprisingly, both of the spleen proteinases have shown the capacity to degrade native collagen. Dr. Mary Burleigh, in our laboratory, made a study of the degradation of both soluble and insoluble collagen. The elastase caused a conversion of the cross-linked β-chains to α-chains, by degrading the non-helical region of the molecule containing the cross-links; there was also further degradation of the α-chains. Cathepsin G also attacked the non-helical regions of the collagen molecule, and both proteinases acted synergistically with specific collagenase, in solution. Both enzymes also had the property of solubilising insoluble fibrous collagen, and it was significant that their action was as rapid with cartilage collagen as with tendon collagen, whereas both specific collagenase and cathepsin BI are several-fold more active against the tendon material.

A study of the specificity of action of the spleen proteinases on the oxidised B-chain of insulin is being made by Dr. Andrew Blow, in our laboratory. The patterns of cleavage show some important differences from pancreatic elastase and chymotrypsin.

Inhibitors. Both the elastase and cathepsin G were completely inactivated by diisopropyl phosphorofluoridate and phenylmethane sulphonyl fluoride, leaving little doubt that they are serine proteinases. Consistent with this was the finding that 4-chloromercuribenzoate, EDTA and pepstatin were without effect. Cathepsin G was much less sensitive to Tos-PheCH_2Cl than was chymotrypsin, but was efficiently inhibited by Z-PheCH_2Br. The elastase was inactivated by Ac-$(Ala)_2$-Pro-AlaCH_2Cl, an inhibitor of both pancreatic and leucocytic elastases. Gold thiomalate strongly inhibited both enzymes at 1 mM concentration and retained some effect even at 5 μM.

The protein proteinase inhibitors of soya bean and lima bean inhibited both of the spleen enzymes. The elastase, but not cathepsin G, was strongly inhibited by turkey ovomucoid, and was much more sensitive to chicken ovoinhibitor than was cathepsin G. Both enzymes were inhibited by α_2-macroglobulin and α_1-proteinase inhibitor from human plasma.

Antisera. A specific antiserum was raised against human spleen elastase by injection into a rabbit of the enzyme adsorbed on beads of Sepharose to which chicken ovoinhibitor had been coupled. The use of the affinity adsorbent carrier had two important advantages in that it provided an additional purification step for the antigen, and also allowed antisera to be raised with the use of very small amounts of enzyme. The antiserum gave a reaction of complete indentity with an antiserum against human leucocytic elastase that was kindly given by Dr.K.Ohlsson, when the spleen elastase was used as antigen.

A rabbit antiserum against cathepsin G was raised by direct injection of the protein, with Freund's complete adjuvant.

Discusion

The spleen elastase was immunologically identical with that of human neutrophil leucocytes. Comparison of its properties with those of the leucocyte enzyme as reported by others show close similarity in regard to electrophoretic behaviour, substrate specificity and sensitivity to inhibitors. We have no doubt that the enzymes are the same, but it is not yet certain whether the spleen elastase is derived entirely from the moderate number of neutrophils in the tissue, or perhaps also from platelets or other cells. The spleen represents an alternative starting material for the isolation of the enzyme in quantities comparable to those obtained from normal blood leucocytes.

The properties of cathepsin G strongly suggest that it is identical with the chymotrypsin-like enzyme of human neutrophils that has been described by Rindler-Ludwig and Braunsteiner (7) and Schmidt and Havemann (8). Moreover, the behaviour of the multiple forms of this enzyme in gel electrophoresis suggests that they constitute the cationic bactericidal proteins present in neutrophil granules (9). Another chymotrypsin – like enzyme named „chymase" has been detected in mast cells of several species (10–13) and this enzyme too, is similar in all essentials to cathepsin G. The availability of a specific antiserum to cathepsin G should allow the possibility of its identity with chymase to be checked.

The work of Dewald et al (9) establishes that both elastase and the enzyme that we have called cathepsin G are located in the azurophil granules of human neutrophils, and to that extent may be described as „lysosomal" enzymes. The broad substrate specificity of the enzymes, embracing all of the major structural proteins of connective tissues, underlines their possible importance in the pathological destruction of tissue elements. Clearly, however, some control must be exerted through the action of the plasma inhibitors, α_1–proteinase inhibitor and α_2-macroglobulin.

REFERENCES

1. Hedin, S.G., J. Physiol. 30, 155-175 (1904).
2. LoSpalluto, J., Fehr, K. and, Ziff, M. In: Tissue Proteinases (Barrett, A.J. and Dingle J.T. eds. North–Holland Publ. Co, Amsterdam and London 1971, pp 263–289

3. Charney, J. and, Tomarelli, R.M., J. Biol. Chem. 171, 501-505 (1947).
4. Jovin, T.M., Ann. N.Y. Acad. Sci. 209, 477-496 (1973).
5. Schumacher, G.F.B. and Schill, W.-B. Anal. Biochem. 48, 9-26 (1972).
6. Takahashi, S., Seifter, S. and Young, F.C., Biochim. Biophys. Acta 327, 138–145 (1973).
7. Rindler-Ludwig, R. and Braunsteiner, H., Biochim. Biophys. Acta 379, 606-617 (1974).
8. Schmidt, W. and Havemann K., Hoppe-Seyler's Z. Physiol. Chem. 355, 1077–723 (1974).
9. Dewald, B., Rindler-Ludwig, R., Bretz, U., Baggiolini, M., J. Exp. Med. 1077–1082 (1974).
10. Lagunoff, D. and Benditt, E.P. Biochemistry, 3, 1427-1431 (1964).
11. Pastan, I. and Almqvist, S., J. Biol. Chem. 241, 5090-5094 (1966).
12. Kawiak, J., Vensel, W.H., Komender, J. and Barnard, E.A., Biochim. Biophys. Acta 225, 172-187 (1971).
13. Vensel, W.H., Komender, J. and Barnard, E.A., Biochim. Biophys. Acta 250, 395-407 (1971).

ISOLATION AND SOME PROPERTIES OF HORSE LEUCOCYTE NEUTRAL PROTEASES

Aleksander Koj, Adam Dubin and Jerzy Chudzik

Institute of Molecular Biology, Jagiellonian University, 31-001 Krakow, Poland

SUMMARY

Horse blood leucocyte granules were found to contain three elastase-like neutral proteases (m.w. 38.000, 24.000 and 20.500, pI 5.3, 8.8 and above 10, respectively). The latter two enzymes were purified to apparent homogeneity by subcellular fractionation, Sephadex G-75 filtration and column chromatography on CM-Sephadex. Their specificity was examined using natural and synthetic substrates. Structure of the active centre was investigated using DFP, isocyanates, photoinactivation with Rose Bengal and iodination with ICl.

INTRODUCTION

Recent studies indicate that azurophil granules from human polymorphonuclear leucocytes contain three groups of proteases: collagenase (1-3), an elastase-like enzyme (4,5) and a chymotrypsin-like enzyme (6-8). Species-dependent differences in the pattern of leucocyte proteases can be expected taking into account studies of human, rabbit (9) and pig leucocytes (10).

MATERIALS AND METHODS

Horse leucocytes were isolated from fresh citrate blood and subjected to subcellular fractionation as described previously (11,12). Almost all neutral protease activity was recovered in saline extracts from the granular fraction. Protease activity was determined with casein at pH 7.4 (11) but in the presence of 2M urea. Esterolytic activities were determined with synthetic substrates of elastase: Z-L-alanine-4-nitrophenyl ester (Z-L-AlaNPE) according to Janoff (13), or with N-acetyl-$(L\text{-alanine})_3$ methyl ester ($AcAla_3OMe$) according to Bieth and Meyer (14). Polyacrylamide gel electrophresis at pH 4.5 was carried out as described by Reisfeld et al. (15), and at pH 7.0 with 0.1 % SDS according to Weber and Osborn (16). Protein iodination was accomplished by the ICl method of McFarlane (17) using carrier-free $Na^{131}I$, 10 mCi/ml, supplied by IBJ (Swierk, Poland). Samples of resorcin fuchsin-elastin and pacreatopeptidase E (elastase I, EC 3.4.21.11) were kindly provided by Dr. W. Ardelt (Warsaw). Octyl and butyl isocyanates were synthesized by Dr. Z. Moskal (Krakow). All other compounds and reagents were high-purity preparations obtained from commercial sources.

RESULTS

Filtration of the leucocyte granule extract on a Sephadex G-75 column resulted in significant purification of neutral proteases and separation of two peaks exhibiting proteolytic and esterolytic activity (Fig. 1). When active fractions were suitably pooled, concentrated and subjected to molecular sieving on a Sephadex G-75 column standardized with ovalbumin, pepsin, chymotrypsin, trypsin, myoglobin and cytochrome C, the molecular weight of leucocyte proteases was estimated as being approximately 38.000 (protease 1 – minor component) and 22.000 (protease 2 – principal component).

Isoelectric focusing of the full granule extract also allowed a separation of two neutral proteases (Fig. 2). The enzyme showing isoelectric point pI=5.3 was identified, as protease 1, while the second peak at pI = 8.8 corresponded to protease 2. However, when protease 2 partly purified on

---

[1] Full description of the experiments will be published elsewhere.

Sephadex G-75 was subjected to polyacrylamide gel electrophoresis at pH 4.5 followed by gel slicing and measuring proteolytic activity, two enzymatic components were always present (designated as protease 2A and 2B). It was established that the faster migrating component showed an isoelectric point above pH 10 and was inactivated during isoelectric focusing.

Separation of leucocyte proteases 2 A and 2B, and their final purification, was achieved by ion-exchange chromatography on a CM-Sephadex column (Fig. 3). Purity of these preparations was confirmed by polyacrylamide gel electrophoresis in the presence of SDS. With suitable protein standards their molecular weights were estimated as 24.000 (protease 2A) and 20.500 (protease 2B). The difference in molecular weight is small and probably for that reason it was not detected during gel filtration on Sephadex G-75. The purification procedure of the proteases is summarized in Table 1. It is evident that proteases 2A and 2B were purified over 100-fold with a combined yield of approximately 15 %.

Purified proteases 2A and 2B digest casein readily at pH 7.4, the activity being doubled in the presence of 2M urea. Protease 2A shows a lower affinity toward casein than 2B, as indicated by K_m values determined from

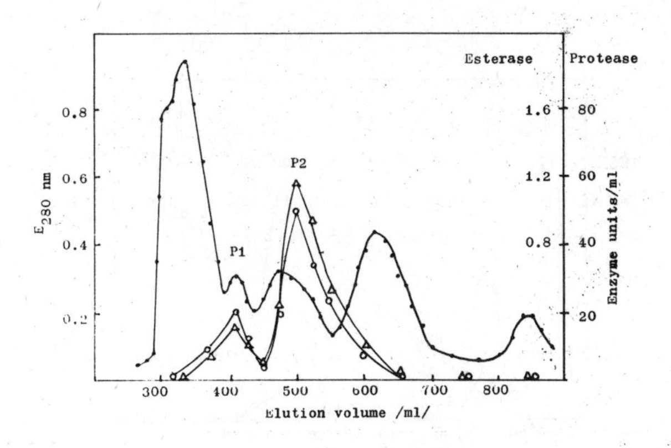

Fig. 1. Sephadex G-75 chromatography of leucocyte granule extract. About 150 mg proteins in 3 ml were applied to a column (3.5x90 cm) of Sephadex G-75 equilibrated with 0.05 M phosphate buffer pH 6.0 containing 0.15 M NaCl. The column was eluted with this buffer, 5 ml fractions being collected. •——• E_{280}; Δ-Δ-Δ caseinolytic activity; o—o—o Z-L-AlaNPE esterase activity. P1 – protease 1; P2 – protease 2.

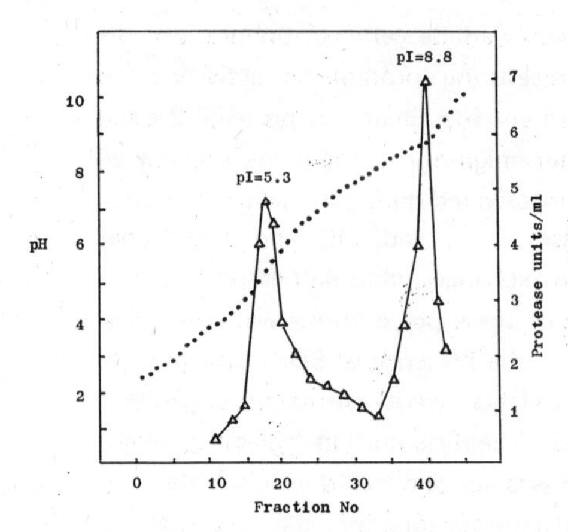

Fig. 2. Isoelectric focusing of leucocyte granule extract. About 15 mg proteins of full leucocyte granule extract were placed in a 110 ml column in continuous sucrose gradient with 1 % Ampholine. After 48 h(300V, 0.5 mA) 2.5 ml fractions were collected for the determination of pH (. . .) and caseinolytic activity (Δ-Δ-Δ).

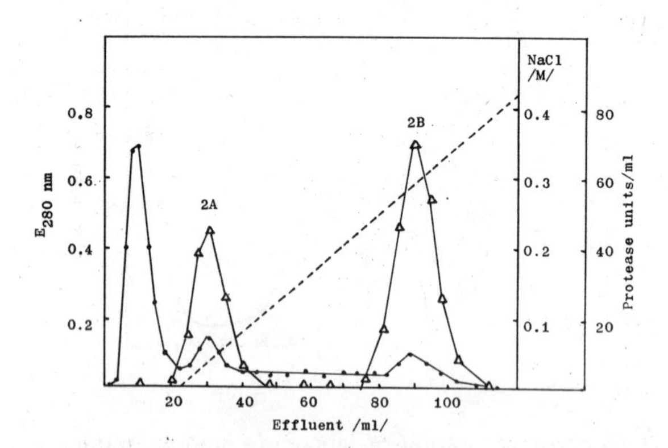

Fig.3. Ion exchange chromatography of protease 2. About 10 mg proteins of partly purified enzyme after Sephadex G-75 filtration were applied to a CM-Sephadex A-25 column (2x20 cm) equilibrated with 0.05 M phosphate buffer pH 7.0. Elution in this buffer in a linear NaCl gradient (●——●); o–o E_{280}; Δ-Δ-Δ caseinolytic activity

the Lineweaver-Burk plot (12.8 and 6.0 mg/ml, respectively). Both proteases exhibit optimum activity against casein at pH 7.4, but even at pH 5.5 or 9.0 they degrade this protein at approximately 25 % of the maximum rate.

When comparing hydrolysis of various proteins at pH 7.4 some further differences in enzymatic activity were detected, since protease 2A preferentially degrades casein and haemoglobin, while protease 2B – casein and fibrinogen (Table 2).

Both proteases are able to digest fuchsin-stained elastin, although at a much slower rate than pancreatopeptidase E. Comparison of esterolytic and caseinolytic activities of leucocyte enzymes and some commercial pancreatic proteases is shown in Table 3. Z-L-AlaNPE is readily decomposed by all enzymes examined. On the other hand, $AcAla_3$ OMe, being a highly specific substrate, is significantly hydrolyzed only by horse leucocyte proteases and pancreatic elastase.

TABLE 1.

Purification of neutral protease 2 from horse polymorphonuclear leucocytes

Purification step	Protein Total (mg)	Protein Yield (%)	Caseinolytic activity Total (units)	Caseinolytic activity Yield (%)	Sp. act. (units/mg)	Purity
Cell lysate	1505.0	100.0	3115	100	2	1.0
Granule extract	141.0	9.40	3807	124	27	13.5
Sephadex G-75	9.22	0.61	1411	45	153	76.5
CM-Sephadex						
peak A	0.89	0.06	218	7	245	122.5
peak B	0.68	0.04	265	8	390	145.0

Protease determined with casein at pH 7.4. Heparin (1000 units per ml) was added to the cell lysate prior to determination in order to reduce the effect of cell sap inhibitors. Paradoxical yield (124 %) in the granule extract is due to removal of inhibitors during subcellular fractionation. Some additional 10 % of caseinolytic activity was recovered in protease 1 peak after Sephadex G-75 column (cf. Fig. 1).

TABLE 2

Degradation rates of various proteins by purified horse leucocyte proteases 2A and 2B.

	Protease 2A		Protease 2B	
Substrate	Specific activity	Relative activity	Specific activity	Relative activity
Casein	170	100	580	100
Haemoglobin	156	92	183	32
Fibrinogen	82	48	398	69
Serum albumin	62	26	108	19

Proteins were digested at pH 7.4 in the presence of urea (2M final concentration). Specific activities are expressed in proteolytic units (ΔE_{280}x10) per mg enzyme protein. Relative activities are calculated assuming specific activity against casein = 100.

Further experiments proved that 2 mM di-isopropylfluorophosphate (DFP) completely blocks both caseinolytic and esterolytic activities of purified proteases 2A and 2B, as well as of the full granule extract. Hence it may be concluded that these enzymes belong to the group of serine proteases.

Brown and Wold (18,19) introduced alkyl isocyanates as active site specific reagents for studying the structure of the substrate binding pocket of serine pr oteases. These authors observed that octyl isocyanate preferentially inactivates chymotrypsin, while butyl isocyanate shows greater efficiency toward elastase. We fully confirmed these observations and we found conspicuous resemblance between elastase and leucocyte protease 2 in the sensitivity to this inhibitor (Table 4).

It is known that pancreatic serine proteases also contain histidine in the active centre. A rather specific modification of the His residue occurs during photooxidation in the presence of sensitizing dyes, such as Rose Bengal (20, 21). We illuminated samples of trypsin, pancreatic elastase and leucocyte protease 2 with Rose Bengal at pH ranging from 5 to 9. Then the inactivation rate constants were calculated for various pH and plotted in Fig. 4. All the experimental curves resemble the theoretical ionization curve of imidazole and this confirms the similarity of the active centre of leucocyte and pancreatic proteases.

Exposure of the enzyme molecule to ICl leads to modification of tyrosine residues, although Cys, His, Met and Try may also be affected (22). Fig. 5 shows the relationship between the number of iodine atoms substituted per enzyme molecule and progressive inactivation of proteases. Significant differences between proteases 2A and 2B in the sensitivity to iodination indicate some differences in the structure of active centres.

TABLE 3

Comparison of esterolytic and caseinolytic activities of horse leucocyte proteases and some pancreatic proteases

	Z-L-AlaNPE		$AcAla_3OMe$	
	K_m(mM)	(E:P)x100	K_m(mM)	(E:P)x100
Leucocyte protease 1	0.714	1.95	1.60	7.52
Leucocyte protease 2A	0.114	0.94	5.55	5.19
Leucocyte protease 2B	0.178	1.76	0.98	5.84
Elastase I	0.133	0.83	1.02	25.00
Chymotrypsin	0.143	0.26	66.70	0.37
Trypsin	0.083	0.19	No activity	

Hydrolysis of Z-L-AlaNPE (0.2-0.025 mM) was determined at pH 6.5 and that of $AcAla_3OMe$ (2.0-0.2 mM) at pH 7.4. The initial velocities were used for calculating K_m from the Lineweaver-Burk plot. Caseinolytic activity was determined at pH 7.4 with urea for leucocyte proteases and without urea for pancreatic enzymes. The ratio of esterolytic: proteolytic activity (E:P) was calculated by dividing the initial rate of Z-L-Ala_3NPE or $AcAla_3OMe$ hydrolysis by caseinolytic activity of a given enzyme solution.

TABLE 4

Differential inhibition of some proteases by butyl and octyl isocyanates

Enzyme	Per cent of remaining activity		
	Butyl isocyanate	Octyl isocyanate	BIC : OIC ratio
Trypsin	86	87	0,99
Chymotrypsin	30	5	6.00
Elastase I	14	86	0,16
Leucocyte protease 2	12	78	0,15

The enzymes were diluted with 0.1 M Tris–HCl buffer pH 7.5 to the concentration of $2x10^{-6}$ – $6x10^{-6}$M and incubated with butyl (BIC) or octyl (OIC) isocyanates at the reagent to enzyme molar ratio 50:1. Then the residual proteolytic activity was determined and BIC:OIC ratio was calculated.

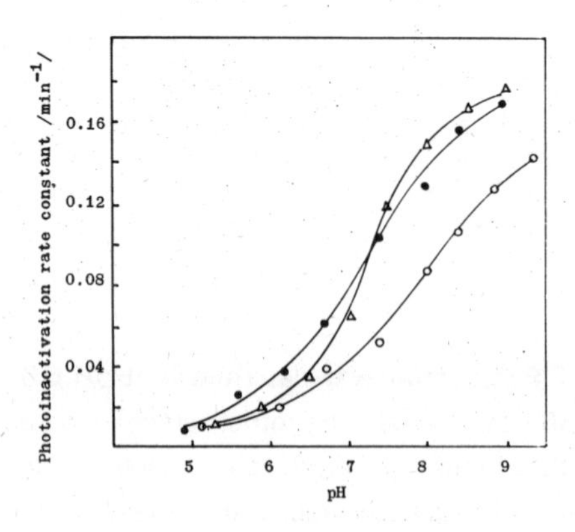

Fig. 4. The effect of pH on photoinactivation of trypsin (Δ-Δ-Δ), leucocyte protease 2 (●-●-●) and pacreatic elastase (o–o–o) in the presence of Rose Bengal (10ug per ml of 0.1 M Tris-acetate buffer).

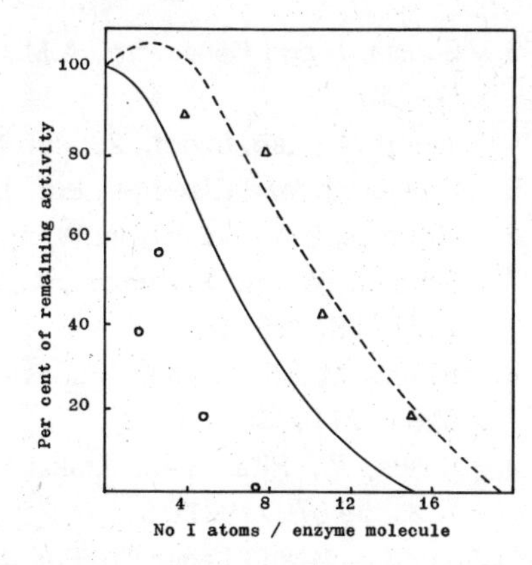

Fig. 5. Inactivation of trypsin (——), pancreatic elastase (–––), leucocytes protease 2A (△) and 2B (o) during iodination with ICl. The curves for trypsin and pancreatic elastase are based on a large number of measurements but experimental points are omitted for the sake of clarity.

DISCUSSION

Our experiments demonstrated that cytoplasmic granules of horse blood leucocytes contain at least 3 neutral proteases differing in principal molecular parameters. They show broad specificity and broad pH optima, and thus may be involved in degradation of various tissue and exogenous proteins in both physiological and pathological conditions. Since they also attack elastin and some synthetic substrates of pancreatic elastase, they should be classified as elastase-like intracellular (lysosomal) proteases. The observed sensitivity to active site specific inhibitors indicates that their active centre includes serine and histidine. At the same time they show distinct structural and catalytic features differentiating them from pancreatic and leucocyte proteases of other animal species.

REFERENCES

1. Lazarus, G.S., Daniels, J.R., Lian, J. and Burleigh, M.C., Am. J. Pathol, 68, 565-578 (1972).
2. Ohlsson, K. and Olsson, I., Eur. J. Biochem., 36, 473-480 (1973).

3. Sopata, I. and Dancewicz, A.M., Biochim. Biophys. Acta, 370, 510-523 (1974).
4. Janoff, A., Lab. Invest., 29, 458-464 (1973).
5. Ohlsson, K. and Olsson, I., Eur. J. Biochem. 42, 519-527 (1974).
6. Mounter, L.A. and Atiyeh, W., Blood, 15, 52-59 (1960).
7. Schmidt, W. and Havemann, K., Hoppe-Seyler's Z. Physiol. Chem., 355, 1077-1082 (1974).
8. Rindler-Ludwig, R. and Braunsteiner, H., Biochim. Biophys. Acta, 379, 606-617 (1975).
9. Davies, P., Rita, G.A., Krakauer, K. and Weissmann, G., Biochem.J., 123, 559-570 (1971).
10. Kopitar, M. and Lebez, D., Eur. J.Biochem. – in press.
11. Koj, A., Chudzik, J., Pajdak, W. and Dubin, A., Biochim. Biophys. Acta, 268, 199-206 (1972).
12. Dubin, A., Chudzik, J. and Koj, A., Przegl. Lek., 31, 440-442 (1974).
13. Janoff, A., Biochem. J.,114, 157-159 (1969).
14. Bieth, J. and Meyer, J.F., Analyt. Biochem., 51, 121-126 (1973).
15. Reisfeld, R.A., Lewis, U.J. and Williams, D.E., Nature, 195, 281-283 (1962).
16. Weber, K. and Osborn, M., J.Biol. Chem.,244, 4406-4412 (1969).
17. McFarlane, A.S. in Mammalian Protein Metabolism (Munro, H.N. and Allison, J.B., eds.), Academic Press, New York and London (1964), vol. I, pp. 297-341.
18. Brown, W.E. and Wold, F., Science, 174, 608-610 (1971).
19. Brown, W.E. and Wold, F., Biochemistry, 12, 828-834 (1973).
20. Westhead, E.W., Biochemistry, 4, 2139-2145 (1965).
21. Murachi, T. and Okumura, K., FEBS Lett., 40, 127-129 (1974).
22. Filmer, D.L. and Koshland, D.E. Jr., Biochem. Biophys. Res. Commun., 17, 189-195 (1964).

NEUTRAL PROTEINASES AND INHIBITORS OF LEUCOCYTES

M. Kopitar, Š. Stražiščar, V. Cotič, M. Stegnar[+] F. Gubenšek, D. Lebez

Department of Biochemistry, J. Stefan Institute, University of Ljubljana and
[+]Institute of Gerontology, IIIrd Clinic of Medicine, Ljubljana, Yugoslavia.

Neutral tissue and cell proteinases have been rather poorly investigated in comparison to acid proteinases. The reason is that it was difficult to obtain stable active samples. A higher neutral proteinase activity from leucocyte cells was obtained from granule extracts (1–4), or from so called reextracts, where the first extract that contains cytoplasm was discarded (5,6), and in the case of very acid extraction of leucocyte cell homogenate (7).

Around 1950 studies (8,9) on the interactions of proteolytic enzymes and inhibitors showed that protein inhibitors in low concentrations block tissue and humoral proteinases. Recently, many data were published on globulin inhibitors of proteinases – $alpha_1$–antitrypsin and $alpha_2$–macroglobulin types (10,11).

The latest investigations in this field have shown that for a successful isolation of neutral proteinases, the inhibitors must first be removed. There are no data concerning the presence of any well-characterized material from animal tissues capable of inhibiting cathepsin D, but there is some evidence of inhibition of cathepsin B by serum inhibitors (12).

Davies with coworkers (13) reported in their study of neutral protinases of polymorphonuclear cells that the cytoplasmic (PGS) fraction caused the strongest inhibition of neutral proteinases. Because the polymorphonuclear cells were obtained from inflammatory exudates, they suggested the.

possibility of inhibitor uptake from plasma. The inhibitory ability of the cytoplasmic fraction toward neutral proteinases was later confirmed by Janoff (14) and Movat (15).

During the isolation of plasminogen activator from leucocyte cells (16), we observed that the elution fraction from Sephadex G-100 that had a molecular weight higher than the plasminogen activator caused inhibiton. That was our first observation of the presence of an intracellular inhibitor of neutral proteinases. In the present paper we have investigated the intracellular distribution of neutral proteinases and inhibitors from pig leucocyte cells, as well as their isolation and characterization.

Leucocytes were isolated from pig blood using dextran (Pharmacia, Sweden) sedimentation of erythrocytes. To a volume of 1600 ml of blood, 100 ml of heparine (Sigma, USA) solution (1 mg/ml in 0.9 % NaCl) and 400 ml of 6 % dextran dissolved in Hank's buffer were added. Contaminating erythrocytes were removed by hypotonic destruction as reported earlier. Microscopic examination revealed 50–60 % lymphocytes and 40–50 % of polymorphonuclears.

Granule and cytoplasmic – postgranular supernatant fractions were obtained from the homogenate of pig leucocyte cells by differential centrifugation in 0,34 M sucrose (Fig. 1).

Granules and cell homogenate samples were extracted with 0.2 M Na–acetate, pH 4.5 and 0.01 M KH_2PO_4 that contained 1.1 M KCl, pH 7.0. The highest amount of nitrogen and the highest neutral proteinase activity was extracted using KCl – phosphate buffer, but when the stability of neutral proteinases of granule extracts, kept at –25 C was followed, it was found that the acid extract retained its full activity up to 4 weeks, whereas it decreased in the neutral extract. Therefore in further experiments only acid extracts were used. Cell homogenate was extracted in the same way as the granules. Beside acid enzyme activity, only a weak neutral activity toward casein was noted.

The time dependence of proteolytic activity of the granule extract and PGS toward neutral substrates is shown in Fig. 2a and 2b. In granule extracts, proteolytic activity increased with time of incubation, whereas in the PGS sample the highest activity was optimally generated after a relatively short period of incubation, only 15–30 min. After 2 hrs incubation of PGS with substrate, a complete reduction in the amount of proteolytic activity occurred. Explanations for the reduction of neutral proteolytic activity of PGS samples with prolonged time of hydrolysis are still unknown. A similar observation has been already noted by Ward (17) in the case of neutral bacterial proteinases. We proposed the possibility that the addition of substrate to the enzyme–inhibitor complex could have resulted in partial and temporary dissociation of this complex.

The effect of PGS on the proteolytic activity on granule extract is shown in Table 1. PGS added to granule extract completely inhibits neutral proteinase activity of granule extract, measured towards all protein substrates, except in the case of casein. With respect to the inhibitory action of PGS on granule extract at neutral pH, the pH stability was determined (Fig. 3). pH stability of PGS inhibitors was 100 %, within the pH range 6–8. Even after 2 hrs of incubation in buffer sulutions, the inhibitor ability was

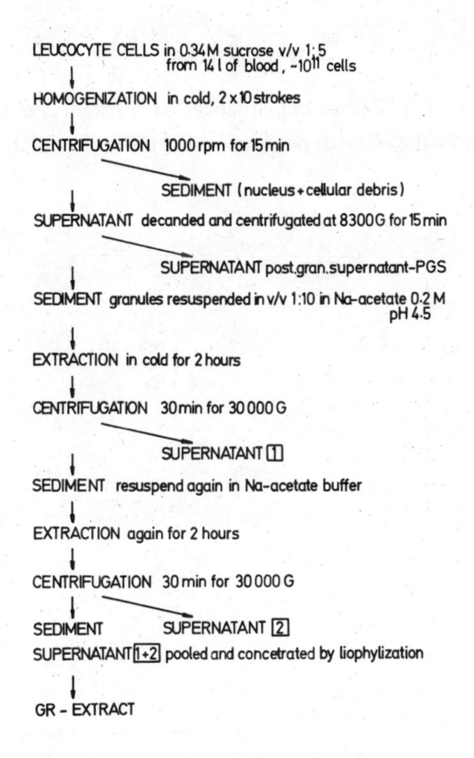

Fig. 1. Shematic diagram of the preparation of granule extract and post-granular supernatant samples.

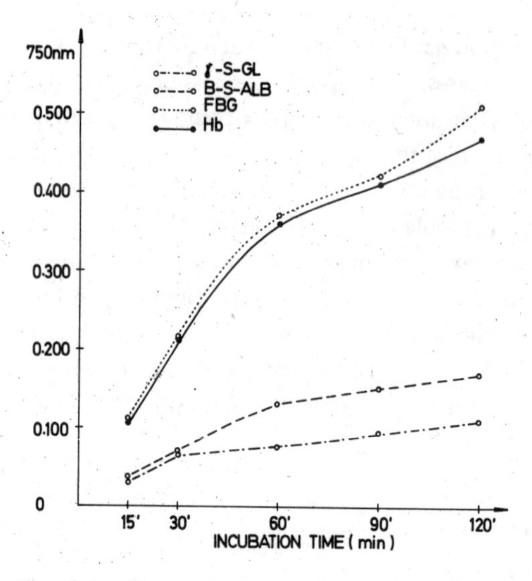

Fig. 2A. Time dependence of proteolytic activity of granule extract. Reproduced with permission from the Europ. J. Biochem.

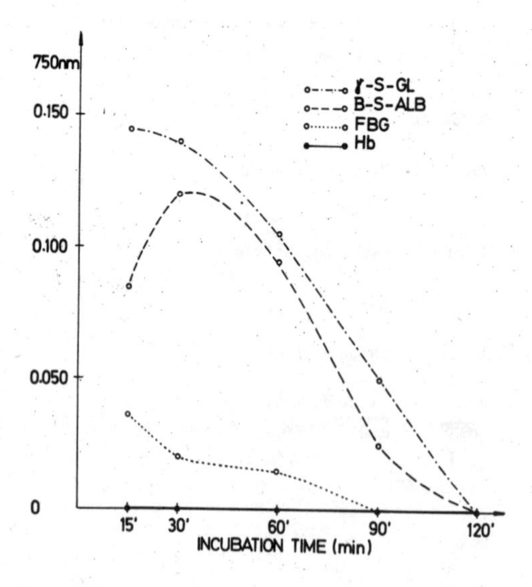

Fig. 2B. Time dependence of proteolytic activity of post-granular supernatant. Reproduced with permision from the Europ. J. Biochem.

TABLE 1.

Inhibitory effect of post granular supernatant (PGS fraction) on granule (GR) extract. γN/ml of GR extract was 820 and of PGS 3500.

Substrate	0.2 ml (GR+H_2O) O.D. 750/2 h	0.2 ml (GR+PGS) O.D. 750/2 h	0.2 ml (PGS+H_2O) O.D. 750/2 h
Hemoglobin pH 3.5	0.382	0.412	0.204
Hemoglobin pH 7.5	0.300	0.000	0.000
Fibrinogen pH 7.5[a]	0.440	0.000	0.000
Plasminogen free Fibrinogen pH 7.5	0.340	0.000	0.000
with plasminogen γ–serum globulin	0.030	0.000	0.000
pH 7.5 α–serum globulin	0.080	0.000	0.000
pH 7.5 Bovine serum	0.160	0.000	0.000
albumin pH 7.5 Casein pH 3.0	0.090	0.070	0.132
Casein pH 7.5	0.800	0.105	0.090

[a] 65 % clottable

completely retained. Maximal inactivation of inhibitory activity occurred within the pH range 3–4. At pH 5.0 a 40 % inactivation is still observed. Therefore, in all metabolic processes whenever a possible pH decrease inside the cells occurs, a reduction of intracellular ability of neutral inhibitors could also be expected. Curve B represents the effect of buffers on granule extract.

In Fig. 4 is shown the extent of inhibition of granule extract on the concentration of PGS, measured by Anson's method on hemoglobin and fibrinogen as substrates. In Fig. 5a, 5b the same effect was given by Astrup's plate technique on fibrinogen with plasminogen and on fibrinogen without plasminogen. By both techniques it is evident that the inhibition depends on the inhibitor concentration.

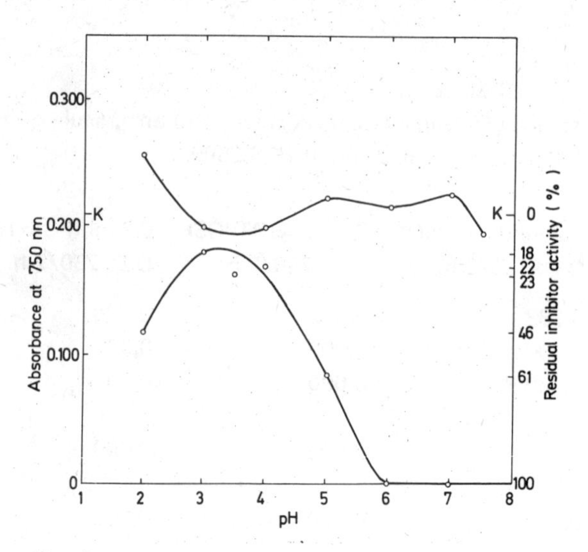

Fig. 3. Effects of pH on post-granular supernatant inhibitory activity (curve A) and on the proteolytic activity of granule extract (curve B). Reproduced with permission from the Europ. J. Biochem.

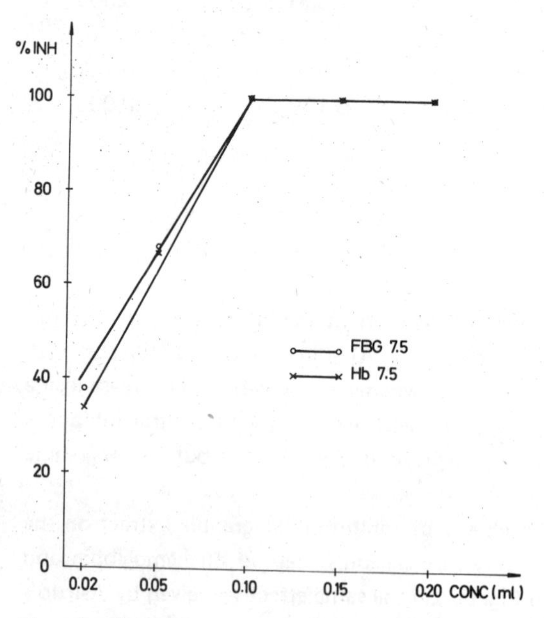

Fig. 4. Inhibition of neutral proteinases (GR—extract) by different concentrations of PGS samples. Reproduced with permission from the Europ. J. Biochem.

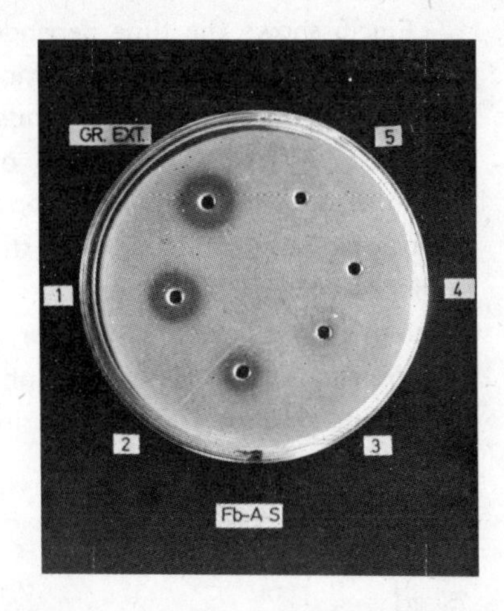

Fig. 5A. Granule extract and mixtures of granule extract and inhibitor (PGS) in ratios 1=10:1, 2=10:2, 3=10:3, 4=10:4, 5=10:5, tested on 0.1 % fibrinogen — 1 % agar plate, containing plasminogen.

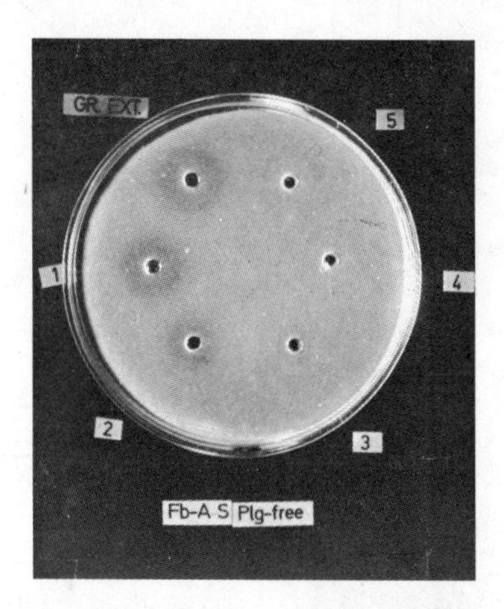

Fig. 5B. Granule extract and mixtures of granule extract and inhibitor (PGS) in ratios: 1=10:1, 2=10:2, 3=10:3, 4=10:4, 5=10:5, tested on 0.1 % fibrinogen — 1 % agar plate without plasminogen.

Fig. 6 shows the time dependence of the inhibition effect of PGS on granule extract, measured on hemoglobin and fibrinogen as substrates. The inhibition effect is not time dependent.

Table 2 gives the results of inhibition of PGS on trypsin and chymotrypsin measured on bovine serum albumin, fibrinogen and hemoglobin, and compared with the inhibition of $alpha_1$ – antitrypsin. The results of comparative inhibition clearly show that the leucocyte inhibitor is different from the serum inhibitor; this finding is in agreement with Janoff's (14) observation in the case of inhibiton of leucocyte cytosol inhibitor on trypsin and chymotrypsin when using synthetic substrates.

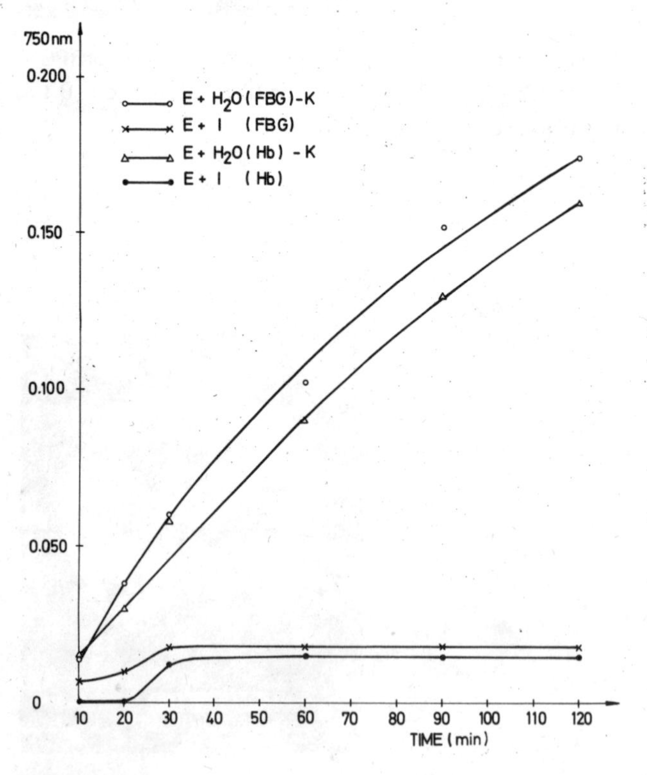

Fig. 6. The effect of inhibition of PGS sample on neutral proteinases (GR extract) at different time of hydrolysis. Reproduced with permission from the Europ. J. Biochem.

TABLE 2.

Inhibition of trypsin and chymotrypsin with PGS fraction and α_1AT. Enzyme activity was measured on the following protein substrates. Time of incubation was 10 min.

Ratio enzyme–inhibitor 1:1	Bov. ser. alb. O.D. 750	Fibrinogen O.D. 750	Hemoglobin O.D. 750
Trypsin + H_2O	0,625	0.640	0.510
Trypsin + PGS	0.430	0.620	0.500
Trypsin + α_1AT	0.310	0.240	0.380
Chymotrypsin + H_2O	0.425	0.620	0.455
Chymotrypsin + PGS	0.285	0.600	0.455
Chymotrypsin + α_1AT	0.350	0.510	0.400

TABLE 3.

Temperature dependence of PGS inhibitors, after 5 min of incubation at various temperatures. The remained proteolytic activity was tested on GR extract with hemoglobin pH 7.5, as substrate. Control value – 0.2 ml GR extract + 0.2 ml H_2O was determined under the same condition, is 0.230.

Temperature ^oC	0.2 ml (PGS + GR extr.) O.D. 750/2 hrs	% of residual inhibitor activity
100	0.225	3
80	0.196	15
60	0.135	40

Table 3 shows the stability of PGS inhibitors at various temperatures. After 5 min incubation at 100^oC the inhibitory ability was completely lost, at 80^oC approximately 85 % of inhibitory activity was inactivated, and at 60° about 60 %.

Gel filtration of a PGS sample on a column of Sephadex G–100 (Pharmacia, Sweden) is shown in Fig. 7. A pronounced inhibitory effect on neutral proteinase activity was obtained in eluted fractions 18–22 and another less pronounced inhibitory effect was found in fractions 30–33. The main inhibitor, which represents 80–90 % of the inhibiton capacity of PGS, was eluted from Sephadex G–100 in the same elution volume as ovalbumin and the remainder with cytochrome C. Unstained gels of fractions 18–22 were sliced into 4 parts (A, B, C, D) and extracted with 0.01 N phosphate buffer, pH 7.0. Concentrated extracts of these segments (A–D) were re-electrophoresed and inhibitor ability was determined. It was found that sample B posesses inhibitor activity.

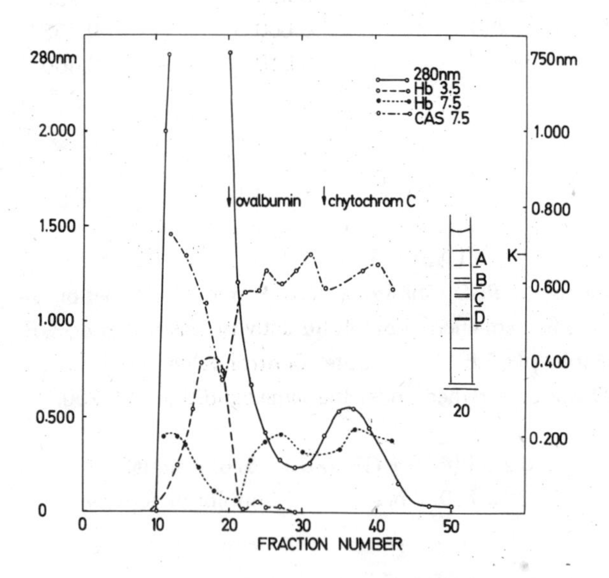

Fig. 7. Profile of gel filtration of a PGS sample (24 mgN) on a 2x50-cm Sephadex G–100 column. The flow rate was 12 ml/h. The inhibitory effect of eluted fractions on granule extracts was determined with hemoglobin and casein at pH 7.5 as substrates. The assay systems were 0.2 ml of granule extract and 0.2 ml of eluted fractions. K, control values from 0.2 ml GR extract + 0.2 ml H_2O. The acid catheptic activity (Hb pH 3.5) was determined from 0.4 ml of eluted fractions. Time of hydrolysis for all substrates was 2 hrs. Reproduced with permission from the Europ. J. Biochem.

By ion exchange chromatography on CM cellulose (Sigma, USA), we obtained electrophoretically rather pure inhibitor (Fig. 8). The elution was performed with 0.05 M Na-acetate buffer, pH 5.7, and after elution of the first protein peak, the buffer was changed for 0.05 M Tris–HCl, pH 7.5, and with a linear gradient of NaCl toward 0.5 M.

The next figure (Fig. 9) shows electrofocussing in a sucrose density gradient containing pH 5–7 Ampholine. The inhibition effect was tested in all three protein peaks. Protein fractions 22–24 showed inhibitory ability.

Inhibitory effects of PGS inhibitors were tested on granule extracts, having high neutral and acid proteolytic activity (shown in Figs. 2,3 and Table 1.), and therefore these samples were separated by DEAE cellulose (Serva, Germany) chromatography in order to determine the types of granule extract proteinases. The column was equilibrated with 0.05 M Tris–HCl buffer, pH 8.0. After elution of the first protein peak, the buffer

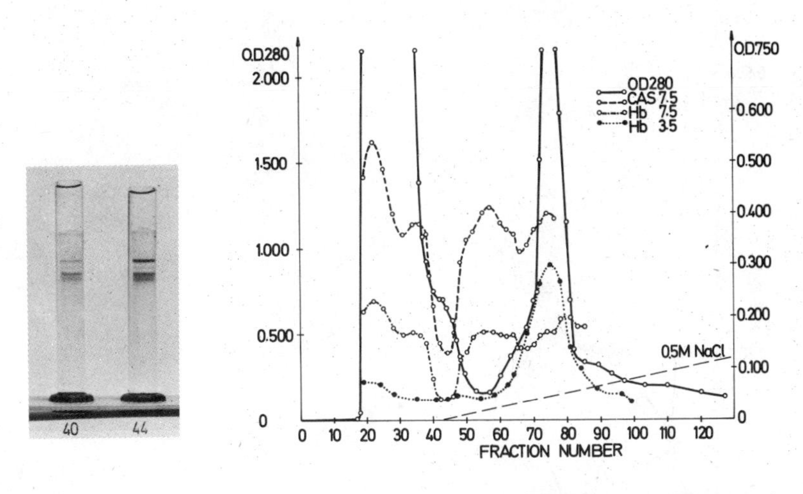

Fig. 8. Profile of ion exchange chromatography of PGS sample (122 mgN) on a 2.5x40–cm column of CM cellulose. The flow rate was 40 ml/h. The inhibitory effect of eluted fractions was tested on granule extract (0.2 ml GR extract + 0.2 ml eluted fraction) with hemoglobin and casein at pH 7.5 as substrates. K, control value from 0.2 ml GR extract + 0.2 ml H_2O. The acid catheptic activity (Hb pH 3.5) was determined from 0.4 ml of eluted fractions. Reproduced with permission from the Europ. J. Biochem.

was changed for 0.05 M Tris—HCl, containing 0.002 M $CaCl_2$, pH 8.0, and with a linear gradient of NaCl toward 0.3 M (Fig. 10). In Fig. 11 are given the electrophoretic patterns of eluted fractions from DEAE cellulose. Experiments based on the determination of the degradation of protein substrates and inhibiton with specific proteinase inhibitors showed that the granule extracts posess 3 groups of neutral and 1 group of acid proteinases. In the first group (fr. 10—13) are present elastase (Fig. 12) and plasminogen activator. It is known that both proteinases are inhibited by DFP (K and K Lab. Inc., USA) and $alpha_1$ – AT (Worthington Biochemical Co., USA), and that plasminogen activator degrades only fibrinogen containing plasminogen, whereas elastase degrades elastin, casein and hemoglobin. The second group (fr. 60—66) of neutral proteinases degrade casein and fibrinogen, but are unstable under the given experimental conditions and were therefore not

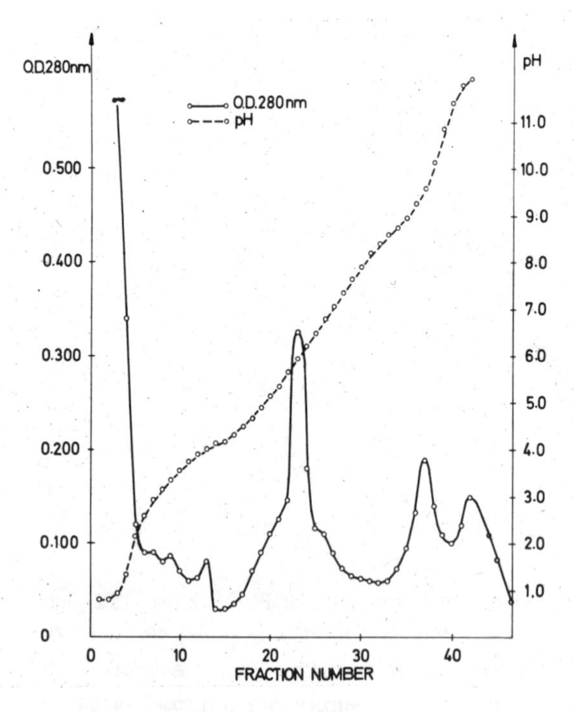

Fig. 9. Electrofocussing of inhibition fractions (40—47) from CM cellulose. Ampholites covering at pH range between pH 3—10, were used in sucrose density gradient.

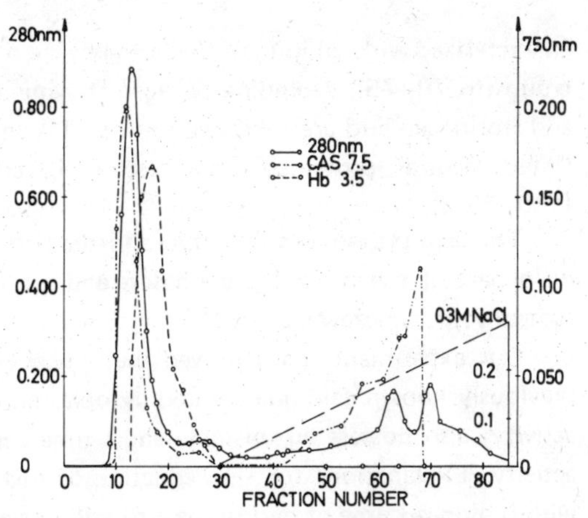

Fig. 10. Profile of ion exchange chromatography of granule extract (47 mgN) on a 2.5x40–cm DEAE cellulose column. The flow rate was 40 ml/h. In the eluted fractions the protein concentration (A 280) and proteolytic activity against hemoglobin pH 3.5 (30 min incubation) and casein pH 7.5 (2 hrs incubation) were measured. Reproduced with permission from the Europ. J. Biochem.

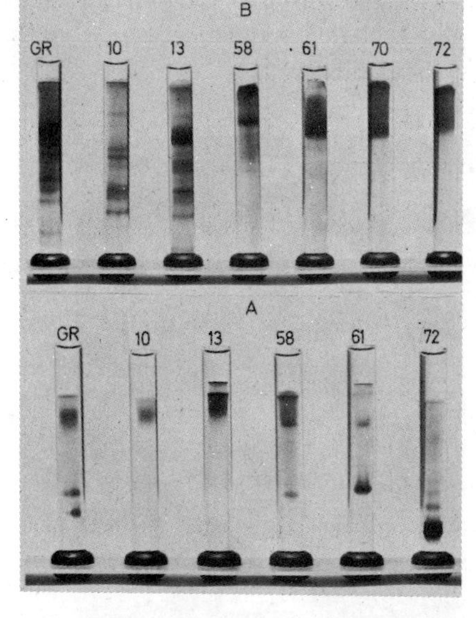

Fig. 11. Electrophoresis on polyacrylamide gel of eluted fractions of a granule extract sample, from DEAE cellulose. A buffer pH 8.4, time 45 min, 7 % gel, B buffer pH 4.4, time 120 min, 15 % gel. Current 5 mA per gel.

characterized with inhibitors. Collagenases are neutral proteinase of the third group (fr. 70–75), degrading collagen (bovine achilles tendon, Sigma, USA) and fibrinogen and are inhibited with EDTA and $alpha_1$–AT. Both types of tested neutral proteinases are also inhibited with leucocyte inhibitor (100–75 %).

The acid proteinases, are eluted in the first protein peak, soon after the main part of the neutral proteinases, and are inhibited only by pepstatin (gift of Prof. Umezawa, Japan).

Our experiments have proved that leucocyte cells, in addition to their previously known acid granule and cytoplasmic activity and granule neutral activity, also possess an unstable short time dependent neutral proteolytic activity. Explanations for the reduction of this neutral proteolytic activity with prolonged time of hydrolysis are still unclear.

The present studies of a leucocyte inhibitor of neutral proteinases were mainly oriented to the determination of its intracellular content. There was no work published dealing with the isolation and characterization of leucocyte inhibitor of neutral proteinases. In our experiments we succeeded in isolating from cytoplasm of pig leucocytes two inhibitors of neutral proteinases with molecular weight about 45 000 and 15 000. The main part of cytoplasmic inhibitory capacity – 80 %, belongs to the high molecular weight inhibitor.

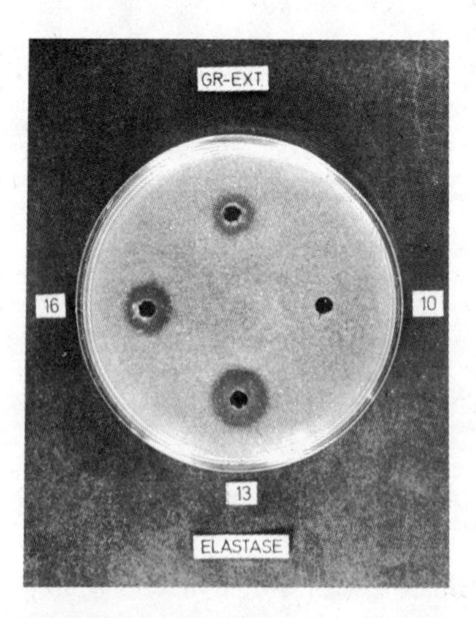

Fig. 12. Determination of elastolytic activity on agar–elastin plates. Diffusion time was 16 hrs at 37°C in termostate.

The comparative inhibition studies made with $alpha_1$ – AT and leucocyte inhibitor have proved that they are different. Also the inhibition tested comparatively with the same two inhibitors on spleen and liver neutral proteinases shows quite different effects. This is especially so in the case of spleen, where the neutral proteinases were inhibited only with leucocyte inhibitor, up to 60 %, but not with $alpha_1$–AT (20).

ACKNOWLEDGEMENTS

We thank Mrs. J. Komar and Mr. K Lindič for their excellent technical assistence. This work was supported by a grant from the Research Council of Slovenia.

REFERENCES

1. Janoff, A., Am.J. Pathol. 68, 579–591 (1972).
2. Lazarus, G., Brown, C., Daniels, J.R. and Fullmer, K.M., Science 159, 1483–1485 (1968).
3. Ohlsson, K. and Olsson, I., Europ. J. Biochem. 42, 519–527 (1974).
4. Koj, A. and Dubin, A., Abstr. 9th FEBS Meeting, August 25–30, Budapest 1974.
5. Anderson, A.J. and Irwin, C., Life Sci.13,601–612 (1973).
6. Fräki, J., Ruuskanen, O., Hopsu–Havu, V.K. and Kouvalainen, K., Hoppe Seyler's Z. Physiol. Cehm. 354, 933–943 (1973).
7. Higuchi, Y., Honda, M. and Hayashi, H., Cell Immunol.15, 100–108 (1975).
8. Grob, D., J. Gen. Physiol.33, 103–124 (1949).
9. Schulman, N.R., J. Exp. Med.95, 593–603 (1952).
10. Metais, P. and Bieth, J., Ann. Biol. Clin. 30, 413–416 (1972).
11. Barrett, A.J. and Starkey, P.M. Biochem. J. 133, 709–724 (1973).
12. Ignarro, L.J., Oronsky, A.L. and Perper, R.J., Clin. Immunol. Immunopathol.2, 36–51 (1973).
13. Davies, P., Rita, A.G., Krakauer, K. and Weissmann, G., Biochem. J. 123, 559–569 (1971).

14. Janoff, A. and Blondin, J., Proc. Soc. Exp. Biol. Med. 136,1050–1053 (1971).
15. Movat, H.Z., Steinberg, S.G., Habal, F.M., Ranadive, N.S. and Macmorine, D.R.L., Immunol. Comm. 2, 547–564 (1973).
16. Kopitar, M., Stegnar, M., Accetto, B. and Lebez, D., Thromb. Diath. Haemorrh.31, 72–85 (1974).
17. Ward, P.A., Chapitis, J., Conroy, M.C. and Lepow, I.H., J. Immunol. 110, 1003–1009 (1973).
18. Unkeless, J., Dano, K., Kellerman, M. and Reich, E., J. Biol. Chem. 249, 4295–4305 (1974).
19. Janoff, A., Ann. Rev. Resp. Dis. 105, 121–122 (1972).
20. Lebez, D., Kopitar, M. and Suhar, A., to be published elsewhere.

NEUTRAL PROTEINASES OF BOVINE LIVER

A. Suhar and D. Lebez

Department of Biochemistry, J. Stefan Institute, University of Ljubljana, Ljubljana, Yugoslavia

The activity of neutral proteinases was found in many tissues and in different subcellular fractions. In the literature there exist reports on the activity of neutral proteinases in homogenates of rat liver, spleen, muscles (1), bone marrow (2), etc. Neutral proteinases were found in the nuclear fractions of different tissues – in the rat liver (3), in the nuclei of calf thymus (4), then in the mitochondrial fraction (5) and in the lysosomal fraction of rat liver (6), in the granules of polymorphonuclear leucocytes (7), in the peroxysomal fraction of rat liver (8), etc. Some neutral proteinases were isolated and partly characterized – from the brain of rats (9), the brain cortex of guinea pigs (10), rabbit skin (11) and elsewhere.

Because of many factors which influenced their isolation (natural inhibitors, relative instability), the properties and function of neutral proteinases are still unknown.

In our experiments, isolation of neutral proteinases from bovine liver was studied. Homogenates of bovine livers were made in physiological solution (0,9 %), 10^{-3}M sodium azide) and after acidification to pH 3.6, centrifuged for 50 minutes at 4000 g. The supernatants were fractionated by precipitation to 20 %, 40 % and 60 % ethanol. The precipitates were further extracted by 0.2 M acetate buffer, pH 4.0. After dialysis of the extracts against water, the proteolytic activity toward 2 % hemoglobin at neutral pH and 2 hrs of incubation at 37°C was measured by the modified method of

Anson (12). The highest neutral proteolytic activity was observed in 20-40 % ethanol precipitate. This fracion was concentrated on an Amicon PM-10 ultrafilter and applied to an equilibrated Bio-Gel P-100 (1.8 x 60 cm) column and eluted with 0.1 M Na-acetate buffer, pH 4.0. The elution diagram (Fig. 1) shows that three proteolytically active fractions, active near neutral pH, were obtained. They were still inhomogeneous as shown by polyacrylamide gel electrophoresis (Fig. 2). One of them (fraction III), which shows the smallest acid proteinase activity, was studied in more detail. The elution volumes of this fraction corresponds to those of chymotrypsinogen A. From this result an approximate molecular weight of 25 000 was calculated. Its pH optimum was at pH 7.0 and its specific activity 300 times higher than the activity in the homogenate (this factor depends on the biological sample). After 1 hr of incubation at 37°C, no loss of activity in

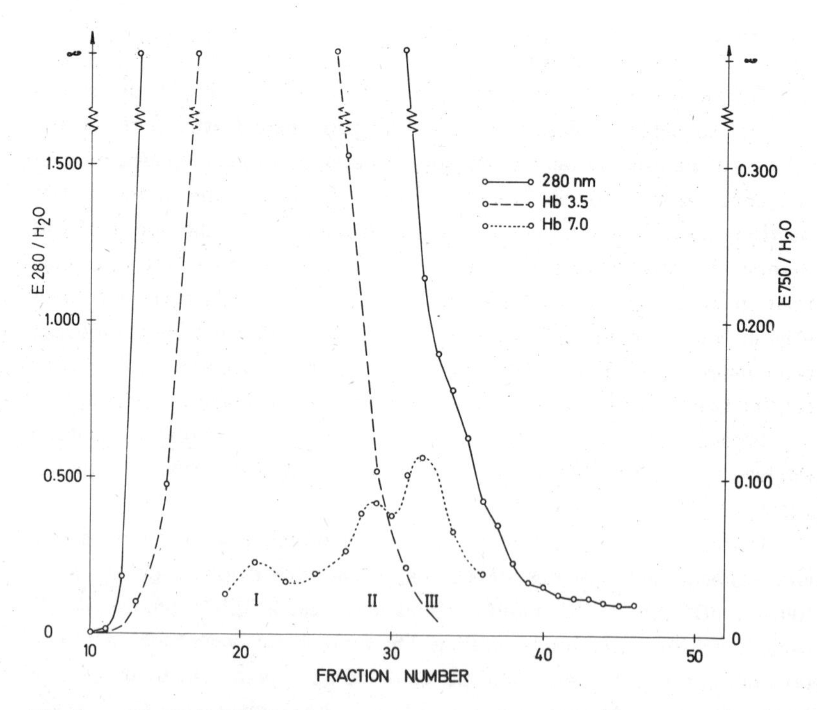

Fig. 1. Bio-Gel P-100 chromatography of the concentrated extract of 20-40 % ethanol precipitate. Dimensions of the column were 1.8 x 60 cm, flow rate was 6 ml/h.

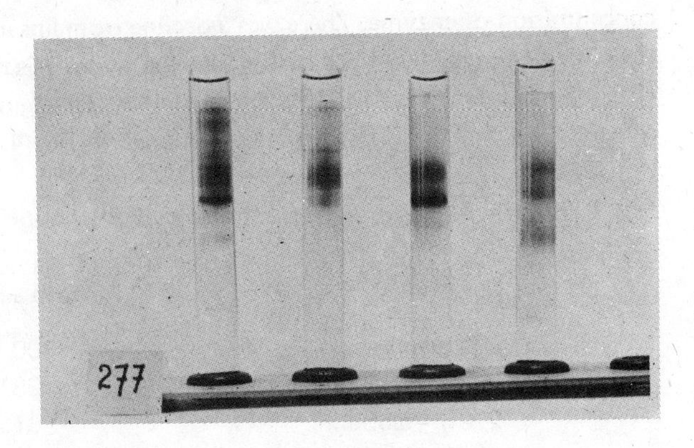

Fig. 2. Polyacrylamide gel electrophoresis at pH 4.4 of separated fractions I,II and III and of the extract of 20-40 % ethanol precipitate.

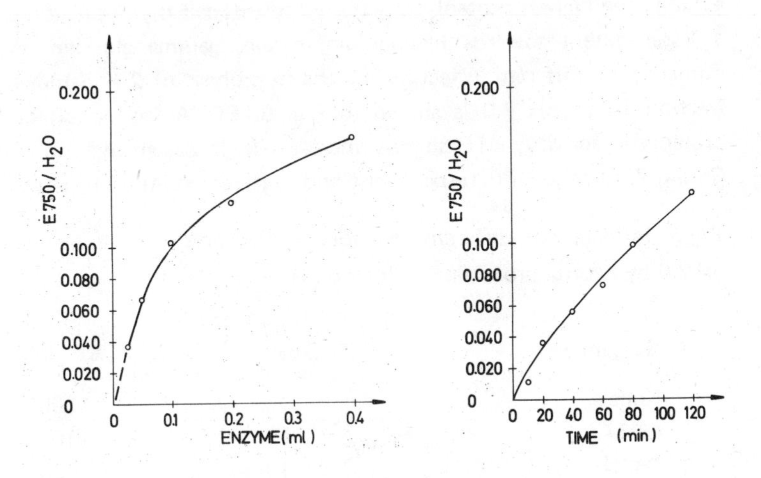

Fig. 3 a. Dependence of hydrolysis rate at pH 7.0 on the substate (2 % hemoglobin) as a function of the concentration of enzyme.

Fig. 3b. Effect of time on hydrolysis rate of the substrate (2 % hemoglobin) at pH 7.0.

the pH region 4-7 was observed. Fig. 3a show the dependence of the hydrolysis rate on the substrate (2 % hemoglobin) as a function of the concentration of enzyme. There was a decline from linearity above 0.1 ml of the enzyme used. The effect of time on the hydrolysis rate of the substrate (2 % hemoglobin) (Fig. 3) was almost linear. The hydrolysis of different substrates by fraction III at pH 7.0 is shown in Fig. 4. Neutral proteinase

Fig. 4. Hydrolysis of different substrates by FR III at pH 7.0

Substrate	Relative activity
2 % hemoglobin	100
2 % albumin	23
2 % γ –globulin	61
1 % casein	74
1 % histones	186
2 % elastin	18
2 % fibrinogen	24

exhibits the highest proteolytic activity towards calf thymus histones and to a lesser extent towards hemoglobin, casein, gamma globulin, etc. The influence of different effectors on the hydrolysis of 2 % hemoglobin by fraction III at pH 7.0 is shown in Fig. 5. EDTA has no effect on the proteolytic activity, DTE slightly increases it. It was strongly inhibited by $CoCl_2$, $ZnCl_2$ and DFP, respectively, and to lesser extent by I-acetamide.

Fig.5. Influence of different effectors on hydrolysis of 2 % hamoglobin at pH 7.0 by neutral proteinase fraction III

Reagent	Conc. (mM)	% of init. act.
None	–	100
EDTA	1	100
$CoCl_2$	1	34
$ZnCl_2$	1	11
DTE	1	124
I–acetamide	1	55
DFP	5	16

The results show that a relatively successful purification was achieved by the described methods for the isolation of neutral proteinases from bovine liver (in our case fraction III). Fraction III, which was strongly inhibited by DFP, can be considered as serine proteinase (13). On the other hand, their inhibition by $CoCl_2$, $ZnCl_2$ and I-acetamide and activation by DTE also indicated the presence of thiol groups (contamination by thiol proteinases?).

Experiments on purification and characterization of all three mentioned neutral proteinases from bovine liver are in progress. Its more detailed characterization will allow us to classify them more accurately.

ACKNOWLEDGEMENTS.

We thank Mrs. S. Košir and Mr. K. Lindič for their excellent technical assistance. This work was supported by a grant from the Research Council of Slovenia.

REFERENCES

1. Umana, C.R. and Feldman, G., Proc. Soc. Exp. Med. 138, 28 (1971).
2. Cotič, V. and Lebez, D., Jugoslav. Physiol. Pharmacol. Acta 1, 141 (1965).
3. Brostrom, C.O. and Jeffay, H., Biochim. Biophys. Acta 278, 15 (1972).
4. Furlan, M., Jericijo, M. and Suhar, A., Biochim. Biophys. Acta 167, 154 (1968).
5. Fitzpatrick, K. and Pennington, R.J., Biochem. J. 111, 29P (1969).
6. Bohley, H., Kirschke, H., Langner, J., Ansorge, S., Wiederanders, B. and Hanson, H., In: Tissue proteinases, North Holland Publ. Co., Amsterdam, London (1971), p. 187.
7. Davies, P., Rita, G.A., Krakauer, K. and Weissmann, G., Biochem. J. 123, 559 (1971).
8. Gray, R.W., Arsenis, C. and Jeffay, H., Biochim. Biophys. Acta 222, 627 (1970).
9. Marks, N. and Lajtha, A., Biochem. J. 97, 74, (1965).

10. Bosmann, H.B., Int. J. Peptide Protein Res. 5, 135 (1973).
11. Lazarus, G.S. and Barrett, A.J., Biochim. Biophys. Acta 350, 1 (1974).
12. Anson, M.L., J. Gen Physiol. 20, 565 (1937).
13. Florkin, M. and E.H. Stotz (Eds.) Comprehensive Biochemistry, Elsevier Publ. Co., Amsterdam, London, New York (1973), Vol. 13.

THE CATABOLISM OF TWO ADRENOCORTICOTROPHIN ANALOGUES FOLLOWING INTRAVENOUS INJECTION.

H.P.J. Bennett, J.R.J. Baker and C. McMartin,

Horsham Research Centre, CIBA Laboratories, HORSHAM, West Sussex RHI2 4 AB, U.K.

We have studied the catabolic fate, following intravenous injection in the rat, of two ACTH analogues, namely corticotrophin – (1–24)–tetracosapeptide (Synacthen ®, and [D–Ser^{1}, Lys^{17,18}] corticotrophin – (1–18)–Octadecapeptide amide (C–41795–Ba). Using the two analogues labelled with tritium in the tyrosine of position two it was established that C–41795–Ba has a much longer plasma and tissue half-life than Synacthen, as indicated by ion exchange chromatography of tissue extracts. This confirms results previously obtained with an isolated adrenal cell bioassay.[1]. There is a marked concentration of radioactivity in the kidneys with both peptides which consists in each case of intact peptide and several discrete fragments. However, the maximum concentration of Synacthen and its fragments occurs after 10 min and constitutes about 10 % of the injected dose whilst C–41795–Ba and its fragments reach a maximum of about 30 % of the injected dose after 30 min which has declined by 60 min. The protection from exopeptidase attack, which the D–Seryl residue at the N–terminus and amidation at the C–terminus confer on C–41795–Ba, considerably enhances its half-life. The relative importance of the kidneys as organs of clearance and degradation is also quantitatively changed by these modifications.

The overall pattern of handling which has emerged is consistent with that observed by other workers for natural ACTH where 20 % of the biological activity was recovered from the kidneys 5 min after intravenous injection [2].

ACKNOWLEDGEMENTS

We thank Dr. D.E. Brundish and Dr. R. Wade of CIBA Laboratories, Horsham, West Sussex, RHI2 4AB, U.K. for supplying us with the tritiated adrenocorticotrophin analogues.

REFERENCES

1. C. McMartin and J. Peters J. Endocr. (in the press). (1975)
2. J.R. Richards and G. Sayers Proc. Soc. Exp. Biol. Med., 77, 87–93, (1951)

AN AUTORADIOGRAPHIC STUDY OF THE RENAL UPTAKE AND METABOLISM OF A SYNTHETIC ADRENOCORTICOTROPHIN.

John. R.J. Baker, H.P.J. Bennett and C. McMartin

Horsham, Research Centre, CIBA Laboratories, HORSHAM, West Sussex, RH12 4AB, U.K.

The cellular and subcellular events in rat kidney following intravenous injection of the radioactively labelled adrenocorticotrophic hormone analogues, [^{3} H–Tyr^{23}] – Synacthen and [^{3}H–Tyr^{23}]–Synacthen (see preceding abstract), have been studied using microautoradiography and quantitative electron microscope autoradiography[1]. The distribution of silver grains 3–7 min after injection is similar for both forms of the peptide and shows that after glomerular filtration most of the label is rapidly resorbed via endocytotic vesicles of the proximal convoluted tubule. After 22 min label from [^{3}H–Tyr^{23}]–Synacthen is twice as concentrated in the lysosomes as in endocytotic vesicles whereas, at 13 and 30 min after injection of [^{3} H–Tyr^2] – Synacthen, label is three to four times more concentrated in lysosomes than in endocytotic vesicles. After 1h label from [^{3}H–Tyr^{23}]–Synacthen is more randomly distributed as determined by the χ^2 test showing similar lysosomal nuclear and cytoplasmic concentrations.

The results indicate that hydrolysis of Synacthen by renal peptidases can occur readily at the C–terminus and contrast with the findings of previous work[2] in which most of the label following intraluminal microperfusion of rat nephrons with I^{125}–albumin was detected within the lysosomes at 1 h. Our findings further indicate that the role of apical tubules is intermediate in the transfer of resorbed peptide from the endocytotic vesicles to lysosomes in the renal proximal tubule.

REFERENCES

1. Williams, M.A. In „Advances in Optical and Electron Microscopy" (R. Barer and V.E. Cosslett, eds), vol. 3, pp. 219–272. Academic Press, London and New York. (1969)
2. Maunsbach, A.B. J. Ultrastruct. Res. 15, 197–241 (1966)

APPENDIX

PRESENT KNOWLEDGE OF PROTEOLYTIC ENZYMES AND THEIR INHIBITORS

A.J. Barrett, I. Kregar[+], V. Turk[+], J. F. Woessner Jr.[++]

Tissue Physiology Department, Strangeways Research Laboratory, Cambridge, Great Britain; [+]J. Stefan Institute, University of Ljubljana, Ljubljana, Yugoslavia; [++] Department of Biochemistry and Medicine, University of Miami School of Medicine, Miami, Fla., USA.

Research in the field of intracellular protein catabolic processes regulated by their corresponding enzymes and their inhibitors has shown remarkable progress during the past decade. This knowledge pertains not only to the nature of these important biologically active components, but also to their distribution, number and structure.

Classification of enzymes and inhibitors is still a difficult task, and intracellular proteolytic enzymes and their inhibitors are no exception. The aim of this appendix is to review the status of our knowledge concerning various proteolytic enzymes and their inhibitors, in order to provide a basis for later discussion, and includes some preliminary data as yet unconfirmed.

The classification of enzymes is basically that of: 1. Enzyme nomenclature in Comprehensive Biochemistry (M. Florkin and E.H. Stotz, eds.) Vol. 13, Elsevier, Amsterdam 1973; 2. „Some remarks concerning the naming and description of tissue proteinases", prepared by Barrett and others, and representing the conclusions of a discussion on nomenclature of tissue proteinases that occured during the 1st International Symposium on Intracellular Protein Catabolism at Reinhardsbrunn, GDR (in the book „Intracellular Protein Catabolism", H. Hanson, P. Bohley, eds., J. Ambrosius Barth Verlag, Leipzig, 1974/6, p. 555).

A. CHARACTERISTICS OF SOME CELLULAR PEPTIDE HYDROLASES

The enzyme names are mostly unofficial, and may be subject to change.

	Source			pH	Mol. wt.	Comments	Reference
	Species	Tissue	Organelle	Optimum	(x10^3)		(if not well known)
Carboxypeptidases (EN 3.4.12)							
Lysosomal carboxypeptidase A	Many	Many	Lys.	5.0		Was cathepsin A	
Lysosomal carboxypeptidase B	Many	Many	Lys.	6.2	52	Was cathepsin B2; cleaves BAA	Otto et al. (p. 282)
Lysosomal carboxypeptidase C	Many	Many	Lys.	5.5		Was catheptic carboxypeptidase C	McDonald & Ellis (1975) Biochem. Biophys. Res. Commun. in the press.
Dipeptidases (EN 3.4.13)							
Ser–Met dipeptidase	Many	Many	Lys.	5.5			McDonald et al. (1972) Biochem Biophys. Res. Commun. 46, 62–70
Dipeptidylpeptidases (EN 3.4.14)							
Dipeptidylpeptidase I (EN 3.4.14.1.)	Many	Many	Lys.	5.5	200	Was cathepsin C, dipeptidyl aminopeptidase I	McDonald et. al. (1971) in ,,Tissue Proteinases (A.J. Barrett, J.T. Dingle, eds.), North Holland Publ. Co., Amsterdam, p. 69.
Dipeptidylpeptidase II	Many	Many	Lys.	5.0		Was dipeptidyl aminopeptidase II	
Serine proteinases (EN 3.4.21)							
Lysosomal elastase	Many	PMNL	Az.gran.	8.5	25-35	Similar to EN 3.4.21.11	Koj et al. (p. 317) Kopitar et al. (p.327)
Cathepsin G	Human	Spleen, PMNL	Az.gran.	7.5	28	Was ,,chymotrypsin-like" enzyme	Barrett (p. 310)
Plasminogen activator	Many	Many		8.5	30-38		Kopitar et al. (p. 327)
Liver neutral proteinase	Bovine	Liver		7	25	Active vs. histone	Suhar & Lebez (p.343)

Thiol proteinases (EN 3.4.22)							
Cathepsin B1 (EN 3.4.23.1)	Many	Many	Lys.	4-6	25		
Cathepsin B3	Rat	Liver	Lys.	6	20-30	Low sensitivity to leupeptin	Kirschke et al. (p. 299)
Cathepsin L	Rat	Liver	Lys.	5	20-30	Low activity vs. BANA	Kirschke et al. (p. 299)
Cathepsin S	Bovine	Lym-phnode		3	14-17	Inactive vs. BAA	Turnšek et al. (p. 290)
Collagenolytic cathepsin	Many	Many		3.5	>25	Inactive vs. heamoglobin	Stražiščar et al. (p. 224)
Carboxyl proteinases (EN 3.4.23)							
Cathepsin D (EN 3.4.23.5)	Many	Many	Lys.	3.2-4.5	42		
Cathepsin D (Bowers)	Rodent	Lympho-cytes		3.6	90,45	High spec. act. vs. haemoglobin	Bowers et al. (p. 230)
Cathepsin E	Rabbit, Bovine	Several		2.5	>100	Activity vs. albumin $>$ haemoglobin	Lebez et al. (1971) in Tissue Proteinases (A. J. Barrett, J. T. Dingle, eds), North Holland, Amsterdam, p. 167.
Metallo-proteinases (EN 3.4.24)							
Tissue collagenase (EN. 3.4.24.3)	Many	Many		7-8	40-65		Woessner (p. 215)
Lysosomal collagenase	Many	PMNL	Az.gran.	7.5	76	Some differences from the preceding	Kopitar et al. (p. 327)
Insulinase	Rat	Many	soluble	8	80	Also active vs. other polypeptides	Ansorge et al. (p. 163)
Proteinases of unknown catalytic mechanism (EN. 3.4.99)							
Cathepsin F	Rabbit, man	Cartilage		4.5	50-70	Resists usual inhibitors; active vs. proteoglycan	Dingle et al. (unpublished)
Pz-peptidase	Many	Many		7-8.5	77	Active vs. Pro-Leu-Gly-Pro. Inhibited by EDTA and pCMB	Woessner et al. (p. 215)

Abbreviations: Az. gran., azurophil granules; BAA, benzoyl-L-arginine amide; BANA, benzoyl-DL-arginine 2-naphthylamide; Lys., lysosomal; PMNL, polymorphonuclear leucocyte.

In order to understand the mechanism of action of these enzymes, it is worthwhile to find out much more about substances which inhibit their action. The classification of proteinase inhibitors is still difficult at present. It is possible to divide inhibitors according to their action on proteolytic enzymes: a) inhibitors of carboxyl proteinases, thiol proteinases, serino-proteinases etc.; b) according to their size as small peptides, (leupeptin, pepstatin), small protein inhibitors (α_1–antitrypsin, ovomucoid) and giant inhibitors (α_2– macroglobulin); and c) there are other possibilities of classification. Therefore we decided to present a glossary of proteinase inhibitors alphabetically. Many well known inhibitors like ions of heavy metals and some organic chemicals are not included. A general survey of proteinase inhibitors and many more not discussed here (if not otherwise stated) may be found in the book entitled „Proteinase inhibitors" (H. Fritz, H. Tschesche, L.J. Greene and E. Truscheit, eds., Springler Verlag, Berlin 1974).

B. GLOSSARY OF PROTEINASE INHIBITORS

β_1– anticollagenase (β_1–AC) a serum protein, MW about 40 000; an inhibitor of human collagenases (D.E. Wooley et al: Biochem. J. **153,** 119 (1976) and Abstract of 5th Meeting of Eur. Fed. Connective Tissue Clubs, Liege 1976)

Antipain – an inhibitor of papain, trypsin, acrosin, etc. isolated from the culture medium of Streptomyces sp. The structure is carboxyphenylethylcarbamoyl-L-Arg-L-Val-Argininal

```
HOOC-CH-NH-CO-NH-CH-CO-NH-CH-CO-NH-CH-CHO
     |           |         |         |
     CH2        (CH2)3     CH       (CH2)3
     |           |        / \        |
  (C6H5)         NH    H3C   CH3     NH
                 |                   |
                 C=NH                C=NH
                 |                   |
                 NH2                 NH2
```

α-1-Antitrypsin - a serum protein that inhibits a much broader spectrum of serine proteases than suggested by the name. The human α-1-antitrypsin has a MW of 54–58,000 and inhibits trypsin, chymotrypsin, elastase, plasmin, etc.

Chloromethylketones – many different derivatives have been developed as inhibitors ofr different proteases. These derivatives usually contain one or more amino acids related to the specificity requirements of the enzyme. However, these compounds can also inhibit thiol proteinases in a nonspecific manner.

Ac-Ala-Ala-Ala-AlaCH_2Cl - inhibits pancreatic and PMN elastases
Tos-LysCH_2Cl – inhibits trypsin
Tos-PheCH_2Cl – ihibits chymotrypsin
Z-PheCH_2Br – inhibits cathepsin G

Chymostatin – an inhibitor of chymotrypsin and cathepsin B1; it has some effect on papain and cathepsin D. Obtained from culture medium of Streptomyces sp.

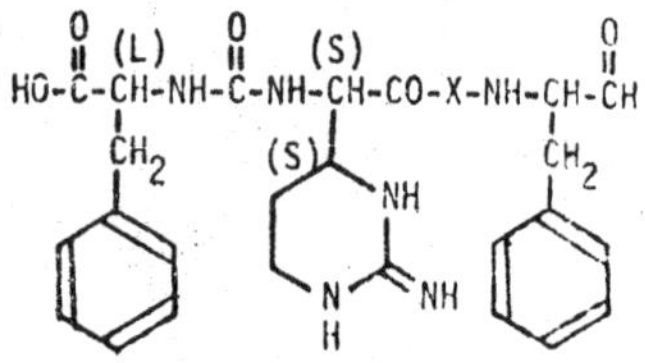

```
     O (L)    O    (S)                 O
     ‖        ‖                        ‖
HO-C-CH-NH-C-NH-CH-CO-X-NH-CH-CH
        |          |          |
       CH2       (S)         CH2
```

X = L-Leu, L-Val, or L-Ile

DFP – diisopropyl fluorophosphate, a compound used for inhibition of most of the serine enzymes. A covalent bond is formed between DFP and the serine residue at the active center.
Diazoketones – in the presence of cupric ions inhibitors of carboxyl proteinases, pepsin, cathepsin D and other bacterial proteinases.
Egg White Inhibitors – avian eggs contain a variety of proteinase inhibitors. Properties are given for inhibitors from chicken eggs.
Ovomucoid – a heat-stable protein of MW 28,000 that inhibits trypsins of various species, but does not inhibit chymotrypsin. $K_i = 10^{-10}$
Ovoinhibitor – a protein of MW 48,000 that inhibits both trypsin and chymotrypsin. $K_i = 10^{-8}$
Papain inhibitor – a heat-stable protein of MW 13,000. It interacts with papain and cathepsin B1 in a 1:1 mole ratio.

LBTI – limabean trypsin inhibitor – a two-headed protein with separate binding sites for trypsin and chymotrypsin. MW about 9000.
Leucocyte inhibitor – an inhibitor of granule neutral proteinases: elastase, plasminogen-activator and collagenase; a heat labile protein of MW 43 000 isolated from pig leucocyte cells (M. Kopitar and D. Lebez: Europ. J. Biochem.**56**, 571 (1975) and in this book p. 327).

Leupeptin — an inhibitor from Streptomyces sp. culture medium that is active against plasmin, trypsin, papain, acrosin, and cathepsin B1. Possibly the aldehyde group of the inhibitor froms a thiohemiacetal with the enzyme thiol group. A number or forms occur; the propionyl form is illustrated:

```
                                                  NH2  NH
                                                    \ //
                                                     C
                                                     |
                                                     NH
                                                     |
                  CH3 CH3        CH3 CH3             CH2
                    \ /            \ /               |
                    CH2            CH2               CH2
                    |              |                 |
                    CH2            CH2               CH2
                    |              |                 |
CH3-CH2-CO-NH-CH-CO-NH-CH-CO-NH-CH-CHO
```

Propionyl Leu Leu D, L-Argininal

α-2-Macroglobulin — a serum protein, MW about 725,000, that is an almost universal inhibitor of protease, inactivating proteinases from all four major groups. The proteases attack bonds in the macromolecular inhibitor, which then undergoes structural rearrangement so as to physically entrap the protease.

Octapeptide: Gly-Glu-Gly-Phe-Leu-Gly-D-Phe-Leu- synthetic octapeptide inhibits cathepsin D and is useful for affinity chromatography, $K_i = 5.10^{-4}$ (F. Gubenšek et al.: FEBS Letters**71**, 42 (1976) and in this book p. 255). Also some other peptides containing D-amino acids — substrate analogues — are competitive inhibitors of carboxyl proteinases (H. Keilova in: Tissue proteinases, A.J. Barrett, J.T. Dingle Eds., North Holland Publ. Co., Amsterdam 1971, p. 45; and others).

Ovoinhibitor, ovomucoid (see egg white inhibitors)

Pepstatin — a pentapeptide from the culture medium of Streptomyces sp. that contains only hydrophobic residues. This binds in 1:1 ratio with proteases of the carobxyl family including pepsin, cathepsin D and cathepsin E. There are some 15 variants of the basic structure; type A is illustrated: (AHMH = 4-amino-3-hydroxy-6-methylheptanoic acid)

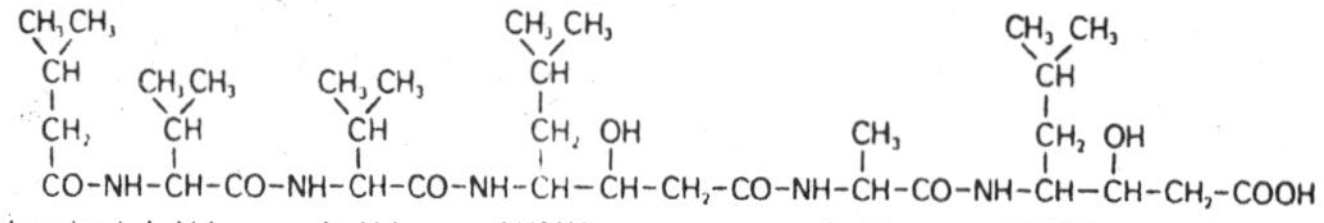

```
CH3 CH3                         CH3 CH3                              CH3 CH3
  \ /                             \ /                                  \ /
  CH     CH3 CH3   CH3 CH3        CH                                   CH
  |        \ /       \ /          |                                    |
  CH2      CH        CH           CH2  OH              CH3             CH2  OH
  |        |         |            |    |               |               |    |
  CO-NH-CH-CO-NH-CH-CO-NH-CH-CH-CH2-CO-NH-CH-CO-NH-CH-CH-CH2-COOH
Isovaleryl L-Val     L-Val        AHMH                 L-Ala           AHMH
```

PMSF — phenylmethane sulfonylfluoride - an inhibitor of serine proteases acting in much the same manner as DFP.

Potato inhibitors: a) from Russet Burbank potatoes, polypeptide of MW 3100, $K_i \cong 1.10^{-9}$, an inhibitor of the pancreatic carboxypeptidase A and B (C.A. Ryan et al.: J. Biol. Chem. **249**, 5495 (1974)); b) from sweet potato, MW about 23 000. Inhibits trypsin at a 1:1 M ratio; c) from Solanum tuberosum potatoes, a heat-labile protein of MW 27,000, inhibitor of cathepsin D, but not of pepsin (H. Keilova and V. Tomašek: Coll. Czechoslov. Chem. Comm. **41** 489 (1976).

Proteinase inhibitors from Ascaris lumbricoides — include inhibitors of pepsin (small proteins with MW of 15,500, $K_i \cong 10^{-10}$) carboxypeptidase A inhibitor ($K_i \cong 10^{-9}$), trypsin inhibitor and chymotrypsin-elastase inhibitor.

Proteinase inhibitors from Scopolia japonica cultured cells — inhibitors of trypsin, chymotrypsin, plasmin, kallikrein and pepsin, but not papain; polypeptides of MW of 4,000 to 6,000, $K_i \cong 10^{-9}$ (K. Sakato et al.: Europ. J. Biochem. **55**, 211 (1975).

SBTI — soybean trypsin inhibitor of Kunitź — a protein of MW 22,000 that inhibits trypsin and chymotrypsin. $K_i = 10^{-8}$
TLCK, TPCK (see chloromethylketones)

Trasylol — tradename for BPTI, the basic pancreatic trypsin inhibitor of Kunitz. Usually preparated from bovine lung, MW 6513. Inhibits trypsin and kallikrein at a 1:1 molar ratio. $K_i = 10^{-10}$

SUBJECT INDEX